90 0587496 X

WITHDRAWN
FROM
UNIVERSITY OF PLYMOUTH
LIBRARY

This book is to be returned on
or before the date stamped below

UNIVERSITY OF PLYMOUTH

PLYMOUTH LIBRARY

Tel: (01752) 232323
This book is subject to recall if required by another reader
Books may be renewed by phone
CHARGES WILL BE MADE FOR OVERDUE BOOKS

New Technologies and Environmental Innovation

New Technologies and Environmental Innovation

Joseph Huber

*Chair of Economic and Environmental Sociology,
Martin-Luther-University, Halle, Germany*

Edward Elgar
Cheltenham, UK • Northampton, MA, USA

© Joseph Huber, 2004

All rights reserved. No part of this publication may be reproduced, stored in a retrieval system or transmitted in any form or by any means, electronic, mechanical or photocopying, recording, or otherwise without the prior permission of the publisher.

Published by
Edward Elgar Publishing Limited
Glensanda House
Montpellier Parade
Cheltenham
Glos GL50 1UA
UK

Edward Elgar Publishing, Inc.
136 West Street
Suite 202
Northampton
Massachusetts 01060
USA

A catalogue record for this book
is available from the British Library

Library of Congress Cataloguing in Publication Data

Huber, Joseph, 1948–
    New technologies and environmental innovation/Joseph Huber.
        p. cm.
    Includes bibliographical references and index.
        1. Technological innovations—Economic aspects. 2. Technological
    innovations—Environmental aspects. 3. Green technology—
    Technological innovations. 4. New products—Environmental aspects.
    5. Product life cycle—Environmental aspects. 6. Industrial ecology.
    I. Title.
    HC79.T4H83   2004
    338'.064—dc22

                                                            2004040570

ISBN 1 84376 799 6

Printed and bound in Great Britain by MPG Books Ltd, Bodmin, Cornwall

UNIVERSITY OF PLYMOUTH

Item No. 9005874096X

Date 2 8 APR 2004

Class No. 338.064 HUB
Coni. No.

PLYMOUTH LIBRARY

Contents

UNIVERSITY OF PLYMOUTH

Figures

Tables

Abbreviations

AOX	Adsorbed organic halogen compounds
BC	Biological control
BOD	Biological oxygen demand (in water)
CC	Combined cycle (steam turbines in combination with gas turbines in power stations)
CFCs	Chloro-fluoro-hydrocarbons
CHP	Combined heat and power (as in DHPS)
CLECs	Cross-linked enzyme crystals
CNTs	Carbon nanotubes
COD	Chemical oxygen demand (in water)
CSP	Concentrating solar power (mirrors, dishes, parabolic troughs)
CVOCs	Chlorinated volatile organic compounds
DHPS	District heat and power station (as in CHP)
DMFC	Direct methanol fuel cell
DNA	Deoxyribonucleic acid (carrier of an organism's genetic information in any cell's nucleus)
DRAM	Dynamic random-access memory
E-car	Electric car
EHS	Environment, health, safety
EPA	Environmental Protection Agency
E-scrap	Scrap from electronic appliances
EU	European Union
FAO	Food and Agriculture Organisation (of the United Nations)
FC	Fuel cell
FDI	Foreign direct investment
FeRAM	Ferro-electric random-access memory
FLOX	Flameless oxidation
FRET	Fluorescence resonant energy transfer
FTO	Flameless thermal oxidation
GM	Genetic modification, genetically modified
GMO	Genetically modified organism
GPS	Global Positioning System (satellite-based)
ICT	Information and telecommunications technology

IGCC   Integrated gasifier combined cycle
ILCA   Innovation life cycle analysis
IPM   Integrated pest management
ISO   International Standards Organisation
IT   Information technology (computing)
LED   Light-emitting diode
LEP   Light-emitting polymer
LISA   Laser in-situ analysis
MCFC   Melted carbonate fuel cell
MEMS   Micro-electro-mechanical systems
MRAM   Magnetic random-access memory
NGO   Non-governmental organisation
NOx   Nitrogen oxides
OECD   Organisation for Economic Co-operation and Development
OLED   Organic light-emitting diode
OTEC   Ocean thermal energy conversion
PAFC   Phosphoric acid fuel cell
PCB   Polychlorinated biphenyl
PCP   Pentachlorphenol
PEM   Proton exchange membrane, or polymer electrolyte membrane
PICs   Products of incomplete combustion
PIN   Personal identification number
PMC   Polymer matrix composites
POP   Persistent organic pollutant
ppm   Parts per million parts
PVC   Polyvinylchloride
R&D   Research and development
$SCCO_2$ Supercritical CO_2
SME   Small and medium-sized enterprises
SOFC   Solid oxide fuel cell
SOx   Sulphur oxides
S&T   Science and technology
TEIs   Technological environmental innovations
TM   Registered trade mark
TOC   Total organic carbon
UN   United Nations
USA   United States of America
UV   Ultra-violet (light/radiation)
VOCs   Volatile organic compounds (benzene, phenol, tuluol, xylol, methyl
       alcohol (methanol), isopropyl alcohol, etc. In total about 50 such
       substances)

PART I

Technological Environmental Innovations (TEIs)

1. Introduction: Upstreaming Environmental Action

This book is about key environmental innovations, i.e. new technologies, products and practices that are of particular importance to solving urgent environmental problems.

One of the challenges of our time still is how to achieve ecologically sustainable living standards with a decent level of affluence for all of the 6–8 billions of people on earth. The approach of technology life cycle analysis explains why a viable answer to the problem needs to be conceived of in terms of ecological modernisation, based on scientific research and technology, controlled by market economies and apt administration, and aimed at structurally changing the industrial metabolism so as to re-embed it into nature's metabolism. Hence the author's long-standing interest in environmentally benign new technologies, referred to hereafter as technological environmental innovations, or TEIs for short.

TEIs are more than just environmental technology such as exhaust-air catalytic converters, or filters in chimneys or sewage water purification plants. This kind of environmental technology is called end-of-pipe technology, applied within the context of add-on measures. They represent a downstream approach in that they come in after the point where some damage or pollution has already occurred, without changing the upstream source of the ecological perturbations in question. Foreseeably, add-on measures will continue to be important in many cases. The far more important portion of TEIs, however, are those new technologies which prevent environmental pollution and deterioration from happening in that they come further upstream in the manufacturing chain and in a technology's life cycle, and change the source of perturbation or avoid it altogether.

TEIs are not necessarily being developed for environmental reasons only. They represent a new generation of innovative technologies that fulfil ecological criteria as much as technical criteria of efficiency, operational safety and reliability. And TEIs have of course to match economic criteria such as price and profitability sooner rather than later as they move along their learning curve. Examples of such TEIs include the following:

3

- replacement of fossil fuels with clean-burn hydrogen, the use of which does not require additional end-of-pipe purification of emissions
- substitution of clean electrochemical fuel cells for pollutant furnaces and combustion engines
- fuelless energy such as photovoltaics and further regenerative energies which make use of sun radiation, geothermal heat, or wind and water currents
- transgenic biochemistry which makes use of enzymes and microorganisms especially designed and bred for various production tasks, thus replacing the conventional high-temperature high-pressure chemistry that is dangerous and poses a heavy burden on the environment
- sophisticated low-hazard specialty chemicals which are, besides other product properties, biodegradable, non-persistent, non-accumulative and non-toxic
- new materials which are simultaneously ultra-light and ultra-strong, thus saving larger volumes of conventional materials and energy
- micromachines and nanotechnology which relieve pressure on resources and sinks compared to larger conventional machines and chemical production
- substituting sonar, photonic and microfluidic analyses for cumbersome conventional methods involving many hazardous ingredients, and thus considerably improving quality and performance of production
- circulatory production processes in which water, auxiliary substances, metals, bulk minerals and fibres are recycled at an optimum rate
- last but not least, overcoming the ecologically devastating practices of today's over-intensified and inappropriately chemicalised agriculture by introducing sound ecological practices in combination with high-tech precision farming and, again, modern biotechnology which makes use of transgenic organisms.

The appendix contains a list of examples from eleven realms of TEIs. It might be advisable at this point to have a look through this list which will give an overall impression of what the empirical basis of this book is about.

To clarify the message straight away: TEIs may help to reduce the quantities of resources and sinks used, be they measured as specific environmental intensity per unit output, or as average consumption per capita, or even in absolute volumes. Overriding priority, however, is given to improving the qualities and to changing the structures of the industrial metabolism. Rather than doing less of something, TEIs are designed to do it cleaner and better by implementing new structures rather than trying to increase eco-productivity of a suboptimal structure which has long been in place. TEIs are about using new and different technologies rather than using old technologies differently.

The attitude does not arise out of taking 'creative destruction' as an ideology, but is derived from the fact that successful systems innovations which represent new technological paradigms and which break new ground tend to have a much bigger learning-curve potential than mature generations of technologies which are approaching the end of their learning curve. TEIs, to give an example, certainly include recycling of organic solvents. Preferable, though, is to replace organic solvents, particularly in open-air applications, with water-based agents or supercritical carbon dioxide. This does not exclude those special cases in which organic solvents remain an unbeatable auxiliary for the time being. In these cases it may in fact be preferable to use organic solvents in a leakage-tight closed-loop process.

An environmental strategy based on TEIs certainly cares about ecological *sufficiency*, i.e. keeping the use of resources and the burdens on sinks within critical limits. This, to be sure, presupposes the availability of some valid knowledge about these limits, not just vague precautionary fears. Given these limits, TEIs will also have to attend to ecological *efficiency*, though this is not an end in itself. If, for example, the ecological problem arises from burning carbon or plutonium, burning it more efficiently does not make much sense. Instead, different fuels or completely different technologies of power generation will have to be developed and diffused. In this sense, an environmental strategy based on TEIs will above all care about the ecological *consistency* of the industrial metabolism as illustrated by the TEI approaches mentioned above (more in chapter I/2.4).

The idea of eco-consistent TEIs is 'fitting in with' in order to avoid anthropogenic destabilisation of geo- and ecodynamics. This is achieved by contriving technological ways and means of maintaining an industrial metabolism that is effectively compatible with nature's metabolism at an optimum level of efficiency, even at large volumes of materials turnover. This contrasts with an oversimplistic notion of 'dematerialisation' that lumps together totally different things, makes no difference between clean and dirty energy, hazardous and harmless substances, and develops no understanding of the major or minor ecological sensitivity of different kinds of materials.

To give just two examples. One is the automobile as the supposed 'environmental enemy #1'. The automobile does indeed have a long list of environmental problems associated with it, ranging from road safety and noise, to land utilisation, soil sealing, and further to the consumption of raw materials on a large scale and to a complex problematic of airborne emissions. All of these problems, however, have proven to be accessible to technological regimes which can keep the environmental impact to acceptable levels, with the exception of CO_2 and some other airborne emissions caused by gasoline burned in internal-combustion engines. So, in principle, there is not too big a problem with being an automobile society, something which most people consider to be an achievement. But there is a major problem with metabolically inconsistent vehicle propulsion, a problem that could already have been solved for the most part if industrial elites had not been so sluggish in developing the necessary awareness and will to deal with that bottleneck.

A second example is energy demand. Intensive use of energy is the physical basis and one of the most significant indicators of industrial development. High and still growing levels of per capita consumption of energy are normally not based on inefficient energy consumption. Efficiency increases right from the beginning of a technology's life cycle, and growth is in turn based on increasing efficiency (II/7.7.1).

So any environmental strategy aimed at 'using less energy', in the sense of bringing down the overall level of energy demand, would have detrimental effects on the ongoing evolution of modern society. What we need instead is a strategy of clean energy which fits in with nature's metabolism even on a large scale in the giga and tera range. This approach certainly entails using energy as efficiently as possible. But the real purpose of increased efficiency along the learning curve of a system, in technical systems as much as in natural organisms, is never to get by on less but to stabilise further growth and development. As can be seen from most of the cases represented in the TEI database of this book, important eco-consistent TEIs tend by themselves to come equipped with much higher eco-efficiencies than previous like technologies have. Mere increases in efficiency, by contrast, tend to occur during the later life cycle stages of mature technologies. The conclusion is obvious: Rather than calling on green saving commissioners, TEIs are calling for green inventors, innovators and investors.

Not long ago, green-minded people used to be at odds with the idea of the greening of industry. Thinking of technological environmental innovation, and expecting science, technology, industry and finance to take a leading role in that greening seemed to be a contradiction in itself. As a Neapolitan colleague once put it, it seemed like entrusting the Mafia with the task of combating crime. Today, green ideas have been diffused throughout society, big science and industry included. The environmental movement has assimilated itself to the general science and technology paradigm such as it is dominating the knowledge base of modern society. Pinning hopes on technology is no longer automatically deemed 'technocratic'. Technology has actually turned out to be the key component of any viable response to ecological challenges.

*   *   *

Environmental policy has much to gain from two basically simple insights. First, solving environmental problems means looking for metabolically consistent and efficient technological solutions, thus TEIs. Second, these solutions are to be found upstream rather than downstream in the manufacturing chain and a technology's life cycle.

Certain scholars of economics, sociology and humanities tend to belittle the role of technology, be that for pretensions to the supremacy of their own discipline, or out of fear of succumbing to a 'technocratic' ideology, while actually missing the pivotal function of technology in modern society which is obvious to everybody else. Some of the classics have clearly recognised the role of technology, among them Karl Marx, Joseph Schumpeter, Jacques Ellul, Lewis Mumford, Daniel Bell, all of them anything but apologists for technology. Social sciences must incorporate a social theory of technology and technosystems development. Otherwise they cannot fully cope with their task of explaining social systems and the place of humans therein. This is all the more true with regard to the environment and human society's relationship with nature.

Ecology refers to the physical exchanges which a population of a given number and a given level of instrumental capacities realises within its living space, i.e. the locations where a population lives and the entire space which is covered by its activities. In particular it includes the natural resources a population makes available and the environmental sinks it employs to get rid of resulting phase-out products. It also includes cooperation and competition with other populations concerning the utilisation of resources and sinks within that living space. Ecology thus deals with the metabolism between a population and its natural surroundings within the geo- and biosphere, how in this process the environment is being transformed, and how populations and their environment co-evolve.

Environmental problem solving needs to be technological in the first place because environmental problems are perturbations of ecological systems of which humankind is part by way of physical operations carried out through its instrumental capacities, i.e. the technology of the time. The origin of ecological perturbations may be geogenic or anthropogenic, or both, as is the case with global climate change. Anthropogenic perturbations are immediately caused by the technological means employed to fulfil certain operative functions. That is why resolving the problem unavoidably includes restructuring or substitution of the technology in place. Also with geogenic perturbations, some kind of technological solution will be needed to face the challenge. Typically these are measures aimed at shielding humans (as well as animals and plants which are considered to be useful) from natural influences, and protecting land and settlement structures.

Ecology is a natural-science discipline which refers to something physical. 'Ecology of freedom' or 'ecology of mind' are unsuitable terms which confuse metabolic operation (exchange of physical substances) with communication (exchange of ideas). It is certainly true that ecological effects which are physically caused by human operations originate ultimately in the value base of humans, in their worldview and certain cultural concepts such as utilitarianism in modern society. In the first instance, however, ecological effects come about through physical operations which modify or transform the state of the environment, for example, by gathering, hunting, building, manufacturing, by extracting, processing and using physical things in the realm of res extensa.

Higher animals and particularly humans have extended the natural instruments of their organism by developing artificial methods and instruments, embedded in more complex social organisation: tools, machines, infrastructures and cooperative arrangements, in brief, technology which serves special operative purposes. Technology thus is the immediate ecological factor in human society besides the fact of humans' sheer biological existence.

The metabolism effectuated by labour and technology needs to be analysed in terms of engineering and natural sciences such as physics, chemistry and biology. At that level ecology cannot be understood in terms of the social sciences, including economics. For example, postulating that 'the problem of anthropogenic climate change is greed' or 'economic growth pressure' or 'distorted prices' or 'compensatory consumption' or 'godlessness' or 'bad government' and 'poor lawmaking', may rightly hint to certain factors which could play a role in a more complex analysis, but it misses the ecological fact that, at the physical source, the problem is an inappropriate way of utilising resources and sinks through technological regimes which are apparently less highly developed than is commonly believed.

In modern society, the metabolism between humans and their environment has taken on the forms of industrial metabolism (Ayres 1993, Ayres/Ayres 1996, Ayres/Simonis 1994). The resulting relationship of industrial society with its environment represents an industrial ecology (Lowe 1993, Allenby/Cooper 1994, Socolow et al. 1994, Frosch 1996). It results in an ecological transformation (Bennett 1976) of the earth's ecosystems unknown in scope and scale to traditional societies. Industrial man is currently causing a per-capita turnover of materials that is 5–10 times the quota of traditional agrarians, in the same way as the agrarian quota in its time was 5–10 times that of primitive hunter-gatherers. Industrialised contemporaries live on 1,320 kg a day, while their stone-age ancestors are said to have lived on just 35 kg (Fischer-Kowalski 1997). Facing the responsibilities ensuing from this opens up the perspective of earth systems management or ecosystems management (Allenby 1999). Today's still rather unfocussed environmental policies represent preliminary steps towards such comprehensive sustainability regimes. Their effective tool will be technology because intended as well as unintended ecological effects are caused by nothing other than powerful technologies in a context of industrialised production.

There is no 'post-industrial' society today and none can be envisaged in the foreseeable future. What is called the service economy, or the information economy, knowledge society, high-tech society or scientific civilisation, remain different aspects of industrial society progressing on its evolutive path. Expecting environmental ease from the sectoral changeover to the service economy has been one of the vain ideas prevalent in environmental discourses. The service society is based on industrial production in ever larger volumes. The supposed paperless office of the future quickly became another chapter in the book of yesterday's tomorrows. Another such futile expectation is that of substituting communications technology for transport of passengers and goods, when in real industrial history transport and communications have always developed in tandem. Industrial productivity is the basis upon which services build up. As a consequence, the service society is much more resources- and energy-intensive than earlier industrial society used to be, in the same way as that earlier industrial society was based on ever more productive agriculture and proved to be much more agro-intensive than any societal formation before.

We can, of course, subdivide the ongoing development of industrial society into different epochs. For example, there are recurrent epochs of technological systems innovation, which introduce new key technologies and so lay the foundations for new industries and markets. These are the so-called long waves – also named, after two of the early researchers in this field, Schumpeter cycles or Kondratievs – occurring within a window of opportunity that seems to open up for about one or two decades every 40–60 years. Thus far they have been mechanisation in the decades around 1800, railroads around 1850, electrification and chemistry around 1900, mass motorisation and telecommunications around 1950, and computerisation, the now somewhat aged new economy, in the past decades (II/7.3.5).

Another such subdivision that could prove to be useful is that of industrial ages, to be measured in centuries rather than decades. So far, two of these could be identified: industrial age I from the beginnings of the industrial revolution more than 200 years ago until the 1960–80s, and industrial age II which has been in the ascendant since around the middle of the 20th century. This can be seen as another long-term stage in the transsecular life cycle of ongoing modernisation of nations in the world system. The defining feature which distinguishes the two is the basis of reference for modernisation. While industrialisation in age I took place in a context of traditional practices of agrarian, crafts and household production which it replaced step by step, industrial age II is now proceeding by modernising the now older modern structures that were created during industrial age I.

Contemporary sociology has chosen to call that process of self-reference 'reflexive modernisation' (Beck et al. 1994). Aficionados of business buzzwords would say industry is re-inventing itself, for example, by substituting the 21st century's specialty steels and a broad range of new materials for the 19th century's cast iron which in turn replaced the wood and stone materials of earlier centuries; and by substituting the 21st century's clean fuels and fuelless energy for the 20th century's oil and plutonium, and the 19th century's coal that replaced the wood, straw and peat of traditional society.

As industrial workers replaced feudal peasants and craftsmen, the industrial workforce of the 19th and 20th centuries is now being replaced by skilled specialists and scientifically educated experts. Production has increasingly become based on scientific methods and advanced technology. The greater part of employment and economic turnover may indeed occur in services, though most of these have taken on scientific characteristics themselves. Apart from this, many services remain directly linked to industrialised production in agriculture, basic industries, manufacturing and crafts. In advanced industrial society each activity is becoming technology-based or technology-intensive, as all technologies are increasingly based on scientific methods of research and development.

There has been little controversy over the expectation that 'post-industrial' society would be high-tech. Hopefully there will soon be no more controversy either on the idea that high technology will be used to environmentally re-embed society. During the last decades of the 20th century, industrial age I has increasingly been coming to its end. At the same time, the high-tech society or industrial age II has been arriving rather than still coming.

*   *   *

The purpose of technology is to enable and empower human operations. Technology is the working medium in the operative subsystem of industrial society. In that function, technology is socially embedded in many ways. To put it in a slightly different way, it is functionally interrelated, co-related with, and co-directional, sometimes counter-directional, to other societal subsystems such as the economic system, the ordinative system of law and administration, the political system, the knowledge system, the value base and further aspects of a society's culture (II/6.2).

Starting from such a systemic-evolutive basis of analysis, we cannot fail to understand that no technical change can occur without concomitant changes in the knowledge base and co-directional inspirations from the value base. Conversely, no sustained change in social ideas can occur without co-directional restructuring of social practices, particularly patterns of work and production, i.e. society's technosystem and co-related institutional and economic control systems. In this sense, a transition to ecologically readapted technologies and practices in, say, agriculture requires that those involved gain certain ecological insights, change their mindset accordingly, acquire more sophisticated knowledge, take decisions on political goals, with whom to cooperate, with whom to compete and whom to fight, how to regulate what needs to be done, and to what ends to invest available money.

Formative factors such as awareness, consciousness, values and knowledge should not be set up in opposition to effectuative factors such as law, financial means and technology, because all of these belong to functionally different subsystems that need to co-evolve. They certainly display individual features, specific 'laws' and constraints of their own, but they do not have an autonomous existence detached from the entire system. Within the whole system, the different subsystems act as a selective and conditioning context to each other, resulting in system integration and functional control.

Functional interrelatedness signifies that the formative subsystems of culture and politics and the effectuative subsystems of law and administration, economy, and production, are all interdependent. For its evolvement, each subsystem needs corresponding evolvements or conditions in other subsystems. Functional interrelatedness thus clearly contradicts any hypothesis of 'factor substitution' (II/6.4). For example, money and capital cannot substitute for work, technology and resources. Substitutional alchemy has been a feature of neo-classical economics, including environmental economics, which was brought back down to earth by ecological economics (Krishnan et al. 1995, Jansson et al. 1994). In much the same sense, environmental awareness cannot substitute for green technology, neither can good consumer behaviour substitute for sustainable production practices, nor administrative procedures and financial policy instruments for those technologies which need to be invented, developed and used. Regulation and financial means can selectively control, but never create and place themselves in lieu of what they control. And seen from the particular angle of ecology, i.e. the geo- and biospheric metabolism between humans and nature, whatever we may believe, think, wish, decide, order and spend money on, nothing will effectively change unless we change technology.

* * *

According to the empirical findings underlying this book, technological environmental innovations (TEIs) can in general be characterised as being upstream rather than downstream, i.e. upstream in the manufacturing chain or product chain respectively, as well as upstream in the life cycle of a technology.

In the environmental literature, the term 'life cycle analysis' is often used interchangeably with 'product chain analysis' or 'eco-balances' that try to gauge the environmental impact of a product from first input by extraction of raw materials to last output by being definitely phased out as waste, the analysis in this sense stretching 'from cradle to grave'. What they represent more precisely, however, is analysis of the vertical manufacturing or product chain. By contrast, another meaning of life cycle analysis, the one preferred here, refers to an innovation life cycle or technology life cycle, i.e. to the evolutive existence of a kind of product or technology, to which the metaphor 'from cradle to grave', i.e. from being born to dying, would apply much better. Life cycle analysis then describes the creative process by which a new technology or product comes into existence, by which it is furthered and diffused throughout a population of adopters, until it finally reaches a stage of retention, or decline and intended phase-out or unintended die-off.

In this sense, technological life cycle analysis includes the stages of structuration of an innovation, from invention or discovery, via successive steps of development to the unfolding of that innovation, followed by later stages of maturation and increasing structural stability before eventually becoming too firmly established to remain adaptive enough for further evolvement. This process of the structural unfolding of an innovation is related to a process of societal diffusion from emergence to take-off and niche saturation. In this process, innovative promoters of a novel item are opposed by the established defenders of previous like items. In terms of evolution theory, the confrontation between challengers and defenders, which can in most cases not be avoided, is a genuine struggle for life, a struggle to gain and maintain a living space, a 'niche' in a society's minds and markets, a position in the divisional structure of society, and to serve a function in its subsystems, which in the case of TEIs are operative production functions that meet ecological requirements and human needs. In the process of unfolding of an innovation life cycle, producers and users of a novel thing have as important a role to play as the inventive originators, and then communicators, investors and regulators who are present throughout the process of innovation.

The discussion of metabolically consistent TEIs in I/4 thus relies on a coherent framework of innovation life cycle analysis as outlined in part II of this book. Even though innovation life cycle analysis builds upon the wealth of empirical research and theorising that has been done in this field during recent years, part II does not represent an all-encompassing recapitulation of innovation and diffusion theories. Rather, it represents a rearrangement and streamlining of particular categories with regard to their specific potential to contribute to the understanding of TEIs, combined with the introduction of a number of new categories and considerations.

If this book were an academic dissertation, part II would have come first, and part I second, because part II lays down the general model and hypotheses of innovation life cycle analysis as the basis on which the more specific discussions of TEIs in part I are drawing. Wherever certain terms or statements in part I may need more explanation in terms of innovation life cycle analysis, the corresponding references to chapters in part II are made in brackets. To scholars of processes of technological innovation and societal diffusion of innovations, part II might be worth reading in its own right.

As will be discussed in more detail in I/2.4 and in II/7.4–8, it is important here to understand that with technologies, as with living organisms, the key features of a novel thing and its life course are determined upstream rather than downstream, for instance with regard to an organism, in its genetic code and in its early days of growth, experiencing and learning; and similarly with regard to technologies, in their conceptualisation and design, in the early stages of research and development, i.e. the early stages of structuration and diffusion. What remains to be determined during later stages consists of incremental changes and modifications of minor importance, after the point of inflection of a learning curve has been passed. Most of the environmental pressure which is caused by producing and using a certain kind of product or technology is determined right at the beginning with the conceptualisation and design of that product or technology. Once in place, there is not much left which can be done with it, aside from some improvements in later new-generation variants of that product, some percentage points of materials and energy savings in the factory, and a few percentage points by being a good consumer. But a truly significant difference can only be made upstream in the life cycle, by changing path on the basis of new technological systems, new products and new practices.

* * *

Beyond and in addition to 'upstream' in the sense of occurring during early stages of a learning curve (of a technology life cycle as shown in Figure 7.3 p. 266), the second meaning is upstream in the vertical manufacturing chain (the product chain as shown in Figure 1.1 next page). The product chain includes the following steps which lead from upstream to downstream:

- Starting from extraction of raw materials, i.e. denaturation of resources in mining, quarrying, agriculture, forestry and fishery
- via processing those materials in successive steps in order to obtain food, fuel, working materials and auxiliary agents
- to producing intermediate products, production machinery and infrastructure, all of which flow into or contribute to
- the production of buildings, vehicles and other short- and long-lived, simple and complex end products, which are in turn going into
- final use or consumption, which is just another form of production, thus the next step in the product chain, in that it uses physical things in order to obtain certain results (as any production consumes resources and sinks, and any such consumption results in metabolic products)
- until being recycled back to the foot of the entire chain as used material or waste, thus on the one hand becoming secondary raw material or fuel for

re-entry into the product chain upwards, or on the other being phased out of human production and released back into natural cycles as re-natured or near-nature materials.

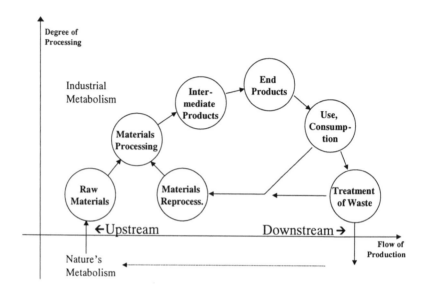

Figure 1.1 The product chain

As a rule of thumb one can say the more products and production processes are placed chain-upwards the more important the potential of their environmental impact is. This is particularly the case with regard to the difference between all of the steps of extracting and successively (re)processing materials, which regularly cause large environmental impact, and the final assembly or otherwise finishing of final products, which causes comparatively less. This insight has implicitly been present in concepts of gauging environmental impact such as the ecological footprint by Rees and Wackernagel (1994) or the materials rucksack by Schmidt-Bleek (1994).

In car-making, for example, the production steps which have high environmental impact are the production of metals, rubber, plastics and textiles from the extraction and processing of the raw materials to their shaping, surface working and coating. The rest, which includes many steps of transport and successive finishing and assembly, is tainted with much less environmental impact. So production steps concerning the original extraction and processing of materials, fuels and food, and the related generation of working energy, seem to be more important targets of TEIs than later steps downwards in the product chain related to the finishing of end products and to final use or consumption.

There is, however, an important exception to this rule: long-lived complex energy apparatus such as motors in cars, jet engines, power plants, heating systems and electric appliances the use of which consumes large quantities of fuel. Food and fuels have a metabolic rate of almost 100 per cent. This makes a big difference to materials which keep their physical structure when used, even if in the production process of materials these are also partly transformed in their physico-chemical structure. Metabolic products of intensive food and feed consumption are highly problematic as we know from environmental and hygienic problems of agriculture and sanitation, although some of that impact has been brought under control. The same holds true for intensive consumption of fuels, particularly carbon and nuclear fuels.

If there are important, unresolved environmental problems to be found in the final assembly, use or consumption of end products, they indeed have to do with the fuels and the energy apparatus involved in using those products. This is most obvious with regard to houses, vehicles and all kinds of electric appliances used in homes and offices. For example, building houses and roads involves vast flows of construction materials and rubble. Most of this, however, with the exception of the production of cement, does not pose a severe environmental problem. By far the most important environmental impact of a building is rooted in its energy design, i.e. in heating and air conditioning.

Energy remains of the utmost importance because there is an energy input into each step along the entire product chain. As a consequence of this another upstream orientation can be formulated: energy-related aspects of producing and using products should be viewed as upstream of any other aspect of product design and production process. This is valid at least for the present age of carbon and nuclear energy. Clean and safe energy must be a priority target in developing TEIs because many things can be done in favour of metabolically consistent and efficient products and processes on the precondition that there is clean and safe energy available.

* * *

A paradigm shift from downstream to upstream in the vertical product chain and technologies' life cycles implies a parallel shift in the emphasis of environmental action and policy (I/5).

First, action has to be refocused onto those industrial operations where large environmental impact actually occurs – in energy, raw materials, agriculture, chemistry and base industries, partly also in building and vehicles, not, however, in distribution, use and consumption.

Second, the key actor groups that have to be mobilised are technology developers, product designers and producers rather than users and consumers. With a focus on products and production, one would not be spending too much time with services, user behaviour and consumer demand. According to the environmental paradox of consumer society, it may be true that environmental effects are ultimately caused by attitudes and the demand of final users, but most of the ecological pressure of producing and using a product is normally not caused end-of-chain in use or consumption, but in the more basic steps of the manufacturing chain, and, seen in a perspective of life cycle analysis, by the conceptual make-up, the technological principles and design of that product. Approaches such as 'sustainable consumption' or 'sustainable household' are undoubtedly useful to some extent. But due to their end-of-chain approach they cannot be particularly effective in changing the industrial metabolism.

Third, demand will continue to play an important role, though the decisive part of it is demand by the manufacturers of end products such as buildings, vehicles, appliances and other more complex goods, and also demand by large retailers such as mail order firms, rather than demand by end users and consumers. Demand is a selective factor in the diffusion of innovations. Demand itself, however, cannot innovate, it cannot 'buy into existence' things which are not on the market yet, with one exception, which is the demand by key manufacturers and large retailers. They are in the position to effectively influence innovation processes and to implement supply chain management, i.e. act as defining chain managers, because they are able to directly exert influence on suppliers along the product chain and have a defining influence on the design and redesign of products. Innovative chain management is none of a user's or consumer's nor of a government's business.

Fourth, Government's business is to implement policies which foster TEIs. Upstreaming environmental policy would induce a shift of emphasis from regulation to innovation. Environmental policy will thus also become a policy of technology development, or will have to cooperate systematically with technological R&D. In consequence there would have to be a shift in the pattern of environmental policy from bureaucratic command-and-control to coordination of national innovation networks, European at the EU level, including suitable and well balanced financial support by granting regular research funds and seed money, as well as providing venture capital and introductory aids in appropriate ways (I/5.2.4).

What has to stay, though, is to set strict environmental performance standards. This remains by far the most effective control instrument for environment and innovation alike – which is not astonishing given the fact that environmental performance standards are of a parametric technical character by themselves.

2. TEIs in Discourse Context

Before going into the details of defining and describing the properties of TEIs, it might aid a better understanding of what all this is about by first seeing environmental innovation within the contemporary discourse context from which it has emerged. Environmental innovation does indeed represent the latest story line in the discourse on ecological modernisation and sustainability. Among the most important story lines in the evolvement of the ecological discourse have been

- the growth debate from 1972 until around 1980
- the discourse on ecological modernisation since the 1980s
- the risk debate since the mid-1980s
- the sustainability discourse propelled by the Rio process before and after the Earth Summit on Environment and Development in 1992.

2.1 THE GROWTH DEBATE

The growth debate focussed on the increase in world population and industrial production, which demand ever higher levels of resource throughput and result in ever more environmental pollution. The debate was launched by the report *Limits to Growth* in 1972, prepared on behalf of the Club of Rome, a network of concerned officials informally linked to bodies such as the OECD and the UN. Forrester and Meadows, the authors, held that if things continued as they were in a business-as-usual scenario, the system would collapse within the next 100 years. The only alternative to a collapse of the system was zero growth. 'The system' was a simple computer model consisting of a handful of variables such as the number of the world population, overall industrial production, soil fertility, resource input, emissions output and the stock of capital (whatever that was meant to represent in reality).

The Forrester simulations have been criticised for being much too simplistic, for confusing economic and physical variables, and for missing important factors, including new knowledge and structural change brought about by technological progress and political reform. In any system there are of course limits to growth at any time, but in dynamic systems there are at the same time also processes which are changing these limits. Any approach claiming to be truly systemic-evolutive would of course have to try to include variables which are able to provide for some adjustment both of the system and in its environment.

Whatever the weaknesses of the report were, it appealed to a wider public who had gone through 25 post-war years of unprecedented growth of industrial production and broad-based affluence in the hitherto utopian consumer society. In parallel, however, the environmental problematic had become conspicuous as had social and economic problems of third world development. The simulation method of computer modelling was the latest thing in science (and at the time not actually very developed), while the message was classically biblical: a doomsday prophecy combined with an appeal to conversion.

There was at the time, however, no clear answer regarding what should be converted. The zero growth hypothesis of the model corresponded to a quite natural psychological reaction: first stop it, or even turn it back, i.e. negative growth or shrinking the economy. Luckily, this was beyond reality. Even if it had been possible, the zero growth run that did not lead to system collapse must have been some programming garbage. The reason is that if contemporary levels of consuming ecologically sensitive resources and burdening on sinks (on the basis of current, environmentally damaging technologies) were unbearable, then zero growth would also be of no use, because zero growth means nothing other than maintaining the current unbearable levels of consumption and retaining the untenable technologies of the time. Zero per cent growth instead of 5 or 10 per cent growth would merely defer doomsday a little longer.

The zero growth proposal nevertheless received some public attention. It was supported by old conservationists and the new ecology movement. They were opposed by the defenders of conventional economic wisdom and supporters of business-as-usual. There were, however, other participants in the discourse who recognised that traditional smokestack industries would indeed have to change structure and direction. There was the beginnings of an exchange of ideas, however half-baked, on selective growth, qualitative growth, organic growth, and decoupling economic growth from energy consumption.

More than a generation ago, any technological perspective was fiercely rejected by the green movement which was then still outrightly anti-industrial. Equally, the then industrial establishment had not yet considered the environmental aspects of technology. As a result, the growth debate contributed much to a growing awareness of the ecological and social 'world problematique', but largely failed to contribute to innovative problem solving.

2.2 RISK DISCOURSE

A similar statement could be made on the discourse over risk. It has contributed much to developing environmental awareness and scientific knowledge
about environmental problems, but has not been particularly creative in coming up with proposals on how to respond to the challenge, except, of course,
by assessing and communicating the challenges posed by various environmental risks. The risk discourse conceptualised environmental problems in
terms of dangers and risk, i.e. the probability of a potential danger to actually
occur and the extent of possible damage caused by this. Again a number of
publications played a discourse-defining role, among them Charles Perrow's
Normal Accidents. Living with high-risk technologies (1984) and Ulrich
Beck's *Risk Society* (1992).

The risk discourse revolves around assessing magnitudes of environmental problems, and assessing the contribution thereto of high-risk technologies as well as analysing the role of different groups of actors who contribute to causing environmental problems. Risk analyses may also stress environmentally benign effects of new technologies, for example the cleanness
of electricity generated by fuel cells. But a genuine concept of effectuative
problem solving or risk containment has not so far been developed in the
context of risk discourse.

The risk perspective triggered a wave of environmental risk assessment
studies in various fields. It has strengthened the role of disciplines such as
toxicology and epidemiology, and given a boost to technology assessment
studies that try to calculate effects of new technologies on the economy,
working conditions, labour market, lifestyles and the environment. After
three decades of research it cannot be denied that there is a certain disenchantment with technology assessment, although such studies as well as environmental risk studies continue to deliver useful results up to the present
day. The centres of concern at the time were the risks of nuclear energy as
well as health hazards by chemicals, for example DDT, or chlorine and its
derivatives, especially dioxins and furanes, or CFCs which are considered to
be the main agents responsible for destroying the stratospheric ozone layer.
In the meantime the centre of risk concerns has shifted to global warming
and climate change.

Since the late 1980s, the chemical industry pioneered risk communication.
The energy business was a bit more reluctant, whilst the agribusiness still
wishes to avoid public debate on its environmental problems even up to the
present day. Risk communication has become an element of a corporate strategy of environmental business management, including disclosure of data on

environmental performance, and trying to establish some regular communication between an industrial corporation and its stakeholders in order to improve image and regain lost confidence.

Similar to the growth debate, a number of scholars have questioned the new emphasis on environmental risks, arguing that what has increased is the perception of risk rather than the degree of risk itself. Among the group was the late Wildavsky (1995). He maintained that there have always been problems of survival, and that previous generations actually seem to have had more problems to be scared about, whereas today's well-off citizens in industrial countries just appear to be more scared about less. As Wildavsky declared himself, however, he did not want to deny that there are environmental problems to be dealt with. His retrospective studies on certain scares such as on DDT were not a reaction to scientific risk assessment, to which he himself wanted to contribute, but were a reaction to the spread of what he considered to be green ideology within the environmental movement and the public and polity in general.

In *The Sceptical Environmentalist*, Björn Lomborg (2001) reacted similarly to what he sees as the environmental 'litany': growing overpopulation, soil erosion and decreasing food yield, increasing undernourishment, depletion of resources, ever growing pollution, dramatically rising earth surface temperature and sea level, continued rapid destruction of rain forests, mass-extinction of species, etc. Lomborg does not deny that most of these issues do represent real problems. But he maintains that according to his figures the extent of those risks is regularly much exaggerated (except for overfishing), and he points out that most often there is not a linear trend, but S-shaped logistical growth curves, some of which are actually levelling out or have even made a turnaround for the better. This can be empirically shown, on the basis of official statistics, with regard to growth rates of population numbers, nourishment and public health, pollution of environmental media in industrialised countries, overall availability of resources and energy, or speed and extent of loss of forests and wildlife.

Such findings continue to contradict, and politically collide with, statistical disclosures from different sources such as the Living Planet Report, published yearly by the World Wildlife Fund International in cooperation with UNEP. Their outlook becomes gloomier every year. One reason for this difference in assessing environmental risks is the use of different methods. The WWF, for example, employs Wackernagel and his 'ecological footprint' (Rees/Wackernagel 1994). This is an aggregate measure that calculates environmental pressures of different kinds in standardised terms of land which equals a certain amount of resources and sinks. Energy engineers do something similar when they calculate energy demand in terms of BTUs (British Thermal Units) or standardised coal equivalents. According to Wackernagel and the WWF, humankind is already heavily overshooting its ecological budget, thus undermining its future, by causing a footprint of 1.2 at the present time, whereas only a footprint of <1.0 (i.e. all of the resources and sinks of the one earth we have) would be sustainable in the long run.

One may wonder about the ecological footprint. By using 1.2 earths instead of the one we have at maximum, it suggests that somehow we are doing what cannot be done. The footprint thus may serve as a good example of the methodological problems of risk assessment. Risk assessment involves modelling which tends to remain controversial, on the basis of assumptions that tend to remain questionable. The footprint of fossil energy, for example, is equated with the area of forests that would be needed if all of the CO_2 were to be absorbed by those forests. Non-forest and oceanic plant life are excluded from the calculation, and the present or some earlier status quo of atmospheric concentration of CO_2 serves as an implicit benchmark – two assumptions that might prove not to be very tenable.

2.3 ECOLOGICAL MODERNISATION

2.3.1 Integrated Problem Solving versus Add-on Measures

The risk discourse succeeded the growth discourse by specifying environmental problems and their causes. Similarly, ecological modernisation succeeded the growth discourse by specifying strategies of environmental problem solving. Ecological modernisation is a term coined in Europe in the early 1980s, but similar ideas have also been discussed and developed elsewhere (Mol 1995, 1997, Mol/Sonnenfeld 2000, Spaargaren 1997, Andersen/Massa 2000, Huber 2000). The modernisation discourse picked up promising aspects of the approaches of selective growth, qualitative growth, and decoupling of economic growth from materials intensity by increasing energy efficiency and resource productivity in general. How could this be done technically in industrial production, and what preconditions need to exist in order to get it done?

Among the key elements of the modernisation discourse during its first stage (from the early 1980s to the mid-1990s) was the distinction between integrated problem solving versus add-on measures end-of-pipe or downstream. By the term 'integrated' was meant 'in-process' or 'product-incorporated'. Sometimes the same distinction has been made by talking of pollution prevention 'at source' versus environmental technologies treating pollution 'after source'. Between the two poles of the spectrum there are various recycling practices of different reach. Examples of integrated problem solving not only referred to increased efficiency, such as increased fuel efficiency in heating systems or internal-combustion motors, but already included technological systems innovation, for example hydrogen as a clean fuel, regenerative energy and benign substitution of hazardous substances such as phasing out asbestos fibres in construction materials or introduction of phosphate-free detergents.

Scientists and engineers obviously have an important role to play in any modernisation strategy, as do managers and investors. Corporate organisation development has been reoriented towards environmental business management; this includes the setting of environmental business targets as well as reporting on a company's environmental performance, compliance with regulations, environmental auditing processes, operational aspects of emissions control and recycling, planning, research and development, marketing, environmental book-keeping and environmental training and motivation of personnel.

Another story-line of the modernisation discourse, which took a sceptical view of industry's endogenous ability and willingness to modernise ecologically, concentrated on the role of regulation and institutional capacity-building. This was called 'political' modernisation. Environmental policy, it was recommended, should more often look for an optimum mix of instruments and make use of liability legislation, negotiated settlements or voluntary agreements and various financial instruments, rather than just relying on bureaucratic command-and-control procedures of planning, application, admission, registering, licensing, or similar (Jänicke/Weidner 1995, Jänicke et al. 2000).

One of the basic aims behind ecological modernisation was to conceive of a way of environmental problem solving that would not fundamentally oppose industrialisation on the basis of science and technology. Instead, ongoing processes of industrial development and modernisation should be ecologically readapted by way of state-of-the-art technology itself. That type of reasoning, geared towards structural reform in general and technological innovation in particular, also influenced the European contributions to the Brundtland report and the pre-Rio process, i.e. the years before the Rio conference on environment and development in 1992. Since then, the strategy of ecological modernisation has become mingled with certain environmental aspects of the sustainability discourse.

2.3.2 Clean Technology

Since the mid-1980s, aspects of integrated problem solving have been at the core of the concept of clean technology (CT) or clean production (CP) (Hirschhorn et al. 1993). Some prefer to say 'cleaner' since, for example, burning gasoline in internal-combustion motors, however efficiently it may be done, will not be clean sensu strictu. CT/CP is certainly among the most important milestones towards a concept of TEIs, although the notion of TEIs goes further, or has become more specific and more differentiated in comparison to the notion of CT/CP as officially defined by the United Nations Environmental Programme UNEP:

> Cleaner production is the continuous application of an integrated preventive environmental strategy to increase overall efficiency, and reduce risks to humans and the environment. ...
> For production processes, CP results from one or a combination of conserving raw materials, water and energy; eliminating toxic and dangerous raw materials; and reducing the quantity and toxicity of all emissions and wastes at source during the production process.
> For products, CP aims to reduce the environmental, health and safety impacts of products over their entire life cycles, from raw materials extraction, through manufacturing and use, to the ultimate disposal of the product.
> For services, CP implies incorporating environmental concerns into designing and delivering services.
> CP ... is a broad term that encompasses what some countries/institutions call eco-efficiency, waste minimisation, pollution prevention ... The concepts of eco-efficiency and CP are almost synonymous. ... CP refers to a mentality of how goods and services are produced with the minimum environmental impact ... (www.uneptie.org/pc/cp, as of July 2001).

A more condensed version reads like this: CP is

> a conceptual and procedural approach to industrial activities that demands that all phases of the life cycle of a product or of a production process should be addressed with the objective of prevention or minimisation of short- and long-term risks to human health and to the environment (Clift/Longley 1995 174).

Definitions of CT reflect the context of the 1980s and early 1990s, particularly risk discourse, and, as described below, that part of the sustainability discourse that revolves around efficiency. A criticism of CT/CP seen from today's perspective would have to stress its one-sided and ecologically non-specified insistence on quantitative input-output efficiency, overall conservation, minimisation or even 'dematerialisation', neglecting qualitative aspects of metabolic consistency as discussed in the following chapter. That would require making certain difficult but necessary distinctions that take into account different degrees of ecological sensitivity of resources and sinks in different places at different times.

Furthermore, although CT/CP explicitly wants to reconcile industrial growth with the environment, it does not have a clear vision of the interrelation between nature and industry. It is partly still enmeshed in traditional ideas of conservation and idealistic non-interference, and has not yet fully accepted the process of ecological transition, i.e. colonisation and cultivation of the earth in accordance with human standards.

2.4 SUSTAINABLE DEVELOPMENT: CONSUMPTIVE SUFFICIENCY, OPERATIONAL EFFICIENCY AND METABOLIC CONSISTENCY

2.4.1 The Meaning of Sustainable Development

When using the term sustainable development reference is made to the meaning this term has taken on in the Rio process and its written documents. The Rio process refers to the ongoing international interaction between new social movements, academia, politics and business that has led to the formulation of environmental policy strategies in the context before and after the United Nations Conference on the Environment and Development in Rio de Janeiro in 1992. The Brundtland report (WCED 1987) belongs to the most important written documents of the Rio process, as well as the Agenda 21 (UNCED 1992).

Controversial as the Rio process may be, for example with regard to the Kyoto protocol aimed at reducing greenhouse emissions, all participants agree upon the threefold mission any politics of sustainability has to fulfil:

1. to promote industrial development and growth of real income, while
2. ensuring ecological sustainability by not exceeding the earth's carrying capacities, and
3. bringing about social equity (also called environmental justice) by creating a better balanced intra- and intergenerational distribution of opportunities to use natural resources and sinks, and by giving access to a fair share of the wealth produced.

Under the aspects of ecology and environmental justice, i.e. income distribution and equitable access to resources as well as an equitable sharing of environmental burdens, the sustainability discourse first revolved around two strategies both of which claimed to represent an 'integrated' innovative answer to the environmental aspects of sustainability: sufficiency and efficiency.

2.4.2 Sufficiency of Needs and Consumption

Sufficiency is an approach that responds to the challenge of limits to growth by self-limitation of needs and consumption. It is aimed at subsisting on less and thus reducing the overall level of consumption of resources, materials throughput and emissions output. Sufficiency represents a continuation of the idea of consumptive self-limitation from the growth debate, for example by adopting a lifestyle of voluntary simplicity.

Sufficiency and further approaches to sustainability can well be arranged according to the IPAT formula given by Paul and Anne Ehrlich (1990):

Ecological *I*mpact = f (*P*opulation, *A*ffluence, *T*echnology)

In terms of the IPAT formula, sufficiency can be achieved in two ways. One is to reduce population growth, if not the absolute size of populations; the other is to reduce the populations' level of affluence. Policies of birth control and population containment have been of some importance in world regions such as China, India, Indonesia, Middle America and Africa. In industrially advanced northern countries, where the point of demographic transition has long been passed and where native populations are actually shrinking (except the US), the ageing of the population has become a matter of concern. Moreover, an unlikely silent conspiracy between feminism (which declares a woman's womb to be exclusively her private affair), the Catholic Church (which declares pregnancy to be exclusively God's affair) and leftists (who consider population to be a right-wing issue) has made population policy an anathema on principle in most western countries.

So proponents of a sufficiency approach have preferred to concentrate on bringing down levels of affluence by demanding people be more self-sufficient. The intention of sufficiency is to remain within supposed limits to growth by doing away with rich-world consumerism in favour of a more modest way of living and a more egalitarian distribution of income, resources and sinks. Supporters of this kind of sufficiency approach to sustainability were NGOs representing certain currents of the social movements of the time, conservationists, third world solidarity groups, certain currents in the churches, and similar.

The sufficiency approach expressly calls for self-limitation of needs and consumption. A clearer notion of that emerged in a study called Sustainable Netherlands by the Dutch section of Friends of the Earth (Milieudefensie 1992). The study became a model for similar approaches in other European countries. Without wishing to oversimplify, one may say that the approach of these studies consists in adding up the resources and sinks available in the foreseeable future and dividing them by the number of living human beings. One thus arrives at per-capita quotas or, in other words, contingents of resources and emissions. Dutch citizens would be entitled to 80 per cent less aluminium, 45 per cent less agricultural land, 40 per cent less water and 60 per cent less CO_2 emission than they have today.

But for whom is this calculation equitable? It represents a crude programme of resource communism to be carried out by a centrally planned austerity regime. It confuses equality with equity. And it does away with any sense of justice which takes achievement into account. This is particularly so in the Netherlands where much of the land was not just there but was wrested from the sea by the hard work of successive generations. The programme does not give due consideration to certain ecological and geogenic facts of life, not to mention the political, institutional and legal preconditions of such a plan. Presumably, an attempt to put it into practice would itself not be very sustainable.

Sufficiency, as much as efficiency and consistency discussed hereafter, can indeed be considered as a strategy of innovation, though in a quite limited sense. Sufficiency represents a special type of social innovation, a cultural revolution, or rather 'renovatio', which attempts to restore value-priorities which are thought to have prevailed in Western culture during the middle ages, or to put in terms of Sorokin, the values of an ideational culture as opposed to sensate, utilitarian culture. It should be made plain what sufficiency means in terms of technology and human action: locally confined, at a low-tech level, slow speed, with little scope and choice. In practice it comes down to a programme of living on much less, at much higher expense.

Already in the mid-1990s, when the sustainability discourse took off fully, the idea of sufficiency was criticised for actually not giving a truly sustainable answer to the sustainability problem and for lacking more far-reaching technological perspectives. Sufficiency does not give an answer to the problem of big numbers and large volumes. It might be a perspective for a world population of several millions of people, not, however, for several billions. Even a hypothetical 50 per cent decrease in volumes under today's technological conditions would clearly continue to be non-sustainable in the long run. Besides, sufficiency tends to be a favourite idea among well educated lifestyle aesthetes. Among the majorities, however, North and South alike, whose attitudes are more than ever utilitarian and consumerist, the idea of 'living on less at higher expense' is completely out of place. So the sufficiency approach does not have enough resonance and support. However sustainable the sufficiency approach might be thought of by its supporters, its goals remain unachievable in social practice.

World population amounts to 6–7 billions at the present time. Most of these individuals persist in their desire for higher levels of consumer affluence. So we face the lasting challenge of providing large volumes in turnover of energy and materials, to be calculated on giga and tera levels rather than yesterday's kilo and mega levels. Seen from the angle of Ehrlich's IPAT formula, technology is more at our disposal than population numbers and levels of consumption are. Population numbers can hardly be changed at political will for a longer time. Also a lower level of consumer affluence would not really help, even if moral suasion or forced austerity could bring it down by half (which, however, is neither realistic nor desirable). This is because it would merely have the effect of doubling the length of the ecological reprieve, whose limits, in the first place, are determined by present-state big population numbers and unsustainable technologies of industrial age I.

2.4.3 Efficiency of Operations: Resource Productivity

The strategy of efficiency, in terms of the IPAT formula, does not care about the factors of population and affluence. It tries to improve on the factor of technology, or more generally speaking, on the structures and infrastructures of industrial production. The efficiency approach draws upon a special element of the modernisation discourse, i.e. expanding the limits to growth by increasing resource productivity in addition to labour productivity, thus producing a desired output on ever less resource input and emissions. Efficiency represents a special type of technical innovation that tries to harness productivity gains, i.e. optimised input-output ratios, brought about by incremental improvements in the learning course of a system's life cycle.

Efficiency gains are real and useful. With regard to ecological sustainability, however, the efficiency approach faces a problem similar to that facing the concept of sufficiency, this even if it were possible to increase industrial eco-efficiency by a factor of between 4 and 10 within a few decades.

First, the approach is wrong in assuming that efficiency gains would translate into absolute savings of resources and sinks. This is an illusion. Increasing per-unit efficiencies are a natural part of a system's learning curve which occur anyway. The desire to push efficiency on within the frame of an established life cycle resembles the vain attempt to push ahead the flow of a river. Not only is there not much need to artificially enforce efficiency gains that will come about anyway, but the function of those learning effects within a system's evolvement is to stabilise ongoing growth and volume maintenance of a system. This is known as the rebound effect: any increase in efficiency translates into further growth of volumes, for example increased motor car efficiencies translating into cars with bigger motors driving faster and more kilometres within an enlarged action radius. This can actually be seen as a general law – the growth-efficiency principle – that holds true in bio-organismic systems as well as in technological systems (II/7.7.1).

Second, when reaching maturity, a system's learning process approaches marginal utility sooner rather than later. Large efficiency gains can be obtained during the rise of new systems, whereas in mature systems an efficiency strategy can no longer lead very far. There are, however, a few exceptions to this. For example, it is possible to obtain considerably increased energy efficiency by better insulating houses. Saving heating energy by modernising the energy design of buildings is one of the rare success stories of an efficiency strategy. Another one is recycling, i.e. replacing virgin materials with secondary materials and thus decreasing the use of natural resources per unit of output, resulting in an increase in materials productivity. Recycling, however, cannot go on forever, because downcycling of materials can to a certain degree be put off, but ultimately not be avoided.

Third, beyond certain limits, efficiency can in some cases be ecologically counterproductive. For example, on-site water recycling may lead to high concentrations of pollutants in the water which in turn need to be neutralised by adding chemicals. If, to take another example, recycling of paper fibres is forced beyond certain limits, it becomes necessary to add chemical binding agents to the downcycled broken fibres in order for them to still be usable. Counterproductive effects of efficiency forced beyond reasonable limits can perhaps best be studied in conventional agriculture, which among modern industries has progressed furthest along its learning curve. Increases in efficiency there have been brought about by an ever increasing intensification of plant and animal production, finally resulting in a type of hyper-efficiency which, seen from an ecological point of view, represents disastrous over-intensification at the expense of both product and environmental quality.

Finally, increasing the efficiency of processes of production and use is beside the point if a process or product is ecologically unsafe and unsustainable by its very nature, for example nuclear fission, fossil fuels, highly hazardous chemicals and unadapted agriculture. If fossil fuels really prove to be the main culprit behind anthropogenic climate change, then it makes little sense to further increase the efficiency of fossil fuels-burning technologies. In this case efforts and investment should be concentrated on innovating clean energy such as solar and hydrogen technologies.

Sufficiency and efficiency, both in their way, are biased towards the quantitative side and do not take into consideration the qualitative side of structural change and of the properties of the industrial metabolism. In technological terms, the sufficiency approach is in actual fact anti-industrial in that it misleadingly tries to handle the problem by cultural restoration, political austerity and economic redistribution. The efficiency approach, by contrast, nurtures a specific technical illusion, which is that unheard of efficiency gains might allow the maintenance of high levels of industrial output by causing physical input to 'dematerialise' – which of course will not work for the three reasons given above: (1) Efficiency increases are limited by technology life cycles reaching marginal utility, (2) efficiency gains are outweighed by rebound growth, and (3) efficiency increases are beside the point when they come with counterproductive effects or represent metabolically inconsistent progress in the wrong place.

All in all, the 'efficiency revolution' represents a rather strange piece of political rhetoric. It is in fact highly conservative because it tries to hold on to outworn and ecologically often untenable systems by incremental low-margin high-cost upgradings. A typical example of this is the use of fossil fuels in internal-combustion engines or thermal power stations, instead of bringing about regime shifts to new systems that are ecologically much less problematic or even benign by their nature.

2.4.4 Metabolic Consistency

The shortcomings of the approaches of sufficiency and efficiency were the starting point of the concept of eco-consistency or metabolic consistency (Huber 1995, 2000, 2001), sometimes also referred to as eco-effectiveness (Rossi 1997, Braungart/McDonough 1998, 2002). The sustainability discourse can in fact not reach beyond certain limits and will repeatedly get stuck, as in the case of the climate convention and the Kyoto Protocol, unless the sustainability perspectives of sufficiency and efficiency are widened and complemented by the environmental innovation strategy of metabolic consistency or eco-effectiveness.

In terms of the IPAT formula, metabolic consistency also starts with the factor of technology. It focuses, however, on the qualitative side of technology, not just input-output quantities within given structures. Metabolic consistency is about how to re-embed the industrial metabolism within nature's metabolism by changing technological structures and infrastructures, thus also the metabolic qualities of products and processes, rather than mere quantity of turnover. For example, energy demand on giga and tera levels may not be an ecological problem as such if the energy were ecologically clean and based on clean fuels, natural kinetic flows, sun radiation and geothermal heat. The consistency approach seeks to change structures by innovating new systems and products, new regimes and practices, in order to make them more compatible with natural systems, rather than reducing turnover within old structures. A product or process, from the outset, ought by itself to be compatible with natural cycles and ecosystem-homeostasis, even at large volumes of turnover, or it ought to be reliably controllable within a closed technical circuit before being reprocessed and released back into the natural environment.

The notion of ecological or metabolic consistency does not in principle include the idea of 'dematerialisation', though it certainly is not an invitation to wastefulness either. In contrast to a conventional understanding of environment-friendly products and processes, a core goal of which usually has been the absolute reduction of energy demand and materials turnover, eco-consistent TEIs are aimed at enabling human systems to grow and unfold, not to shrink and vanish. Empirical life cycle analyses of energy demand clearly indicate further increases in energy demand, partly through the growth of demand in newly industrialising countries, partly because the whole world system is entering into a new long-term follow-up cycle of energy growth. What is actually needed is not less energy but more, and what we therefore need is metabolically adapted energy, i.e. from sources that are compatible with nature's metabolism. A sound living system does not shrink, and if it is about to enter a final stage, this occurs on the highest level of

structural unfolding and quantitative turnover which the system in consideration is able to achieve (II/7.7.1).

Ecological or metabolic consistency must not, however, be misunderstood
as an invitation to unrestrained exploitation of natural resources. Truly eco-
consistent technologies do not need much denaturation of sensitive resources.
Eco-consistent energy, for example, can be kinetic or radiant clean energy,
i.e. fuelless energy, or hydrogen fuel derived from water, requiring neither
that landscapes be excavated or turned upside down nor the destruction of
ecosystems over wide areas in order to extract fuels which then pollute sinks
when being burned. So, with regard to the resource input into production
processes, metabolic consistency equals ecosystem compatibility. Metabolically consistent technologies are capable of using ecosystems without destroying them. Metabolic consistency means fitting into geo- and biospheric
dynamics without creating severe perturbations.

When put into a context of industrial metabolism, the idea of materials recycling also can be reframed in a more comprehensive way so as to encompass the cycles and cycling of all materials in anthropogenic *and* geogenic
materials flows. This leads in turn to the notion of industrial ecology, which
refers to the entire metabolism of physical input-throughput-output caused by
modern society (Socolow et al. 1994, Bourg/Erkman 2003). Restructuring
the industrial ecology, however, is not just about materials recycling. It also
includes, for example, product design for environment (I/4.1).

In retrospect, those more far-reaching approaches can be seen as the beginning of a new story line in the environmental discourse, i.e. the environmental innovation discourse (Klemmer et al. 1999, Hemmelskamp 1999,
Hemmeskamp et al. 2000). The discourse on environmental innovation, and
thus TEIs, is now gaining momentum as the sustainability discourse seems to
have exhausted itself in so far as it has revolved around consumptive sufficiency and operational efficiency, whilst neglecting metabolic consistency.

Sometimes, eco-consistency is understood in a foreshortened way as a
strategy that wants to mimic nature. Bionics has indeed much to contribute
to a consistency approach, for example applying the lotus effect through
mimicking the surface of the lotus leaf which is self-cleaning by not allowing
airborne particles to take hold on the surface. Scientists and engineers who
have been successful in mimicking nature in individual cases deserve our
admiration. Even so, in most cases our technological intelligence does not
appear to be advanced enough yet to equal nature's ingenuity. Moreover, a
strategy of metabolic consistency is different from bionics in that it involves
artificial systems that do not exist outside the anthroposphere. A strategy of
metabolic consistency intends to apply any useful tool which modern science
and technology may offer in order to restructure the industrial metabolism. In
many cases this will continue to include completely new artificial techno-

systems and materials with no precedent outside the anthroposphere. Humans are after all creative creatures.

A blatant misinterpretation of metabolic consistency is to accuse it of representing a technocratic attitude which is looking for a technical fix. As a matter of fact, leaving all ideas and ideological preferences aside, any sustainable answer to ecological problems will have to include technology and will ultimately be carried out by technology. Sufficiency too comes with certain technological implications, i.e. downscaling to smaller size and much lower production capacities, dismantling big industrial corporations and transnational industrial cooperation, retardation and de-specialisation of production which would all result in fewer options, lower levels of productivity and accordingly low living standards. Some sufficiency protagonists would even like to restore pre-industrial elements by opting for a small farmer economy with only a local reach to replace continent-wide or global agribusiness. Similarly they propose a shrinkage of transport, a bigger role for crafts compared to factories, and an extended role for unpaid work in the informal economy in order to compensate for shrinking paid employment. Such is the, not even hidden, industrial and technological policy side of a sufficiency approach.

Ecological problems are perturbations of a population's metabolism within its living space. Environmental problems of industrial society are perturbations of the industrial metabolism within the earth's ecosystems all of which have become part of the human living space or are at least being touched by it. The industrial metabolism, rather than relying on traditional physical labour, is actuated by technical work, i.e. work that has been technologically reshaped or which is in some way carried out by machines and technical energy. Furthermore, technisation of industrial work has proceeded increasingly on the basis of scientific methods. Therefrom, ecological re-adaptation of modern society can in the final analysis only be achieved by structural change of the industrial metabolism through scientific and technological innovation.

Technology, though, is a societal medium. It belongs to the *operative system*, representing one of three effectuative subsystems of society, the others being the *economic system* of price-regulated markets and monetary and financial processes, and the *ordinative system* of public government, administration and private management based on law and on specified authorities which have the power to issue directives. The price-related impulses of the economic system and the law-based impulses of the ordinative system are control factors, which control the operative system from the outside and complement endogenous techno-evolutive conditions (II/6.2).

Moreover, all of the three effectuative systems are in turn directly controlled by formative factors from cultural and political subsystems of society, particularly the knowledge base (cognition), the value base (evaluation or judgement), formation of styles (expression) and processes of will-building and decision-making (conation). Most sociologists today describe technology as being socially embedded, i.e. as an integrated societal subsystem interconnected with other functional subsystems. Technological innovation does not come as a solitary event. It necessarily happens in co-evolution with co-directional processes in other subsystems. New technology thus involves new knowledge and new or reinforced attitudes and worldviews, particular styles of working and living, promoters and institutional carriers, legal and regulatory conditions, investment and markets with effective demand, all in addition to technical feasibility itself. Understanding technology involves some extended understanding of society, and conversely modern society cannot be understood without understanding technology.

2.4.5 Sustainable Technology: Improving Metabolic Consistency at an Optimum Level of Efficiency

In technological terms, the sufficiency approach was not enthusiastic about innovation. It would prefer to stick to what already exists and reduce that to much lower levels of production capacity and volumes of turnover. The strategy of efficiency, by contrast, is in favour of technical innovation, at least in principle. It tends to be conservative, however, in that it prefers to concentrate on increasing the efficiency of existing products and processes by incremental innovations in later stages of a long-established life cycle. Such innovations tend to be of comparatively minor importance, although incremental improvements of major importance may occur as well. Normally, efficiency increasers are rather risk-averse when it comes to changing path. They are averse to structural change which demands that we give up old habits and venture into the less familiar or even into the unknown. Taking some entrepreneurial risk, by contrast, is exactly what a strategy of metabolic consistency calls for. Any strategy of metabolic consistency will have to be based on generic TEIs that open up new paths by introducing new systems or system components, upstream rather than downstream in the technological life cycle and the product chain as discussed in the introductory chapter.

The concept of ecological consistency of the industrial metabolism, brought about by TEIs that are run at an optimum level of efficiency according to their stage of structuration, can be seen as the core element of an advanced programme of ecological modernisation (Andersen/Massa 2000a and b). The concept of TEIs indeed represents a further step in the unfolding of the ecological modernisation discourse. It is not only broader but also more specific and selective than the concepts of integrated problem solving and clean technology previously were. This is because such an advanced understanding of ecological modernisation, without wishing to preclude valid options, comes with clear priorities and does not include just anything.

Instead, a clear ranking order emerges which puts consistency before efficiency, and efficiency before sufficiency. Excluded from this are enforced efficiency (it does not make sense to push ahead the flow of a river) and progress in the wrong place (it does not make sense to be efficient in doing things that are by their very nature unsustainable). Also excluded is enforced sufficiency which would turn friends of the earth into enemies of the people.

Generic innovations of systems and components which improve ecological consistency come first, for example clean-burn technologies, biotechnology, eco-consistent agriculture, and many more as will be described in I/4. Consistency certainly does not neglect efficiency. In actual fact the two accompany each other along the learning curve of a technology's life cycle. Nor do the two of them deny the moral virtues of a modest way of living. There are, however, different degrees of feasibility as well as a clear factual precedence, as opposed to normative preference. First and foremost, it is essential that technological systems are or can be made metabolically consistent on quite a high level of turnover. This is necessary because we have to start from the fact that 6–8 billions of humans expect material living standards on quite a high level. That is why first priority must be given to systems innovations which are metabolically consistent.

Mere efficiency-increasing measures come second. They are not considered to be unimportant, particularly in heating/cooling, where efficiency gains can still be considerable. Increased efficiency, however, develops naturally in the course of a technology's learning curve. Efficiency ought to be optimised according to the life-cycle stage, not forced to the maximum, nor indeed pursued at all in cases of metabolic inconsistency. It should not be a priority to push efficiency increases further than they will naturally reach by themselves, because cost-benefit relations tend to be non-optimal in the majority of such enforced cases. In earlier stages of a system's evolvement efficiency gains can be obtained more easily and to a larger extent than during later maturity and retention where the marginal utility of further efficiency increases declines. So it is important to be sure about the remaining potential for further reasonable efficiency increases. That is why mere efficiency-increasing measures should no longer be pursued at all in mature technologies near or beyond marginal utility as is the case, for example, with gasoline-fuelled internal-combustion engines in cars.

Ecological and economic utility are not always identical. Increasing the efficiency of processes for cleaning exhaust gas from 90 to 95 per cent is in most cases wasted effort both in ecological and in economic terms. In other cases ecological and economic utility are not identical. The predominant practices in contemporary industrialised agribusiness clearly demonstrate how an increase in economic efficiency can represent ecologically devastating over-exploitation. So the most important thing indeed is to be sure about the metabolic consistency of the target in order to avoid progress in the wrong place. It is pointless to strive for greater efficiency in technological systems or products that are metabolically inconsistent by their very nature, for example nuclear fuel which will remain highly poisonous for many thousands of years. That efficiency comes second to improving consistency is a matter of fact rather than of ideological preference.

Sufficiency- and efficiency-oriented changes in consumer practices come third. As explained by the ecological paradox of consumption (I/5.1), environmentally concerned consumer practices can in most cases only make a marginal contribution. Voluntary sufficiency comes last, because it has in fact very little to contribute ecologically, whatever its mental and cultural achievements may be. Enforced sufficiency, i.e. bureaucratically imposed central planning of resource consumption, must be ruled out. It would be completely at odds with the political consensus of a society which gives priority to civil liberties rather than to the rule of statism and corporatism.

Nature, though it obviously economises in its physical and biological intra-systems routines, is lavish, productively creating and wastefully destroying with regard to external systems dynamics. So rigid adherence to over-generalised sufficiency principles of minimisation of everything ought to be ruled out. Over-generalised sufficiency principles appear on the left side in the following semantic profile:

Small scale	versus	large scale
Decentralisation	versus	centralisation
Local	versus	global
Short distance, low speed	versus	long distance, high speed
Natural	versus	artificial
Non-interference with nature	versus	high-impact interference
Minimisation of materials	versus	turnover in large volumes.

Undoubtedly these dimensions are often related to certain ecological problems, particularly under conditions of carbon-based energy and highly hazardous practices in agriculture, chemistry and basic materials processing. But rigidly maintaining a one-sided position at either pole will result in some sort of disorienting ideology which can give no useful answer to any practical problem. Given the numbers of people and the high living standards they expect, some in-between tendency which leans towards the right side of the spectrum is unavoidable – and will have to be rendered possible by metabolically consistent TEIs.

Against this background we cannot sign up to a number of simplistic concepts derived from the sufficiency approach, particularly that of dematerialisation, de-chemicalisation, or that of traffic avoidance. Dematerialisation is far too unspecified, i.e. metabolically unqualified from a consistency point of view. The idea needs to be specified at least in three ways: as decarbonisation of the energy base, as disuse of hazardous high-impact chemicals outside industrial closed-loop applications, and as recycling of materials such as water, auxiliary process materials, metals, other minerals and fibres in those cases where recycling makes sense. Positively speaking, decarbonisation relates to the vision of clean energy, i.e. clean fuels and fuelless energy. Disusing hazardous chemicals relates to benign substitution, the introduction of biotechnology as well as sophisticated mechanical processes replacing hazardous chemical ones. A generalised perspective of de-chemicalisation is simply unfeasible, besides being undesirable. A sensible vision, by contrast, would be that of low-impact chemistry, or green chemistry as it is called in the writings by Anastas, or chemistry of restrained reach as conceived of by Scheringer (2002). Traffic avoidance is a similarly unqualified goal, just as are all other unspecified minimisation slogans. Traffic can be optimised, but avoiding it would be tantamount to disrupting modernity itself which is based on interactive dynamics at high speed over long distances. What the idea reasonably comes down to is clean vehicle propulsion, vehicles designed for disassembly and recyclability, and optimisation of traffic and its modal split. These and further perspectives of metabolically consistent and efficient TEIs are explained in more detail in I/1.4.

In a recent paper commissioned by the EU research directorate (EU Commission Community Research 2001), a group of authors has tried to 're-invent' sufficiency by putting forward ideas under that heading which would have to be described as aspects of metabolic consistency and socio-technical use-efficiency. The previous definition of sufficiency was extended in that EU paper so as to encompass elements such as new energy systems, clean production methods, new industrial products; i.e. technological innovations that would pertain to a modernising approach of metabolic consistency, or practices of marketing and leasing, for example car-sharing, or leasing of household durables and clothes. These represent examples of efficiency increase in that they optimise the use of products to fuller capacity and increase the circulatory efficiency of materials. Even ecological modernisation is declared in the paper to be an aspect of 'sufficiency', although ecological modernisation has always been identical with a combined strategy of metabolic consistency and operational efficiency according to the logic of innovation life cycles (Andersen/Massa 2000b, Huber 2000).

The authors of the EU paper equate sustainability with a catch-all notion of 'sufficiency'. It ought to be clear just how confusing such a 're-invention' of the sustainability discourse would be. It falls behind already achieved distinctions between different approaches to sustainability, i.e. (1) consumptive sufficiency as a strategy of limiting and possibly bringing down levels of consumer affluence, whereas (2) operational efficiency aims at increasing output-input ratios, and (3) metabolic consistency tries to improve on the physical qualities of the industrial metabolism in order to reintegrate and better fit in again with the metabolism of earth's ecosystems. Without such distinctions and a clear ranking of priorities among them one would have to deal with a mere hotchpotch of many various activities and measures without making any difference regarding their cost and benefit, potential and importance.

3. Environmental Innovation Discussed in More Detail

3.1 AN EXPANDABLE DEFINITION

According to Klemmer et al. (1999 25) environmental innovations are

> a subset of innovations that lead to an improvement of ecological quality. This is working definition of environmental innovation; it encompasses any innovation which serves to improve the environment, regardless of any additional – economic – advantages.

Particular reference is made to new products, materials and production processes that economise on energy, resources, land, emissions and waste; also to managerial procedures, institutional arrangements, regulatory instruments, as well as to changes in people's normative orientations, consumer behaviour and lifestyles. Taking this working definition as a starting point, it can be further specified in three ways:

1. Innovations should be more precisely differentiated according to systemic function (II/6.2), particularly

- innovations in the operative subsystem, i.e. technologies and technological regimes
- effectuative controls innovations of economic and ordinative (i.e. legal and administrative) content
- formative innovations of cultural and political content.

This includes a clear understanding of how ecological effects come about: only by technology and work, i.e. physical activity transforming matter, thus causing metabolic effects. Effectuative controls innovations, and formative innovations with regard to value orientations, normative preferences, will-building and decision-making, are not ecologically effective by themselves, but gain sense and purpose only by controlling and effectuating technical and behavioural choices. Seen from an ecological point of view, the focus is on the industrial metabolism in the first instance, i.e. on transformation of materials by technology and work in activities of production, or consumption respectively, hence on technological environmental innovations (TEIs). Economic and ordinative controls need to be specified according to TEIs. By contrast, ideas, value orientations, the knowledge base, communication styles, will-building and decision-making are formative factors which lay the foundations of any effectuative function. In a certain sense they can be said to represent impulses prior to laws, money and technology, but – to use that metaphor of the information age – no formative 'software' can in fact be put in lieu of effectuative 'hardware'. The idea of clean energy can inspire and orient action but cannot be a substitute for technology which makes the idea of clean energy become reality.

2. Within the category of operative innovations, and as discussed in the introductory chapter, technological upstream innovations are much more relevant than changes downstream life cycles and production chains. To put it in terms of the sustainability discourse, ecological impact for the better can in effect only be brought about by major technological innovations upstream in the product chain and technology life cycles, whereas the contribution of environmentally-oriented user and consumer behaviour, by comparison, is of educational rather than immediate practical importance (I/5.1).

3. The above definition of environmental innovation is still one-sidedly oriented towards ideas of 'efficiency revolution' and 'dematerialisation' in the sense of economising on everything as much as possible regardless of whether it makes ecological sense. The definition fails to explicitly stress the utmost importance of generic technological systems innovations that contribute to ecological consistency of the industrial metabolism.

In economics and organisation theory there is a distinction between 'orientation to efficiency' versus 'orientation to innovation' (Clark 1987 132). In economic terms this is about the difference between cost saving within maintained old structures versus improved quality and new qualities through some sort of significant structural change. Thus, the crucial point in the recent discourse on environmental innovation is indeed TEIs, i.e. generic systems and component innovations that improve metabolic consistency at an optimum level of efficiency.

3.2 ENVIROTECH AND TEIS

Environmental technology (envirotech) and technological environmental innovations (TEIs) should not be confused. Envirotech can be seen as belonging to the whole set of TEIs, representing, however, just a certain part of it. Typically, envirotech is add-on (such as end-of-pipe filters, catalysers, downstream waste and sewage treatment) or involves specially designed instruments for environmental monitoring, supplied by the envirotech industry that has become a multi-sector business of its own. TEIs that are not envirotech, by contrast, are product- and process-integrated and reach far beyond envirotech. TEIs can in principle occur with any technology having a relevant ecological impact.

Classifications of envirotech, and TEIs in general, mostly start from that difference between add-on measures and integrated solutions as introduced since the 1980s in the first stage of the discourse on environmental modernisation. Tables of types of environmental innovations, as given, for instance, in Kemp (1997, Kemp et al. 2001) read as follows:

- End-of-pipe processes, pollution control technologies
- Recovery and recycling
- Waste management and sewage water treatment
- Clean-up technology (remediation)
- Product-integrated innovations, i.e. cleaner products
- Process-integrated innovations, incl. logistics, i.e. cleaner processes
- Energy-saving products and processes
- Measurement and monitoring
- Organisational measures (e.g. environmental audits of production sites).

With special regard to the chemical industry, Becker/Englmann (2001) distinguish end-of-pipe innovations such as multi-step effluent treatment from production-integrated innovations such as new syntheses, new catalysts, new solvents, and changes in input materials.

Obviously, the defining criterion is indeed whether an action represents add-on measures or integrated problem solving. Recycling represents something in-between. Organisational measures belong to a further group – that of controls innovations – and also include regulatory innovations such as standards setting, administrative procedures of registering, licensing, applying for, data reporting, disclosures, etc., and economic measures such as subsidising, taxing, emissions trading, etc.:

Technical add-on measures
- End-of-pipe processes, pollution control technologies
- Waste and sewage treatment
- Clean-up technology (remediation)
- Recovery and recycling

Integrated technological problem solving
- Cleaner energy
- Product-integrated innovations, cleaner products
- Process-integrated innovations, incl. logistics, cleaner processes
- Measuring and monitoring (technical controls measures), incl. eco-auditing

Non-technical effectuative controls measures
- Organisational measures such as corporate environmental management
- Regulatory innovations such as standards setting, administrative procedures of registering, licensing, applying for, data reporting, disclosures, etc.
- Economic measures such as subsidising, taxing, emissions trading, etc.

What is still missing here is a further specification of integrated technology. 'Integrated' was a demarcation against add-on and downstream. Framed in more positive terms, however, it can refer to quite different things. The clean technology approach, as discussed above in I/2.3, was a first attempt to become positive about the concept of integrated problem solving. The clean technology discourse, however, has not yet recognised a basic distinction between clean technologies: namely whether they increase efficiency or improve consistency or achieve both at the same time.

The notion of TEI includes technological principles and processes as much as materials and products. It also includes the rules, practices and co-operative routines of a technological regime (II/6.3). For example, the practice of crop rotation in agriculture is based on such a regime rule. Crop rotation is neither a device, machine nor infrastructure, nor a seed nor other material. It is an operative procedure incorporating know-how, i.e. knowledge about how to make effective and efficient use of things. So-called management rules of ecological sustainability ('Keep within carrying capacity, especially regenerative capacity', etc.) also represent regime rules, albeit at the overall level of general guidelines.

Still more far-reaching aspects of a technological regime, however, for example aspects of organisation in a broad meaning, referring to institutional capacities as well as various ordinative and economic controls functions, are not included in an appropriate notion of TEIs. Those regulatory and money-related factors are part of TEIs' selective context conditions. Regulation, on the other hand, tends to have an all-effectuative reach, i.e. it is not just about administrative procedures, but also includes highly specified technical and environmental standards that define industrial processes, materials and products in operative terms. These standards, as far as they define operative procedures, properties of materials and emissions, and the design of products in a specified way, are a subset of technological regime rules and thus also pertain to the notion of TEIs.

In terms of ecological discourse lines, TEIs can be characterised as being the operative tool-set of technological regimes aimed at ecological modernisation, i.e. structural change towards benign or at least strain-relieving effects on resources, sinks, ecosystems and the biosphere. TEIs eliminate, or reduce, or help to control environmental hazard (risk). TEIs create metabolic consistency and optimise eco-efficiency. TEIs can be add-on as well as integrated, although integrated solutions are in most cases preferable from an ecological point of view.

In II/7.3, key technologies are characterised as being pervasive innovations in that they serve as a starting point for a greater variety of new technical processes, products and practices in a broad range of applications. Typically, key technologies are the basis of long waves of industrial innovation, such as steam power, rail, electricity, motor cars, semiconductors and computers, microengines and nanotechnology. There are only a few key technologies of such structural importance at any time. With TEIs it is the same. One would have difficulties in finding many examples of key environmental innovations; 'key' in the sense of being pervasively key to many follow-up applications. Examples of TEIs fulfilling the criterion of being 'key' would perhaps include photovoltaics, photonics and nanotechnology, and certainly include hydrogen replacing fossil fuels, as well as electrochemical fuel cells big and small replacing conventional furnaces and combustion engines.

There is, however, one outstanding field of technology today which represents a Schumpeterian Basisinnovation indeed, one which is laying the foundations for many new businesses and markets. It will forseeably play a big role in the next long wave which will succeed today's 'new economy', and is fundamentally transforming technologies in medicine, agriculture, chemistry and materials. That key innovation is transgenic biotechnology. Besides clean energy, it is biotechnology that has emerged from the database of TEIs as today's seemingly most important TEI which is 'key' to very many things.

Biotechnology will no doubt have to be considered to represent an important field of TEIs, full in the knowledge that such an assessment will certainly remain controversial for another one or two decades to come. As tends to be the case with true key innovations of major structural importance, there is strong conservative opposition to biotechnology, including opposition from fundamentalist activists who bedevil transgenics; this notwithstanding the fact that there may be some applications that will indeed turn out to be untenable, as again is the case with any important innovation.

Beyond the notion of key innovation sensu strictu, general usage tends to call any new technology of some importance a key innovation, for example, in the field of TEIs, fully automated waste separation, membrane technologies in chemical production, lightweight ultra-strong compound materials, water-based non-toxic dyestuff, thermopaints, radiation treatment of water, transparent insulation or GPS-based precision tilling.

Furthermore, and this is an additional meaning specifically related to environmental innovation, there are key environmental problems; they are key in that they create an urgent need for action, and TEIs differ according to their potential for dealing with key environmental problems. Wind generators, for example, represent high tech, but they are not a key technology; environmentally, however, they contribute to solving the urgent key problem of airborne pollution and anthropogenic climate change, though their share in doing so is relatively small.

In theoretical principle, a key environmental innovation could be incremental, but in fact it will almost certainly constitute a generic innovation at the system or component level, or a redesigned new product generation, and these innovations may perhaps even be disruptive (II/7.3.4–5). Similarly, a key environmental innovation can in principle be add-on, as it happened to be in the case of a number of epoch-making technologies such as sewage water treatment or exhaust-gas catalysers. More often, however, and more systematically, it will be upstream problem solving which improves metabolic consistency and increases efficiency.

Any technology comes with some kind of environmental impact – a damaging or benign impact, to a low or high extent. A TEI is new technology, but a new technology in itself does not automatically constitute a TEI. New technologies may even be overtly non-TEI such as ABC weapons, or vast arrays of high-frequency towers, or deep-sea fishery fleets of world-wide reach armed with giant fishing nets and equipped with high-tech capabilities to detect shoals of fish.

A key technology that was originally thought to be of some environmental importance is information and telecommunication technology (IT). So far, the expectations have proven to be exaggerated or even wrong. Applications of IT in environmental measuring, monitoring and control certainly represent additions to the database of TEIs, but IT in general cannot be classified as a TEI. In contrast to original expectations, IT neither substitutes for physical transportation nor does it save electricity, or paper or any other materials. Efficiency gains of IT have been transmitted into rebound effects of further growth in physical turnover, both in mobility of goods and passengers as well as volumes in telecommunication and data processing. The IT industry itself, furthermore, is characterised by a number of common environmental problems related to hazardous chemicals, toxic waste and waste water.

3.3 LOOKING FOR ECOLOGICAL CRITERIA OF ENVIRONMENTAL INNOVATION

3.3.1 EHS Requirements

Which are the ecological criteria that qualify a new technology as representing a TEI? In the chemical industry, there is a set of such criteria known as the industry's EHS requirements. They crystallised during the 1980s. EHS is the acronym for Environment, Health and Safety. Originally, the order of the words was the other way round (safety, health and environment) with that original order indicating some kind of historical sedimentation. Concerns about operational safety arose right at the beginnings of the chemical industry in the 19th century, later on followed by growing insights into health hazards. Environmental concerns in today's sense are more recent, not having arisen before the late 1960s and 70s. The now accepted order of the terms also represents an order of comprehensiveness. Environmental concerns subsume health issues as well as operational safety, as health care subsumes operational safety when employees come into more or less direct contact with substances. Environmentally adapted processes and products could thus be characterised as being operationally safe, i.e. not dangerous in production and use, not detrimental to health, and not harmful to the environment with regard to depleting resources, overburdening sinks, and damaging biosphere and ecosystems. The development of the EHS requirements as an explicit basis of what was later called 'environmental management' took place within the context of the risk discourse and attempts at technology risk assessment.

3.3.2 Environmental Problems and Urgency of Action

TEIs differ according to the kinds of environmental problems they address. Table 3.1, without claiming to be comprehensive, gives an overview of current environmental problems. TEIs, it can be said at the outset, are technologies that contribute to environmental problem-solving in one or several of the problem categories as listed in Table 3.1.

The Society of Environmental Toxicology and Chemistry SETAC, a global body working on product chain analyses (eco-balances), has compiled a list of important environmental effects. Products and related production processes are considered to be the more ecologically adapted the less they contribute to the problems listed as follows:

- Global warming
- Ozone depletion
- Human and environmental toxicity
- Acidification
- COD discharge (chemical oxygen demand in water)
- Photochemical oxidant formation (for example ozone near ground)
- Space requirements, nuisance (smell, noise)
- Occupational safety
- Final solid waste being hazardous or non-hazardous
- Effect of waste-heat on water.

A similar list by the International Standards Organisation (ISO), compiled for purposes of environmental auditing, contains

- Abiotic resources (limited availability)
- Biotic resources (sustainable vs. non-sustainable use)
- Land use
- Global warming
- Stratospheric ozone depletion
  (continued on next page)

- Photochemical oxidant formation
- Acidification
- Eutrophication
- Ecotoxicological and human toxicological impacts (OECD 2001c 30).

According to Ayres (1993 172), technologies and the industrial metabolism can be called ecologically sustainable to the extent they fulfil the following criteria:

- no anthropogenic contribution to climate change
- no net increase in the acidity of the environment
- no net accumulation of toxic heavy metals, radioactive isotopes, or long-lived halogenated chemicals in soils or sediments
- no net withdrawal of groundwater
- no net loss of topsoil
- nor further loss of wetlands, old-growth forest, and biological diversity.

Such listings are bound to change and have added items to them in the course of time, depending on the state of ecological knowledge and environmental policy priorities. Twenty to thirty years ago, pollution problems of air and surface water had top priority. The urgent need for action they represented could be seen and smelled. Much of that has been dealt with in industrial countries, bringing down levels of sulphur, lead or soot in the air, or chemical oxygen demand in rivers. This, however, has not been an answer to growing amounts of toxic waste or greenhouse emissions. Moreover, it has created some problem deference in that it has replaced air- and water-borne pollution by problems of toxic waste resulting from add-on treatment of exhaust gas and sewage water.

These examples point to unsettled problems of relative urgency and relevance to the present. The above listings are not explicit about whether they represent a ranking, although this might implicitly be suggested in the lists of SETAC and Ayres. Any such ranking, necessary as it is for policy agenda setting, will remain a problematic endeavour, as epidemiologists, toxicologists and researchers into eco-balances are quite aware. How can drug deaths be compared with deaths through traffic accidents? (Although we do know, at least, that far more deaths are caused by smoking and bad nutritional habits.) For the time being, nevertheless, the above listings may suffice. They can be seen as being compatible and complementary, and they represent broad-based up-to-date international expert opinion.

One may have reservations about prevailing expert opinion because in principle it is subject to the same mechanisms of large group dynamics and issue cycles as public opinion is. Although scientists are not expected to replace truth by majority vote, truth is a tricky concept, relating to multiple

Table 3.1   Current environmental problems

Air, climate (atmosphere)
Anthropogenic global warming and intensification of natural catastrophes, ozone near ground, loss of stratospheric ozone shield, smog, acid rain, air pollutants
Water (hydrosphere)
Pollution and eutrophication of surface water, contamination and scarcity of drinking water, contamination of groundwater, loss and lower levels of groundwater (karstification, desertification), marine pollution by permanent releases of phosphor, oil, tar, acids, alkaline solutions, heavy metals, etc.
Soil (pedosphere, lithosphere)
Soil contamination through agriculture and diffuse airborne immissions, acid soil, existing industrial waste deposits, erosion, karstification, desertification
Landscapes and biocoenoses
Destruction by mining, quarrying; soil sealing, land denaturation and spoiled, disfigured landscapes through urban sprawl
Flora (botanical biosphere)
Reduction of crop variety by monocultures, loss of biodiversity, deforestation, forest damage by acidification of environmental media
Fauna (zoological biosphere)
Extinction of species, incorrect animal medication, inappropriate and over-intensified animal farming, experimentation on animals
Radiation
Exposure to radioactivity, high-frequency electromagnetic radiation ('electric smog')
Further spoiling of human environment
Noise, foul smelling air
Health hazards to humans and other organisms
Hazardous substances in the food chain, water and air, resulting in manifold symptoms of poisoning, illnesses and physical complaints.

---

approaches which produce different results, with the results often being probabilistic rather than clearly determined. So the state of scientific knowledge at any given time may indeed tend to be expressed as a majority vote of experts, the outcome of which voting may be (slightly) different next time. Scientists are of course not going to vote on the laws of gravity. But when it comes to critical loads and limit values in ecosystems, and to environmental priorities, the boundaries between science and policy become blurred. This cannot be other, it simply has to be seen. It is also necessary to accept that in a knowledge society there is no sensible alternative to expert opinion which, without doubt, will remain disputable and has to be justified in the same way as policies in general have to be justified.

There may also be some 'northern' bias in the above listings. Not only do environmental problems change over time, they particularly change according to the stage of industrial development (WDR 1992 12). In the early stages, there are typical problems of urbanisation, hygiene and overexploitation of forests and grazing land, perhaps in combination with wasting of scarce water. Later on, additional fuels and materials are made available and the problem focus shifts towards industrial pollution. The next stage is that in which advanced industrial countries now find themselves. The problems involved in this stage have so far preliminarily been dealt with by downstream measures, with the result that more complex problems of general resource mobilisation, diffusion of hazardous substances, and perturbed ecosystems dynamics come to the fore. So the ranking of urgencies might indeed be different in different world regions. Even so, advanced TEIs will for the most part have to come from advanced northern countries where almost all of the world's technological R&D still is carried out. (There are some recent exceptions in biotechnology where there are now important R&D capacities dedicated to food and agriculture in a number of developing countries.)

In the UN-related development discourse, which takes into account 'southern' priorities, there is some talk about the 'big six', referring to six ecological problems considered to be of particular importance (GlobeScan various years):

- Climate change
- Shortage of drinking water
- Deforestation
- Poverty, food security
- Loss of biodiversity
- Population growth.

A survey conducted by the Scientific Committee on Problems of the Environment of the International Council for Science among 200 environmental experts in 50 countries resulted in a quite similar list (UNEP 2000 334):

- Climate change (51%)
- Scarcity and pollution of freshwater (29%)
- Deforestation, desertification (28%)
- Poor governance (27%)
- Loss of biodiversity (23%)
- Population growth, changing social values (22%)
- Soil deterioration (18%)
- Chemical pollution (16%)
- etc. (more than 30 thematic issues in total, mentioned by 3–15% of the experts).

It should be stressed that neither list of environmental problems raises concerns about depletion of ores and fossil fuels. In SETAC's list, maintenance and availability of environmental media as basic resources (such as surface water, groundwater, topsoil, wetlands, forests, wildlife and biodiversity) does not even get a mention. Whereas ISO and UN experts obviously keep a keen eye on resources, to other experts there seems to be an implicit consensus that metabolic effects are more immediately urgent than scares about the availability of input resources, with the exception of drinking water in certain poor countries.

It is true, indeed, that in general there is no shortage of solar and geothermal energy, water, minerals and microorganisms on earth, representing a potential for replenishment of whatever we can imagine. The conservationist movement of the late 19th century was wrong in expecting reserves of iron ore soon to be depleted, and so was the ecology movement of the 1970s when it forecast that oil would soon run out. Today there are still huge reserves of fossil fuels, and the cause of the most devastating metabolic effects is burning rather than mining them. Therefore, it might seem that, rather than availability of resources, the most pressing environmental bottlenecks are caused by ill-placed metabolic products, i.e. dislocated and transformed materials that badly fit in with natural cycles, particularly emissions that pollute sinks and thereby deteriorate the quality of land, water, soil, air and the health of living organisms.

Such considerations, plausible as they seem to be at first sight, may nevertheless turn out to be somewhat short-sighted. Land, water, soil, air and living organisms function simultaneously as sinks and resources. Soil, water and air represent fundamental resource inputs as well as major sinks. Beyond critical thresholds of impact, forests, animal species and landscapes would not recover, at least not in a way acceptable to contemporary opinion. There has already been much irreversible erosion of topsoil and growing aridification and desertification around the globe. It takes a very long time for groundwater tables to be restored. Wildlife is seriously endangered in many places, and in fishery, forestry and agriculture the extent and structure of biodiversity is being reduced to suboptimum levels which threaten ecological stability in manifold ways. Environmental deterioration of this kind, even if it does not result in a doomsday-like breakdown, will eventually result in unintended megaeffects on the dynamics of geo- and biosphere.

The above listings can be regrouped into three broad categories: (1) global warming and climate change, (2) dissemination and accumulation of pollutants and hazardous substances, and (3) problems of maintenance and availability of natural resources, also including classic issues of conservation of nature. A similar compilation was given by Graedel (1994, Graedel et al. 1994). There, again, global climate change comes first, followed by environmental accumulation of pollutants and hazardous substances, and thirdly over-exploitation of resources and loss of biodiversity.

There is one important aspect still missing here which has to be taken into consideration, i.e. the diffusion of genetically modified organisms (GMOs). Changes in organismic microstructures by conventional breeding and farming practices as well as by modern transgenics constitutes a problem category of its own in that genetic engineering and released GMOs are interfering with and possibly altering important aspects of the dynamics of biosphere – altering, though not necessarily and certainly not one-sidedly for the worse.

Certain problems not explicitly appearing in this aggregation, for example noise, radioactivity, electric smog or toxic waste, would pertain to the category of pollution in the sense of representing health hazards to the flora and fauna in the biosphere. Non-toxic waste, by contrast, and in contrast to public opinion, is of ecological relevance only insofar as it may touch sensitive resources such as natural fibres or metals. Littering or not littering streets may be a question of being a considerate citizen, and using or not using plastic bags may to some persons be a question of style, but neither one is of particular ecological relevance. Neither is dog droppings, though it represents a not unimportant health hazard which, amazingly enough, is tolerated by most authorities (representing another example of environmental patchwork awareness and inconsistent risk assessment).

In summing up, we can identify four broad categories of environmental problems:

(A) *Pollution,* i.e. the dissemination and accumulation of pollutants and hazardous substances such as soot, oxides of sulphur, nitrogen and other acid formants, volatile organic compounds, halogenated compounds, phosphate, nitrate, heavy metals, etc. in the environmental media of air, water and soil, as well as in organisms, thus also in the food chain, as well as in utility goods of everyday use, causing problems of acidity, toxicity, carcinogenity, mutagenity and similar problems.

(B) *Natural resources,* i.e. problems of maintenance and availability of natural resources such as water, groundwater, ores, other minerals, fuels, natural fibres, topsoil, wetlands, forests and wildlife, caused by or resulting in erosion of soils, deforestation, desertification, eutrophication of lakes and rivers, groundwater depletion, high sea overfishing, urban sprawl, all resulting in destruction of landscapes and natural habitats. The topic of natural resources obviously includes classic issues pertaining to the conservation of nature.

(C) *Geodynamics, especially climate change* or global warming caused by net release of CO_2, methane and other greenhouse gases, i.e. the thermal physics and chemistry of the atmosphere, and its geocybernetic ramifications in interaction with oceans and continents. Anthropogenic climate change, according to present evidence, is a factor that significantly contributes to altering important aspects of the dynamics of geosphere. Also part thereof are large territories covered or not covered by forests, rainforests and deserts.

(D) *Biodynamics, especially the spread of GMOs* (genetically modified organisms) and transgenic products, altering biocoenoses of bacteria, microorganisms, plants and animals, including their natural succession and the extent and structure of biodiversity. Also part thereof are significant alterations of natural habitats and organisms brought about by conventional practices.

At this level of aggregation, the problem categories obviously do not represent priorities or urgencies. It might be said that geo- and biodynamics are more complex than mere single items which relate to pollution and natural resources, whereby pollution and a deteriorating resource base are contributive factors to unstable, crises-prone geo- and biodynamics. Even so, the categories give a useful orientation on where today's important TEIs are to be found, or have to be researched and developed respectively:

(A) The category of pollution and hazardous substances relates to energy as well as to agriculture, chemistry and other materials-processing industries.

(B) Natural resources relates to regimes of resource utilisation, urban development, land use and practices in agriculture, forestry and fishery.

(C) Anthropogenic climate change relates to (today's fossil) energy.

(D) GMOs and transgenic products relate to agriculture and chemistry.

Apparently, the most relevant TEIs are to be found in the realms of

- energy
- agriculture
- chemistry and neighbouring basic industries such as metallurgy, cement production, tanning, pulping and paper-making, all of the latter being nothing other than applications of materials chemistry
- technological regimes of utilisation of natural resources.

3.3.3 Upstream versus Downstream Position of Activities

All of the industries in the above-listed realms, and related technologies and products, are positioned upstream in the product chain. While (according to the above categories A–D) a ranking of the urgency of environmental problems remains somewhat arbitrary, product chain analysis and technological life cycle analysis, by contrast, are in no doubt about priorities: upstream is prior to downstream:

- production is prior to final use or consumption of products
- processing and reprocessing of materials is prior to the use of products in service businesses and households

- generic innovation of new systems or components is prior to their incremental improvement in later stages.

The structure of demand and supply along the manufacturing chain neatly fits this picture. User or consumer demand selectively controls industrial production, in the same way as industrial production controls demand for natural resources. There is, however, an important difference. Industrial manufacturers (or materials consumers respectively) directly and in material substance determine the demand for raw materials, i.e. natural resources. But control of manufacturers exercised by user demand relates to aspects of functional utility rather than aspects of material substance. It has rightly been pointed out in the context of environmental product policy and circulatory production policy that what users actually demand is utility rather than materials. Users have need functions. For example, users need to prepare a written document, need to communicate, want to know about news, or want to be entertained. For that purposes they use instruments such as pen and paper, computers, telephones and TV sets. To run these appliances, users need electricity, but they are not particularly interested in the material details of power generation. Users want reliable cars, they do not care too much about construction materials and propulsion technology. They even do not necessarily want cars, they want instant individual mobility, and it happens to be the automobile that most often fulfils that need best. In the rooms of a building, people want to feel comfortable, they do not want to be bothered with the technicalities of heating and air conditioning. How these functions are fulfilled technically, and which kind of environmental impact they cause, is for the most part left to the suppliers or manufacturers.

Action is most urgently needed in those core fields of industrial production that are ecologically most sensitive and chain-influencing. Those core fields today are energy, agriculture, biotechnology, chemistry and materials processing, followed by some further manufacturing sectors such as construction, vehicles, industrial machinery and utility goods (household and office appliances, textiles, furniture). TEIs in agriculture could, for the time being, be ranked as being even more urgent than innovations in chemistry and basic materials processing, because in the latter fields some improvements have already been achieved whereas the greening of agriculture has not made much progress so far.

3.3.4 Metabolic Production Function

Each one of the successive steps in the manufacturing chain – more generally speaking, the entire chain of production and consumption from first steps of extraction to final steps of phase-out – can be described by a transformational or metabolic production function that includes a number of factors, partly representing input, partly output.

Output consists of the desired product or service. Often there are byproducts that can be used as an input in further steps, and always there are emissions in some form of solid waste, effluent water, gases, noise or radiation. The term emission thus includes any output factor that is neither product nor byproduct and leaves the anthroposphere as an immediate re-entry into nature's metabolism.

$$\text{Output} = \text{Product} + \text{Byproduct} + \text{Waste/Emissions}$$

Products and byproducts, though, are also bound to become waste sooner or later. From a viewpoint of ecological book-keeping, a product can be considered as a kind of run-through position, running from original denaturation of resources, i.e. input into industrial metabolism, via many processing steps of throughput to final output from industrial metabolism, i.e. renaturation into sinks. A product can thus also be regarded as a temporary artificial sink.

As to the input factors, they can be said to represent some specific combination of materials, energy and information:

$$\text{Product} + \text{Byproduct} + \text{Waste/Emissions} = f \text{ (Material, Energy, Information)}$$

Information in the sense of knowledge, know-how, skill and experience is certainly different from materials in that it represents 'intangible assets'. In a metabolic equation, intangibles can be left out as can any form of mental energy such as will or motivation. Physical energy then, by which a production process, literally speaking, is fuelled, also comes down to using some materials, especially fuels. Even if it is fuelless energy, such as wind power, some device and infrastructure made of various materials is needed to make that fuelless energy available.

Products are equivalent to materials, and production processes are equivalent to the transformation and processing of materials through the use of process materials and physical energy of some kind. Any product (which is an operative, technosystemic item rendering a service to humans or fulfilling a societal function) represents, in ecological terms, an array of materials. Hence the idea of conceiving of environmental policy as the task of managing materials flows. For humans there are basically three types of materials: food, fuels and use materials of any kind.

The above equation symbolises a metabolic production function. This is not an economic function. An economic production function is actually an oxymoron. In economics, there can be financial input-output functions, or price-related profitability functions, or transactional demand and supply functions. But any appropriate production function is operative and metabolic; it consists of physical variables, not monetary ones. Ecology indeed calls for a reconsideration of the issue of which factors are endogenous and exogenous to the economy (II/6.4).

Furthermore, such an equation states that, seen from an ecological point of view, production and consumption are interchangeable terms: any production or product automatically entails the consumption of a specific array of materials, just as any consumption of materials unavoidably results in the production of an array of transformed materials. Any production process consumes pre-products as it produces products and byproducts. In the same way, any consumption or final use of products is itself a production process, resulting in products, byproducts and emissions of various kinds.

A particular production function represents a particular step within the manufacturing chain. The interplay between the factors in a production function is present in each step along the chain. Input materials, i.e. intermediate or pre-products, represent the link to prior steps in the chain, whereas products and useful byproducts are connecting that step to subsequent ones. Energy input and emissions output, by contrast, are not in the same way integrated in the product chain. Rather they are accompanying the entire chain, energy at the input side, emissions at the output side. As a conclusion from this one can say that energy should in general be considered to be 'upstream', i.e. prior to any other manufacturing step, as emissions should in general be considered to be 'downstream', i.e. subsequent to all steps of production/consumption – keeping in mind that each step along the product chain represents a metabolic production function.

3.4 CLASSIFYING TEIS

Designing a taxonomy involves systematics and classification. On which defining aspects of a technology should a taxonomy of TEIs be based? Having gone through the above considerations one can think now of a number of such aspects:

- the kind of environmental problem to which a TEI contributes a solution
- the field of science and technology which a TEI belongs to
- the industrial sector or a TEI's position in the manufacturing chain
- the functional cluster, i.e. the need function which a TEI fulfils
- the metabolic function which a TEI fulfils.

3.4.1 ... by Field of Science and Technology

In contrast to the biological classification of organisms, there is no common general systematics of technologies. Textbooks vary from scientific foundations of technologies to applicative purposes of technologies. The scientific aspect seems to be more authentic, but outside academic faculties there is no general systematics of science either.

In a conventional way, and fitting aspects of ecology, technologies can be characterised as being predominantly (a) mechanical, (b) physico-chemical or (c) biological.

Typical mechanical technologies (a) include the shaping and putting together of solid materials, for example spinning and weaving, pressing or cutting to size or assembly of parts. Methods of putting things together, however, also include physico-chemical methods such as welding and gluing in addition to mechanical methods such as sewing, clipping, fitting together or screwing. Important fields of mechanical technologies are:

- Civil engineering above and below ground (e.g. construction, mining, quarrying)
- Mechanical engineering (machines, plants)
  · Mechanical materials processing (spinning/weaving of fibres, cutting, bending, boring, grinding, putting together of solid materials, etc.)
  · Assembly and disassembly, mechanical separation technologies
- Micromechanics
- Motors and engines engineering (actually a combination of mechanics and physico-chemical technologies)
- Hydraulics, i.e. pressure mechanics on the basis of fluids
- Mechanics on the basis of gases, which normally includes steam, i.e. pressure.

(b) Typical physico-chemical technologies are burning of fuels, melting of ores, founding of metals, glass, plastics, metal coating, cracking of oil molecules, pulping of wood, varnishing, painting, spraying, or similar. Such processes usually need high temperature or high pressure, or the use of chemotechnical aids such as acids, organic solvents, or other. Important fields of physico-chemical technologies are:

- Energy technology and electrical engineering
  · Mechanical kinetic energy (wind, water)
  · Electrical chemistry
  · Combustion technology, thermal technology
  · Nuclear energy
  · Solar energy
  · Electrical engineering, electronics (outside computers and other ICT)
- Chemistry (synthetic and analytical, e.g. production of synthetic materials and substances, analytical chemistry)
- Materials physics and chemistry (physico-chemical technologies of processing and transforming natural and synthetic materials)
- Information and telecommunication technology (ICT)
  · Semiconductor technology, computer electronics
  · Optoelectronics
- Photonics (laser, optics)
- Acoustics
- Nanotechnology (abiotic or inorganic).

(c) Typical biological technologies include growing of plants and animals, fermentation of biomass of any kind, enzymatic reactions (enzymes are biological catalysts). More generally speaking this covers any process that makes use of metabolic (anabolic and catabolic) capabilities inherent to genomes, living cells, micro-organisms, plants and animals:

- Conventional farming and food biotechnology (or macrobiology)
  · Agricultural technology (growing and breeding of plants and animals, including forestry and fish farming)
  · Conventional food and biomass processing
- Modern biotechnology (or microbiology)
  · Microbial biotechnology
  · Molecular biotechnology
  · Cell technology
  · Genetic engineering (transgenics, genetic modification)
- Nanobiotechnology.

Today, biological technologies are attracting particular attention because in the transsecular shift from traditional to modern production the introduction of modern biotechnology represents a field of novel key technology which is increasingly shaping the present and future of industrial production.

Aside from this particular feature, however, many TEIs do not properly fit the above structure of technologies. They often combine elements from different categories. In particular they may include mechanical as much as physico-chemical aspects. And they can of course be relevant in quite different fields of industrial application, e.g. energy TEIs that are simultaneously applied in electricity generation, vehicles, construction, and office and household appliances.

A taxonomy of technologies can certainly profit from drawing heuristic analogies to the biological taxonomy of species. There is nevertheless a different underlying logic in many respects.

3.4.2 ... by Industrial Sector or Position in the Manufacturing Chain

New technologies, particularly of the generic systems type, constitute new industries and markets. The structure of industrial sectors, which has developed over time as reflected in official statistics, can in a certain sense be seen as a historical sediment of technological innovation, representing branches on the tree of traditional and modern technology. Economists tend to perceive industrial sectors as a structure of businesses and markets rather than clusters of interlinked productions and products. With regard to systematics, it does not make much difference. Insofar as production lines cross industrial sectors or markets, they reflect the macrostructure of division of labour, i.e. the vertical manufacturing chain or product chain, which also includes some horizontal connections as well as certain aspects of metabolic function. The UN standard industrial trade classification (SITC/ISIC Rev.3) mirrors this structure which is in the same way present in national statistics worldwide:

A Agriculture, hunting and forestry
B Fishing
C Mining and quarrying
D Manufacturing
   · Food, beverages, tobacco
   · Fuels, lubricants, and related materials
   · Oils, fats, waxes
   · Chemicals and related products
   · Manufactured goods classified by material (i.e. processed or refined materials)
   · Machinery
   · Vehicles, transport equipment
   · Miscellaneous manufactured articles, commodities not classified elsewhere
E Electricity, gas and water supply
F Construction

G Wholesale and retail trade, repair businesses
H Hotels and restaurants
I Transport, communications
J Financial intermediation
K Real estate
L Public administration and defence
M Education
N Health and social work
O Other community social and personal services
P Private households
Q Extra-territorial organisations and bodies.

The structure clearly separates production (A–F) from services, households and administrative bodies (G–O). Within production, it reflects the vertical chain from raw materials to finished end products, which are for use in production itself as much as in services, households and administration.

Here again that basic insight from the consistency approach imposes itself, stating as a rule of thumb that most of the environmental impact of a product and its use is caused chain-upwards, as it is predetermined in the early stages of a technology life cycle. That is why, for example, airlines and car drivers are not among the metabolically important actors, even if they burn huge amounts of fossil fuels and thereby emit huge amounts of greenhouse gas. They do not have much choice. The choice of structural change is in the hands of the creators (researchers, developers) and producers. The users' concerns may enter the manufacturing chain by way of chain management and cooperating with producers in an innovation network. Even so, sector-related TEIs, in most cases include sectors which produce physical goods or provide process energy, not, however, airlines, car leasing, tourism, banking, insurance, because the latter are simply users, not producers of materials, energy sources and physical goods. Hence the policy message, somewhat simplified but to the point: do not bother users and consumers, whether corporate or private. Ecologically it will not lead very far. Instead, cooperate with the inventors, developers and manufacturers of energy apparatus, materials and physical goods, the more so since, at least for the time being, they exert a defining influence on the creation and use of technologies and products (I/5.1.2).

Many TEIs actually belong to just one industrial sector, or to put it differently, have a clear position in the manufacturing chain. Others, however, do not. They are of relevance to a number of sectors, particularly TEIs relating to energy and biotechnology, or TEIs in the field of measuring, monitoring and technical control.

3.4.3 ... by Need or Use Function

Another structure, which also touches on various sectors, is the structure of needs or use functions (II/7.3.8). This relates to a recently much debated, or at least much quoted, idea of shifting from product-orientation to purpose-orientation. This is usually meant to represent a structure of needs such as nutrition, clothing, housing, cleaning, heating, lighting, mobility, etc. Some of these clusters are cross-sectoral in that they include final use processes as much as certain production processes. Typical clusters, some of which have been discussed in environmental research and literature, are:

- Food
- Health
- Clothing
- Shelter
- Housing
  - Building
  - Heating/cooling (indoor air conditioning)
  - Indoor living (furniture, household goods, household and office equipment...)
  - Washing, cleaning, painting, ...
- Transport (mobility)
- Telecommunication
- Physical information handling (writing, documenting, sending, shipping, publishing, data processing and storage via print, audio, video, film media)
- Education, teaching and learning, knowing
- Leisure, entertainment, sports, tourism
- Public security and defence
- Other ones.

We, it has been said, do not actually want to have floor lamps or ceiling lights for their own sake; what we really need, i.e. the function to be fulfilled, or service to be rendered, is lighting. We want clean clothes, not washing machines; not heating systems, but decent room temperatures; not cars, but mobility however it is achieved.

Complementing product-orientation by purpose-orientation, and the approach of clustering need functions, can be helpful in reshaping product stewardship and producer-consumer relations (for example leasing a service instead of buying a product, producer-organised take-back, and reprocessing of used products). In the present context, however, the approach has proven not to be very appropriate for several reasons.

One reason is that, enlightened as the approach may seem to be from an engineering and planner's point of view, it remains debatable on socio-cultural grounds of style and behavioural preferences. Furthermore, and most importantly here, focussing on final purpose or the needs of private consumers creates an overt downstream orientation whilst what technological environmental innovation needs to do most urgently is to look upstream life cycles and the vertical manufacturing chain.

It became clear that clustering according to need function is in most cases neither informative nor relevant. In too many cases TEIs cannot be ascribed clearly enough to a single need cluster. In addition to the need clusters of final users, further functional clusters relating to producer 'needs', i.e. requirements of production, have to be added. In particular the question arises of whether specific TEIs under consideration serve as energy of any kind, or working material of any kind. The distinction remains in any event rather insignificant, since anything in the industrial metabolism represents some aggregate state of matter and energy.

3.4.4 ... by Metabolic Production Function

The factors in a production function as discussed above in I/3.3.4 could be interpreted for taxonomic purposes as representing

on the output side

- final products
- recycling and reprocessing of byproducts and used products
- treatment of waste and further emissions

on the input side

- product materials
- process materials
- process machinery and infrastructure
- fuels
- energy apparatus
- management of natural resources and environmental media.

One could think of classifying TEIs according to these categories. Here again, however, similar aspects arise as with categories of science and technology, or industrial sector. In quite a number of cases the categories fit well, in other cases they do not. There are overlaps, and metabolic aspects in every step of production are of course not an alternative to, but are additional and complementary to, the sectoral position of those steps within the vertical production chain.

3.4.5 ... by Realm, Domain, Family, Kind and Case

As a result of the preceding reflections on environmental problem solving and possible classification of TEIs, and certainly as a result of empirically collecting and analysing examples of TEIs, a structure has gradually emerged which combines all of the approaches discussed in this chapter, and has the position of a specific TEI within the product chain and its metabolic function clearly at its core. Eleven realms of TEIs have been clustered:

1 Energy (including fuels, furnaces, motors and engines in propulsion systems, fuelless energy, as well as generation and distribution of electricity)
2 Natural resources (raw materials)
3 Agriculture
4 Chemistry and chemicals
5 Materials processing
6 Building, settlement structures
7 Vehicles, transport
8 Utility goods (machines, office and household appliances, furniture, textiles, leather)
9 Materials reprocessing, recycling and waste management
10 Emissions control
11 Environmental measuring and monitoring.

These realms represent the basic framework for classifying TEIs. Subdivisions at the next lower level are called domains, followed by families, these in turn by kinds, the latter often being exemplified by cases. The overall structure thus consists of:

- Realms, and subrealms if applicable
- Domains and subdomains, in some cases also cross-domains
- Families, and subfamilies if applicable
- Kinds and cases.

The systematics of TEIs in the appendix has been compiled on the basis of that structure.

4. Trends and Visions of TEIs

4.1 SELECTION OF TEIS AND DATABASE

This chapter gives examples of TEIs and explores major new paradigms, trends and ensuing scenarios of technological environmental innovation, some of which serve as conceptual lead visions. The presentation of TEIs is structured according to the eleven realms that give rise to the systematics of TEIs as explained in the previous sections.

This structure emerged in the course of empirical research work aimed at creating a taxonomy of TEIs. Since 2001 it has been carried out in connection with the Key Environmental Innovations group of the German Federal Research Ministry's Initiative on Sustainability and Innovation. An explorative databank has been created which now numbers 305 datasets on TEIs, many of which in turn contain several kinds or cases of the TEI in consideration, so that a total of about 500 examples of TEIs are represented.

The datasets include information on a technology (i.e. on products, materials, processes and practices), its structural impact, the stage of structuration and diffusion, on rival like technologies, competitiveness and adoptability, as well as ecological properties and environmental improvements which have been achieved or can be expected.

The databank has been fed by a continuous survey of innovations as they were reported in articles in the following sources from the beginning of 2000, partly earlier, through to March 2003:

- *MIT Technology Review*
- *The Economist* and *The Economist Technology Quarterly*
- *VDI Nachrichten* (official weekly of the Association of German Engineers)
- *Industry and Environment* (by UNEP Cleaner Production Programme)
- *The IPTS Report* (Institute for Prospective Technological Studies, Seville, in cooperation with the European Science and Technology Observatory Network)
- *Journal of Industrial Ecology* (published at the MIT)
- *DBU aktuell* (National Foundation for the Environment)
- *Biotech Brief*, Deutsche Industrievereinigung Biotechnologie, Frankfurt, (official industry publication)
- *Transgenics Newsletter*, Öko-Institut, Freiburg/Berlin (alternative expertise)
- Ecodesign examples, www.sustainability.at.

Information tapped from further publications has been collected on a more occasional basis and also fed into the datasets. Any information on TEIs given hereafter, if not indicated otherwise, is based on these sources as they have become established in the datasets. It is assumed that the above listed sources of information can be trusted to be reliable. One would otherwise have had to check original sources. Such a project could then not be carried out within a reasonable period of time, aside from the fact that funds available do normally not allow for such methodological tightness.

It should be pointed out in this context that the purpose of this work is systematic and conceptual rather than to fulfill a need to specialise in in-depth assessments of this or that single technology. It certainly involves the ability to make a difference between innovations of major and minor importance, but this does not need to be based on absolute precision of figures and forecasts. As said by Johann von Neumann, an Austro-Hungarian mathematician who spent his life in America and contributed to quantum physics and game theory, it is better to be approximately right than exactly wrong.

According to the criteria discussed in I/2.4 and I/3.3, technologies qualified for entry into the databank if they improved on metabolic consistency or significantly increased eco-efficiency. Examples could thus be expected to represent at least one of the environmental innovation strategies as listed in Table 4.1, all of which have played a certain role in the environmental discourse of the recent past.

That most of the examples chosen in this way turned out to be upstream rather than downstream in both their life cycle and the product chain was not a prerequisite but emerged as the most distinctive feature of the resulting selection.

Industrial symbiosis (second line in Table 4.1) means combined production processes between different companies where any output is going to be a useful input in some consecutive process.

A zero emission process, or pollution prevention, is achieved by process-integrated emissions prevention resulting in zero pollutants (Pauli 1998).

Circulatory economy means continuous recycling and reprocessing of materials, thus decreasing the use of natural resources per unit of output, resulting in an increase in materials productivity. Recycling, however, cannot go on forever, even if clean energy were available, because downcycling of materials can to a certain degree be put off, but ultimately not be avoided.

Sustainable resource management refers to so-called sustainability management rules as formulated in the Rio process in various ways, regarding consumption rates of exhaustible and regenerative resources, critical emission loads, carrying capacity of ecosystems, including biodiversity, but being different from pure conservation of nature and landscape.

Table 4.1   Selected approaches to technological environmental innovation according to the realms where they most apply

Approach \ Realm	Energy incl. propulsion and heating	Raw materials. Natural resources	Agriculture, forestry and fishery	Chemistry and chemicals	Basic materials processing and reprocessing	End-prod.: Build., vehicles, utility goods	Emissions control, waste processing
Industrial symbiosis / Zero emission process / Circulatory economy	×	×	×	×	×	×	×
Sustainable resource management	×	×	×	×	×	×	
Cleaner technologies	×	×	×	×	×	×	
Benign substitution				×	×	×	
Product design for environment				×	×	×	
Bionics				×	×	×	
Add-on purification technology							×

Clean technology, as characterised in I/2.3.2, was a predecessor concept of TEIs from the 1980s, focussing, however, just on production processes and eco-efficiency.

Benign substitution refers to any substitution of non-hazardous substances and materials for hazardous ones. This is of particular relevance with regard to anything which has to do with chemistry, materials processing and chemicals (see for example in Anastas/Williamson 1998, Anastas/Warner 1998, SubChem 2002).

Product design for environment, or ecodesign, includes a number of design principles such as energy efficiency, the use of safe, non-hazardous and non-toxic materials, the choice of materials anticipating their recyclability, an assembly process anticipating disassembly and recovery of components and materials, and optimum durability of components and parts (Paton 1994, Tischner et al. 2000, UNEP/IE 1997, Kazazian 2003). It started from the idea of the recycling-oriented design of cars, TV sets and other more complex products, and eventually evolved into a broad concept of redesigning whole product groups and material flows, phasing out unsafe, toxic or ecologically otherwise hazardous materials and replacing them by better recyclable and ecologically more neutral or even benign ones. For this to make sense, the process in turn needs to be part of a strategy of management of materials flows throughout the product chain from prime recovery to reprocessing.

Bionics refers to designing products and processes in a way that mimics forms and principles found in nature. There is constructive bionics and material bionics. Constructive bionics mimics natural principles of construction, form-giving and shaping. Material bionics wants to replicate the physico-chemical properties of naturally occurring materials and substances.

4.2 ENERGY: FROM CARBON TO HYDROGEN AND FUELLESS ENERGY

4.2.1 Paradigms and Metabolic Classes of Energy Technology

The trends which can be identified in the realm of energy revolve around the choice of fuels. The fundamental choice determining everything else is whether to stick to conventional fuels, i.e. fossils and uranium, and try to make them cleaner and safer, or whether to convert to alternative sources and forms of energy such as biomass, or hydrogen as a clean fuel, or fuelless energy such as solar, wind, hydropower and geothermal energy.

In terms of life cycle analysis this leaves us with three types of innovative dynamics (II/9.4). One is a succession struggle (new against old) between the defenders' paradigm of clean coal, or perhaps better clean carbon, and clean nuclear energy respectively, and the challengers' paradigm of decarbonisation and denuclearisation. Secondly there is a retentive rivalry among the defenders (old against old) between carbon and nuclear, and thirdly an innovation contest among the challengers (new against new) to establish themselves as the dominant future paradigm.

Pioneers of renewable or regenerative energy have so far behaved as if they were of the same breed, sharing a common paradigm and serving the same interest. It is time to recognise that they are actually of at least three breeds and that combining their potentials and reconciling their diverging paradigms will not in every case be an easy option. The reason is that in addition to the criterion of 'exhaustible vs. renewable/regenerative' there are two further metabolic criteria that make a big difference. These are the criteria of 'polluting vs. clean', which is at least known, if not given due attention, and 'fuels-based energy vs. fuelless energy', which has barely been given due consideration at all. Taking into account all three criteria gives a more appropriate picture of the arena as shown in Table 4.2.

Table 4.2   Energy according to metabolic class

	Energy based on fuels i.e. chemical and nuclear	Fuelless energy i.e. kinetic and radiant
Exhaustible, non-regenerative and polluting	Coal Mineral oil Natural gas (methane, methanol) Methane hydrate Uranium	– none –
Regenerative but exhaustible (limited availability at a particular time) and polluting	Wood Energy plants, plant oil (e.g. biodiesel/methylester) Waste biomass of any kind Biogas Ammonia	– none –
Inexhaustible or renewable/regenerative and clean	Hydrogen (H_2) Deuterium Tritium Silane (?)	Light, sun radiation Solar thermal Solar hydrogen Wind Hydropower Tidal flows, wave power Geothermal heat and power 'Free energy' (?)

The categories in Table 4.2 give a robust orientation:

1. From an ecological point of view, fuelless energy is preferable to fuels because any extraction and processing of fuels is costly in terms of resources, energy and money, and may entail denaturation or even destruction of ecosystems, bad use of land, dislocation of excavated materials and groundwater, and also significant airborne emissions even before the fuels are burned; in other words, fuels come with very large ecological footprints or materials rucksacks. Fuelless energy comes free of all that. Energy apparatus, such as solar panels or geothermal pipes, certainly also constitute some demand on materials (resources), but this also holds true for power plants, furnaces and motors of fuels-based energy.
2. Clean energy is preferable to polluting energy because emissions from clean fuels can be released without causing environmental problems whereas emissions from polluting fuels, such as carbon-containing fuels or uranium, require laborious and expensive purification, if they can be purified at all.
3. Regenerative fuels are preferable to exhaustible fuels, even if a regenerative fuel may be available only in limited volumes at a time and be somewhat

polluting. This, however, is a second-best fall-back option if inexhaustible fuelless or clean-fuel energy are not sufficiently available.

In consequence, there are five metabolic classes of technological paradigms within the spectrum of fuels-based/exhaustible/polluting versus fuelless/regenerative/clean energy, representing actual empirical trends, though of different reach and potential, and representing the preferred guiding visions of their followers:

- Cleaner and safer nuclear energy
- Cleaner carbon energy, bringing about coal age II
- Biofuels
- Clean fuels, i.e. hydrogen in the foreseeable future, bringing about a hydrogen economy or the hydrogen age
- Fuelless energy, most commonly symbolised by a smiling sun heralding the coming of the solar age.

Not included are scenarios of mere energy saving in the sense of 'dematerialisation' by way of 'efficiency revolution' for reasons explained in I/2.4. A clean energy future will in all probability not include using less energy. Energy is something like the physical currency of human operations. Nothing works without spending on energy, i.e. exchanging energy for operative services of any kind. The use of energy can be optimised, and is being optimised simply for economic reasons. But, according to the growth-efficiency principle (II/7.7.1), the purpose and result of economising on energy is to maintain a high level of energy consumption, or to put it differently, a high level of availability of operative services. We need to minimise environmentally undesirable impacts of energy use, but there is no need to minimise the use of clean energy for ecological reasons. If one had the strange idea of considering the behaviour of celestial bodies in a utilitarian way, one would have to conclude that our sun lavishly spends on energy and that the earth's energy supply is thus extremely 'wasteful'. If we continue to economise on energy, and also on clean energy – a thing we are certainly bound to do – then we will have to do so for economic reasons.

4.2.2 Safe Nuclear Energy: Still in the Realms of Fancy

There were some attempts to relaunch nuclear energy as the spearhead of decarbonisation. Nuclear plants, for sure, do not emit airborne carbon or nitrogen compounds. But they emit solid radioactive waste. Hazardous nuclear waste which stays radioactive for many thousands of years, rather than operational safety, turned out to be the Achilles' heel of nuclear fission.

Cut back by industry and abandoned by a growing number of governments, nuclear energy cannot be expected any longer to be a virulent field of innovation. Yet there are a couple of incremental improvements and a few new generation designs. One is the pebble bed reactor. Its prototype dates back to the 1960s. The fuel it uses is uranium dioxide which is encased in billiard ball-sized graphite 'pebbles' that remain below melting point. The graphite encasement also contributes to making long-term storage of spent uranium less unsafe. A pebble bed reactor is cooled with helium gas instead of water, allowing higher, more efficient temperatures, and yet it does not need a containment dome and a regional evacuation plan because of the inherent safety of the process. It can also be run in smaller units contributing to some decentralisation of electricity generation. Such a reactor is now in operation in Koeberg near Cape Town, South Africa. It represents obvious progress in the safety and efficiency of nuclear plants. However, it does not give an answer to the problem of nuclear waste.

A design of late is called CAESAR, i.e. Clean And Environmentally Safe Reactor. This wording might be somewhat euphemistic. The idea is to reuse spent uranium-238 fuel rods in the reactor, instead of disposing of them. In a CAESAR reactor there would be no enrichment with uranium-235, just uranium-238. Steam would serve as a moderator, i.e. ensuring that passing neutrons attain the necessary speed to split uranium-238 atoms. In any such fission process 'delayed neutrons' are also created. In a conventional process, these are slowed down so much that they cannot re-enter the reaction, whereas in a CAESAR the delayed neutrons would remain speedy enough to again hit a uranium-238 nucleus. As a result, spent fuel rods of pure uranium-238 are no longer waste products but could be perpetually reused as fuel for several decades. In addition there would be a practical side-effect: Since the process would run for decades without intervention, the core of the reactor could be sealed. So nuclear proliferation could be controlled more effectively, because in order to produce plutonium, the reactor's core needs to be accessible, not sealed. For the time being, however, CAESAR is just a new design, there is no prototype yet. And while it certainly could postpone the problems involved in end-storage of nuclear waste, it would ultimately not do away with them.

CAESAR is in a certain sense reminiscent of the fast breeding reactor which can be 100fold more efficient in making use of uranium-238 than normal reactors. While this was an answer to the problem of finite reserves of uranium, it offered no answer to the question of nuclear waste, even if breeding reactors do produce relatively less of it. The darker side of breeders is the production of uranium-239 for use in nuclear warheads. In contrast to a fast breeder, CAESAR could be sealed. But who could prevent seals from being broken?

There have been rumours time and again of some technology of transmutation of nuclear waste that would shorten the half-life of its radioactivity to a couple of decades instead of many millennia. That would indeed represent a solution to nuclear waste. Shortening the half-life of nuclear waste might actually re-open a window of opportunity for nuclear energy and could give it a new boost, assuming that concerns about safety of nuclear plants would not outweigh the obvious advantage of building upon well established foundations of nuclear science and technology.

An approach pursued by the US Department of Energy, the University of Nevada and the Los Alamos National Laboratory, New Mexico, is to extract plutonium and other long-life elements (about 1 per cent) from the spent nuclear fuel, and then to bombard these same elements with high-speed neutrons from a particle accelerator, thereby splitting the nuclei into smaller elements that either are not radioactive or decay within decades. As a result, nuclear waste, which under present conditions has to be stored in underground repositories for thousands of years, would not be such a burden any longer. The quantity that needs to be stored under ultra-safe conditions would be reduced to one hundredth of what it is today. However, it is uncertain whether this approach really would work in practice. Building a demonstration facility might take 20 years and cost $4–7 billion. Nuclear waste would be significantly reduced, though not completely eliminated. Operating nuclear fission plants would continue to be a high-risk technology.

Have similar statements to be made on nuclear fusion, hot and cold alike? Fusion, i.e. fusing atomic nuclei, is the opposite of nuclear fission. There can be no explosive chain reaction. It represents an attempt to replicate technically on earth what causes the sun to shine under natural conditions of tremendous heat and corresponding high pressure. In a fusion plant, a 'burning' (fusing) plasma of hot, ionised gas consisting of the hydrogen isotopes deuterium and tritium would be created. Deuterium is heavy hydrogen (^{2}H), tritium superheavy hydrogen (^{3}H). Such a process would set free ten times the energy that was put in. Fusion would thus qualify as pertaining to the class of so-called free energy technologies which, by definition, have a greater energy output than energy input.

Nuclear fusion would be an inexhaustible source of power because its fuels, deuterium and tritium, are not scarce and regenerate or can be regenerated within brief periods of time. Fusion would be safe power in that there is no risk of explosive chain reactions, though running a plant at very high temperature and pressure always poses a risk in itself. Finally, and again in contrast to fission, hot fusion would be clean power in that it comes with no long-life radioactivity. In principle there is no doubt that hot fusion is technically feasible. Whether it can be done in practice continues to be as much the province of science fiction as a number of further options that are not yet available in any foreseeable future.

Research into nuclear fusion has been conducted since the 1950s, without ever yielding the spectacular results that were expected, but having cost the taxpayer dozens of billions of dollars. Since the early 1980s both politicians and the public have become reluctant to support the funding of further fusion programmes, though research activities are still being carried out on a relatively large scale. For how long can something continue to be the next 'big thing' without turning into the latest has-been candidate? Whereas nuclear fission remains an ecological non-option, hot fusion remains in the realms of fancy.

Hot fusion has in cold fusion a spurned relative who is being denied official credentials because it entails electrochemical processes that seem to be at odds with the established thermodynamic paradigm. Cold fusion, also called low-energy nuclear reactions, or chemically-assisted nuclear reactions, is said to occur in electrochemical cells consisting of metals such as palladium loaded with heavy hydrogen. Like hot fusion it is said not to create damaging radiation or radioactive byproducts, and its obvious advantage would be its working at ambient temperature and pressure.

Cold fusion was reported in 1989 by Stanley Pons and Martin Fleischmann at Southampton University, UK. Other researchers, however, failed to replicate reported results. The approach was disgraced. Similar reports, however, have now emerged again, for example by Rusi Taleyarkhan, Oak Ridge National Laboratory, Tennessee. Cold fusion would represent infinite amounts of clean energy. For the moment, however, and for the couple of decades we can and have to plan ahead, cold fusion too remains just another would-be source of clean nuclear energy.

This is not to dismiss grand visions of future energies. According to our present understanding of nature, everything in the universe consists of matter, energy and information. So literally everything can in principle be considered to contain energy of some kind and be a potential source of energy. If humankind continues its evolution, it will almost certainly make new sources of energy available which we cannot even dream of today.

4.2.3 Cleaner Carbon Energy: A Big Weak Option

4.2.3.1 Why fossils and methane hydrate are unsustainable
The best known carbon-containing fuels which have been of some importance include wood (fresh biomass), peat (rotten biomass), brown coal, hard coal, mineral oil (fossil biomass) and natural gas (from rotten and fossil biomass). The listing represents a ranking order of pollutiveness of carbon fuels in terms of soot, sulphur, CO, CO_2, NO_x, VOCs, organic micro- and nano-particles and many other noxious and hazardous substances. Fresh biomass, wood, peat and brown coal score worst. Oil then achieves a somewhat better performance, which is yet surpassed by natural gas (CH_4). Natural gas can today be bought for \$10 a barrel, comparing with petroleum at \$20. Natural

gas has practically no sulphur and zero aromatic pollutants (no benzene) which taint other petroleum products. Cars that run on gas, compared to gasoline cars, emit 25 per cent less CO_2, 80 per cent less CO, 20 per cent less hydrocarbons, 60 per cent less VOCs, and particle emissions are next to none since 99 per cent of natural gas consists of methane. Compared, however, to the benchmark of cleanliness, i.e. zero hazardous emission, methane nonetheless performs badly.

Methane hydrate, often described as burning ice, is another source of natural gas that is now being explored. Frozen water under a degree of pressure forms special rigid structures (clathrates) which act like a container hosting methane molecules without chemical bonding. The source of the methane is bacterial breakdown of plankton and algae. If temperature rises and/or pressure eases, the rigidity of the water structure dissolves and the gas is released. One litre of frozen water releases up to 163 litres of gas. Methane hydrate forms in colder waters below a depth of 300 metres and under the permafrost of more northern and southern latitudes. If the big energy corporations are allowed to exploit methane hydrate, it might be extracted on a regular basis in about 15 years. In the interim, suitable extraction methods and infrastructures would have to be developed.

Reserves of methane hydrate are gauged at more than twice those of hitherto known gas, oil and coal combined. This serves to confirm the assumption that the crucial restriction to the use of fossils is not that we were to run out of them soon. The bottlenecks are ecosystems and sinks: mining and extracting of fossils destabilises and destroys ecosystems, and hazardous emissions from burning fossil fuels overburden natural sinks, which in turn adds to the pressures on ecosystems. Extraction of methane hydrate has the same problematic effects as drilling and mining. If, moreover, gas – natural gas from subterranean cavities or methane hydrate from beneath the permafrost and the open sea – is extracted on a large scale, large amounts of the volatile gas will unavoidably leak from holes, pipes, pumps etc. The greenhouse effect of unburned methane in the open air is 30–60 times that of CO_2.

Defenders of carbon fuels are still trying to play down anthropogenic climate change. They have recently gained unexpected support by the solar wind theory as worked out by the Danes H. Svensmark, E. Friis-Christensen and others. It states that the most important control factor of climate on earth are thermodynamic activities on the sun's surface transmitted to the earth by solar wind. That in turn would control the formation of clouds and the level of temperatures on earth. Strong empirical correlations indicate that the solar wind theory does seem to have some truth to it. The authors do not deny, though, that the carbon cycle is another, if less outstanding, control factor of earth climate, so that there is indeed an additional anthropogenic greenhouse effect, no matter to what extent the sun or the earth's volcanoes, oceans and forests might by themselves cause ongoing geogenic climate change. Literally adding fuel to the flames is irresponsible in any case.

Even if there was no climate change at all, carbon fuels will never represent an environmentally benign option because their production is accompanied by large ecological footprints, i.e. it causes strong pressure on ecosystems and heavy pollution of sinks. Furthermore, seen in retrospect from a possibly not too distant future, it might look pretty silly to burn, i.e. destroy organic molecules (non-recyclable for humans) created by photosynthesis (not replicable by humans yet) instead of retaining them as part of valuable natural stocks of resources, and using them as components in the production of feeds and use materials in various industries. This corresponds with a widespread view which chemists and other scientists have long held.

4.2.3.2 Extracting and burning fossils at lower impact

The response of carbon defenders has been manifold attempts to make carbon environmentally less harmful. Cleaner carbon has the following aspects:

- minimising the ecological impact of extracting raw materials
- beneficiation of fossil fuels
- burning the fuels more completely, thus more cleanly
- keeping control of resulting emissions.

As for minimising the ecological impact of extracting raw materials, new practices of resource management have been devised, among them renaturation of disused strip mines. This includes the reconstruction of certain layers of overburden, already planned for in the mining process itself, artificial addition of topsoil, controlled succession of plants in the process of recultivating the area, and perhaps controlled flooding of holes and low-level areas. Turning a dead moon surface into a greening landscape of lakes and peninsulas with bathing beach and marina is not so bad after all, even if nature there will never be again what it was before. Underground pits, though, cannot be restored, but may literally undermine the foundations of those who live on top when the pits flood. Groundwater aquifers in any case take many decades to re-establish themselves.

Another approach is single shaft recovery, a method of exploration and production of oil and gas (E&P technology). It significantly reduces the impact on sensitive surface ecosystems. Eco-impact of deep sea drilling can also be reduced by multi-directional wells aided by seismic imaging of oil fields which results in much higher precision of drilling and extraction.

Multi-phase pumping can then prime and transport mixtures of oil, gas and further solid components over long distances. Conventional technology needs expensive separation of components on-site which are then distributed in different pipeline systems (which is expensive), whereby gas has to be burned off (a loss one would like to avoid). Multi-phase pumping, by contrast, retains everything until arrival at the point of refinery. It also renders possible – certainly a controversial advantage – the exploitation of small oil and gas fields, fields at sea ground and beneath ice, as well as a longer exploitation of fields. Multi-phase pumping is an alternative to drilling rigs. Investment costs are 40 per cent less than with comparative technologies.

Such innovative practices which reduce undesired impact and increase efficiency are certainly welcome. As long as fossils are being used, those methods ought to be applied, because recovery of fossils, as with uranium, or metal ores and other minerals, always collides with interests aimed at conservation of nature. One should be aware of the fact, however, that more circumspect methods of exploring and extracting fossils do not change the ecological consistency of the industrial metabolism involved. They may contribute to reducing the problems with fossils, but they do not solve them.

With processing and beneficiation of fossils it is quite similar. Huge oil refineries can be run and certified according to best-practice standards of environmental management. They nevertheless remain an environmental problem. Beneficiation of carbon fuels renders the fuels less harmful, but they continue to be harmful on the reduced level. Coal can be processed to have a much lower content of sulphur, mercury, less humidity, or less ash content. Burning coal nevertheless implies acid sulphur emissions, toxic mercury emissions, etc., which have to be dealt with.

Coal can be aged artificially under high temperature and pressure resulting in a cleaner-burning feedstock with up to 60 per cent more heat output per coal unit. Similarly, any carbon base, particularly coal and gas, can be transformed into some synthetic fuel, a carbo-synfuel. Synfuel includes a number of conversion processes such as upgrading of coal to gaseous and liquid hydrocarbons with improved fuel characteristics. The conversion is done at high temperature and pressure by a reaction with steam (hydrogen) in the presence of air (oxygen), with or without a catalyst.

Coal liquefaction is done by pyrolysis, by solvent extraction (for example Exxon donor solvent), by catalytic liquefaction (H-coal) or by indirect liquefaction (resulting in Fischer-Tropsch-methanol). The Fischer-Tropsch process dates back to 1923. The alternative now sought is that of a suitable catalyst as the basis of a modern gas-to-liquid (GTL) technology which would convert natural gas into petroleum at normal temperature and pressure. Candidates are catalysts of mercury (40 per cent yield) and platinum (70 per cent yield). It would render commercially viable all of the remote places where natural-gas fields can be found. In Alaska and Siberia there are reserves with a span of 50–200 years compared to mineral oil of 40–160 years. Synfuel processes can also be applied to the recovery of petroleum crudes from heavy oils and tar sands, or to the retorting of oil shale.

Synthetic gasoline was invented for military purposes in Germany, a country with reserves of brown and hard coal but short of petrol. Today, synthetic gasoline has been heavily subsidised in the US since the 1980s by synfuel tax credits in order to make the country less dependent on foreign oil imports. The US has by far the highest per capita consumption of petrol and a rather poor performance with regard to energy efficiency.

Liquefaction of coal and gas would make sense under conditions of scarcity of petrol with no alternatives to the use of petrol available. There are much bigger reserves of coal than of petrol, and still bigger reserves of gas if those huge reserves of methane hydrate are included. So synthetic gasoline opens up the prospect of a prolongation of the oil age. All these carbo-synfuels nevertheless remain carbon fuels, and thus clearly do not represent a step towards decarbonisation. From an ecological angle this represents a typical example of progress in the wrong place. Carbo-synfuels could only be seen to be of some advantage if they are used for producing methanol as a transitional step towards introducing fuel cells and a hydrogen energy base. Carbo-synfuels do not make sense if there are clean, high-quality, less cumbersome and cheaper alternatives available.

The next avenue for cleaner carbon is that of burning the fuels more completely, thus more cleanly, in power plants, blast furnaces, industrial production processes, in car motors, aircraft engines and in heating systems. There are a number of new generation burning processes particularly in thermal power stations. One is the integrated gasification combined cycle IGCC. This

> converts coal into gas which is then cleansed and burned in a combustion turbine. The advantage of IGCC is that it allows generating companies to capture carbon dioxide from the exhaust more easily, while producing electricity more efficiently than is possible with other clean-coal methods. Best of all, the process can be retrofitted on to existing power plants (*Technology Quarterly*, Sep. 2002 5).

Combined cycle means a process by which a fuel serves a double use in that it simultaneously drives a steam turbine and a hot gas turbine. This can be done by burning oil or gas in a furnace which directly produces steam, driving the steam turbine, and by pressurising the hot exhaust air from the furnace, driving the hot gas turbine. Both turbines are linked to the same electricity generator, thus reinforcing its capacity. Gross energy efficiency of combined cycle power plants is about 55 per cent, making a difference indeed to the range of 20–35 per cent of former coal plants.

A previous new generation process which significantly increased burning-efficiency was pressurised-fluidised bed combustion. Air is pressurised at 12 bar, then injected into the furnace, where coal is suspended together with the 'bed' material (which is sulphur-absorbing ash and limestone, thus rendering add-on desulphuration unnecessary). Coal particle concentration in the air is just 2–3 per cent thereby resulting in a burn-up rate coming close to

100 per cent. Temperature in the furnace is only 850°C resulting in a low emission rate of nitrogen oxide (NO_x). The remains are neutralised in the furnace by injecting ammonia so that there is no need for add-on denox processes. Add-on de-dusting, however, is still necessary.

Energy efficiency can furthermore be increased by using the heat from power generation for heating, or using the heat from heating for electricity generation. Both variants constitute a combined heat and power system. Cogeneration of heat and power has long been conventional practice in big power stations which deliver long-distance heat, and is now also being applied in district heat and power stations (DHPS), which deliver short-distance heat and power, typically in a neighbourhood, a factory, a large office edifice, a hospital, a hotel, or similar. DHPS usually run on natural gas. New generation DHPS generators which can operate at variable revs/min are even 50–100 per cent more efficient than more conventional ones running at a constant 1,500 revs/min (to achieve 50 Hz mains frequency).

These are remarkable examples of increased energy efficiency and integrated emissions control. But we are still dealing with the burning of fossils, emitting CO_2 in large quantities, creating hazardous waste in the form of ash and filtering cake from de-dusting which have to be dumped or reprocessed, and which put manifold pressures on ecosystems when denaturalised as raw materials in large volumes all over the globe, on and below ground, and beneath the sea. So, despite these increases in efficiency, there is no real change in ecosystem compatibility and metabolic consistency. As a consequence, the rebound growth in which this will end up (II/7.7.1) continues to be carbon-based, and carbon-related pressures on the environment continue to be unsustainable despite these major incremental innovations in components of the carbon energy paradigm.

4.2.3.3   Total emissions control: CO_2 sequestration

As if somehow they suspect this, the defenders of coal are now trying to ensure total control of all emissions from burning processes, including CO_2, by integrated solutions as much as by end-of-pipe measures. CO_2 is considered today as the crucial question concerning carbon, notwithstanding the fact that the problems of carbon are obviously more complex and extend beyond emissions of CO_2. Whereas a technical breakthrough in nuclear energy which would solve the problem of hazardous nuclear waste is not in sight, such a breakthrough is in principle possible for fossil energy by way of CO_2 sequestration. This is certainly not suitable in small or mobile sources of CO_2 emissions such as in cars or home heating, but it is technically and economically feasible in large central power plants where it makes sense indeed, for example in an integrated gasification combined cycle IGCC as noted above.

Carbon sequestration can be done in various ways all of which include capture and subsequent storage or fixation of CO_2, at source or after source. Capture is the more difficult and expensive part. For example, CO_2 can be 'scrubbed' from exhaust gas by passing it through a chemical reagent such as mono-ethanolamine. Scrubbing is suitable in large plants where exhaust gases come in large volumes and high concentration. It can be retrofitted on to existing power stations or similar big industrial emitters.

Steam reformation is an approach which can be implemented by the integrated gasifier combined cycle. Reformation is the process by which most of today's hydrogen is produced. Steam reformation is done by making a carbon-containing fuel, or CO_2, react with oxygen and steam. This results in synthesis gas composed of carbon monoxide and hydrogen (the latter coming from the steam, i.e. water). Assisted by a suitable catalyst, the CO in the gas can be transformed into CO_2 which re-enters the process. In addition more hydrogen is produced. The hydrogen can be burned in a conventional gas turbine, or in a motor or in a fuel cell. Originally, IGCC plants were designed for ammonia production, not for removing CO_2 from the exhaust. Combining that with further processes into polygeneration of synfuels, hydrogen and electricity together certainly represents a major component in a scenario of cleaner carbon, apart from possibly bridging the transition to the hydrogen age.

Capture of CO_2 after source can be achieved by catching it in large 'wind fences' soaked with dissolved calcium. The CO_2 can then be leached and fixed by one of the methods below. The calcium solvent can be reused. It remains unclear, however, how many such fences of what size would be needed in order to have an impact.

Interim storage of CO_2 is possible by liquefaction or by dry freezing of captured CO_2. For final storage or fixation, liquefied or dry-freezed CO_2 can then be handled in a number of ways, for example by dumping it into deep sea below 3000 m where it is hoped pressure will be high enough to prevent CO_2 from surfacing again. Theoretically, deep sea dumping would have the biggest potential of re-absorbing CO_2 at the level of thousands of gigatons of coal (GtC). It remains uncertain, though, whether the method is actually reliable, while its high costs are in no doubt.

CO_2 can also be stored by injecting it into deep rock layers, with a re-absorbing potential by the hundreds of GtC. Disused mining galleries or depleted oil and natural gas reservoirs could also be used. These places are well known and available at no additional cost of exploration. A particular variant of this way of storage is injection into coal seams which are too deep to mine economically. CO_2, re-gasifying there, is absorbed on to the surface of the coal, thus locked up safely. As a welcome byproduct, methane in the seams is often displaced by the injected CO_2. That methane can be captured and used.

The snag is that this method has the lowest re-absorbing potential of all (just by the tens to hundreds of GtC). A comparatively better performance can be achieved by injecting CO_2 into deep saline aquifers, also called geological sequestration, with a re-absorbing potential of hundreds to thousands of GtC. CO_2 dissolves in the salt water to a certain extent. It also reacts with silicate minerals present to form carbonates that remain stable for many millions of years. The method has been pioneered since 1996 by StatOil, Norway, because of a tax on carbon emissions. As an idea, CO_2 petrifaction is actually quite convincing. Captured CO_2 would have to be compounded with magnesium silicate, resulting in carbonate, i.e. rocks of dolomite. This would do artificially what nature does over millions of years.

Finally, one could get rid of CO_2 by 'sea fertilisation'. Fixation of CO_2 would in this case be done by farming of marine microalgae (plankton) with iron the algae need but indeed often lack for their growth. When growing, the algae incorporate CO_2 by way of photosynthesis, and when dying, they take the absorbed CO_2 with them down to the sea bed. For the moment, however, this is hardly more than a bold idea. Whether it works, and what such an artificial blossoming of plankton would do to marine life, is not clear at all. It might help to restore fish schools. It might also help to ultimately kill them.

CO_2 sequestration, if it were operational on a regular basis, would reduce but not eliminate anthropogenic climate change. The reason is, first, that most methods do not really solve the problem once and for all but rather postpone it for an uncertain period of time. We lack the experience to be able to say. Furthermore, possible methods of CO_2 sequestration need to be applied in central plants. Apart from the cases of calcium fences and sea fertilisation, which lack credibility, they do not cover exhaust from decentral processes such as combustion in car motors or individual heating systems. In consequence, if emissions of CO_2 are to be avoided, these processes would have to switch to burning hydrogen, produced in central coal-based thermal plants with CO_2 sequestration – thereby ushering in nothing other than a transition scenario in which carbon would be phased out and hydrogen introduced. Consistent with this perspective is the consequence that CO_2 sequestration significantly raises the cost of fossil energy which in turn levels the commercial playing field for regenerative clean fuels and fuelless energy so that these can become competitive earlier on their learning curve.

The various components of a strategy of cleaner carbon represent an impressive demonstration of ingenuity. So carbon still seems to be a big option. And yet it is a weak option, firstly because of the environmental pressure (fuel footprints) which still remains to a large extent, secondly because carbon-related technologies are approaching late stages of maturation and retention in their life cycle (II/7.4). Seen like this, cleaner carbon resembles the last blossoming of a dying tree, activating once again all of the means and measures available to a mature system. Technologies based on coal, oil and gas still have the big advantage of being deeply rooted in the culture, institutions and technical structures of contemporary industrial society, including national innovation systems (II/10.2). Carbon-based energy technologies are thus highly connective and diffusable (II/8.2). This big advantage, however, also represents the biggest disadvantage of carbon, i.e. its being caught in long-established path-dependence (II/7.8), definitely locking in people's careers and professional skills as well as the fate of companies and their capital. Better to say good bye in time and turn to new horizons...

4.2.4 Biofuels: Another Weak Option

To expect energy from biomass to be one of the new horizons in energy would be erroneous. Historically, burning wood represents the oldest energy horizon we know of since the domestication of fire by primitive cultures. Biofuels look like an evasive manoeuvre by path-dependent carbon burners, forming a conservative defenders' coalition with a hyper-intensified low-quality agriculture which is looking for substitutional business opportunities.

Biomass is nothing other than fresh organic carbon compounds in contrast to fossil carbon. Consequently, biomass energy derived from direct burning or from conversion into liquid or gaseous biofuel which is then burned, entails the same or similar environmental problems as burning fossils. In particular it makes much worse use of land than does burning fossils, and it pollutes exhaust air with toxic and otherwise hazardous particles and substances. Just as natural substances can be highly poisonous, biofuels are quite 'dirty' in every respect. Fresh wood and other greenery have the worst emission performance of all carbon fuels, entailing expensive measures of fuel beneficiation and emissions control. The misperceptions about biofuels probably stem from some ideological bias in favour of anything labelled 'bio' or 'natural'. To promote biofuels and solar energy in the same breath is actually confusing. The possible contribution of energy from biomass is a case for reconsideration.

The big argument in favour of biofuels has been the neutral CO_2 balance of plant biomass. Fuels from plant biomass only release as much CO_2 as they have absorbed during their growth. As new plants grow, so the story goes, they re-absorb the CO_2 released by burning biofuels, which will then be released again when the newly grown plants are burned, etc. This indeed is an advantage in comparison to fossils, though a number of factors make the advantage much smaller than it looks at first glance.

CO_2 may be an important factor, but is actually not the only important environmental aspect of carbon fuels, as CO_2 is not the only factor of anthropogenic climate change. Energy from biomass also includes energy from burning or converting animal waste and animal food waste into biogas. This is not included in the calculations of the neutral CO_2 balance of biomass. The balance goes into deficit with the inclusion of the negative CO_2 balance of all of the working energy, machinery, agrochemicals and further materials which have to be invested in growing plants and tending animals before some of their physical energy content can finally be harnessed. If this advance energy were also to be covered by biomass, this would reduce the availability of energy from biomass accordingly, or would lead to severe problems of mass production of biomass respectively, as discussed below.

The potential of energy from biomass appears to be huge in theory. Two hundred billion tons of biomass grow annually on the planet, representing one hundred billion tons of hard coal units, about eight times the current global energy demand. This, though, is a naive consideration. How can this widely dispersed biomass be sensibly collected? And should it really be done in the face of urgent ecological requirements of reforestation and augmentation of the stocks of biomass on earth? Energy from biomass has difficulties regarding large-scale collection. More importantly, biomass represents highly valuable stocks of resources too good for just being burned.

A scenario in which biofuels could make a significant contribution to the energy supply would be large-scale plantations of energy crops, in Europe sugar beet, sunflower, rape, wheat and potatoes. Various gasoline-range hydrocarbons can be derived from methanol and ethanol, which are alcohols produced from crops, for example bioethanol from sugar beet or from the starch of potatoes, and methanol from rape as di-methyl ester, known as biodiesel, for use in tractors with robust motors and an exhaust plume which one had better not get too close to. Methanol from sugar cane has been used as a petrol substitute in Brazilian cars since the 1980s. Interestingly, the programme has of late quietly been cut back. In New Zealand methanol is derived from natural gas by a catalytic process, providing one third of domestic petrol demand.

The production of biofuels, however, creates relatively large amounts of problematic waste which cannot all be processed for further use, for example lignin from wood, or inorganic salts which need laborious and expensive treatment before they can be used in some way. Moreover, what biofuels may save in terms of net CO_2 emissions is counterbalanced by acidic airborne emissions of a similar or even larger per capita amount. It is certainly not very eco-consistent to replace one evil by another.

A number of technology assessments of biofuels and fuel crops, carried out or published by environmental authorities, were unsupportive (BMU

1999), mainly because of pollution problems which accompany the production and use of biofuels, and because of bad use of the land on which the crops are grown. Astonishingly, these findings are not being considered (Jensen 2003), perhaps because of a combination of those bio-biases and agricultural lobbying for biofuels which promise, or already have produced, another highly subsidised agroindustrial business.

If systematically planned for and produced on a large scale, fuel crops perpetuate all of the unsustainable features of conventionally industrialised agriculture such as intensive use of chemical fertilisers and pesticides, possibly combined with incorrect irrigation and waste of scarce water resources (certainly depending on the region), degradation and loss of topsoil, monoculture and loss of biodiversity. And yet, even if driven to the environmentally devastating extreme, yield per hectare cannot be high enough to make bio-fuels competitive and give farmers a satisfactory income. All studies on the subject have concluded (in a rare example of consensus among experts) that biofuels would not be competitive even at energy price levels that were twice as high. This is the main reason why Brazil has stopped its bio-alcohol programme. The cost of other regenerative energies can be expected to come down because of considerable learning effects and economies of scale; agriculture, however, as an already highly intensified mature industry, no longer has that potential. Biofuels would probably have to be subsidised forever.

Bad use of land in this case also includes an ethical dimension. If a significant part of the energy supply were to come from fuel crops, the land used for this purpose would have to be widely extended. This would lead to conflict over mutually exclusive types of land use: food fields or fuel fields? Food for all, or energy for the rich at the expense of the poor? Forests which endure at a high level of biodiversity, or biologically poor industry forests which are pushed up to be chopped down and burned or gasified?

This may not bother citizens in more developed richer countries, but is an issue of outstanding importance at the global level. Besides being an ecological question, and a question of availability of food of high or low nutritional quality, it also poses an ethical question which must not be answered by the power of money alone. Fertile land has become scarce on the globe, and using it to grow plants that are going to be burnt is as immoral as destroying excess food in the face of millions of starving people simply in order to prevent prices from fluctuating.

Biofuels, moreover, contain a complex chemical mixture of many things. Except for sulphur, biofuels are polluting and come with a number of ingredients which can be disruptive of technical processes. This is why biofuels need comparatively more, and more expensive modification, i.e. measures of fuel beneficiation and emissions control, and in many cases also modified furnaces and motors. Methanol, for example, is highly poisonous (and this is one reason why it might not be sustainable in fuel cells either).

Burning biomass would seem to be more acceptable where it represents the final step of making use of waste or residual byproducts which are left over after all possibilities of use and recycling of organic matter have been exhausted. This, however, is a niche application, an exception to the rule, particularly applicable in rural areas or in bigger wood-processing factories. Waste wood, lean wood, straw, also in a processed and dried form of pellets, briquettes, wood chips, chaff, or similar, can thus be burned in especially designed local heat and power plants. Burning those fuel surrogates in standalone houses is again not recommended unless a lot of equipment and infrastructure is added. In certain regions, enthusiasts have tried to set up regional networks (i.e. markets) for fuelwood in pursuing a strategic idea of decentral heat and power based on biomass. They consider this to be a clever alternative to the disposal of residual wood, straw etc. It may not be such a good idea. It might be better to satisfy energy requirements by local fuelless energy, and leave those residues to the soil and its micro flora and fauna, for example by ploughing them under as a measure of soil melioration.

Quite similar aspects also hold true for biogas. Biogas, like natural gas, is primarily methane (CH_4) from various sources, particularly from manure and slurry fermenters on farms, or captured as dumping site gas or seam gas. The possible volume of biogas worldwide is estimated in a range of tens to several hundreds of million cubic metres a year, which is not much, but should not be spurned either in a local context. Biogas can be used in gas motors of combined heat and power plants. In the German town of Rotenburg organic wastes from private households, distilleries and farm slurry are co-fermented in a municipal biogas plant which feeds a district heat and power station. Any other residual bio-materials, such as organic waste from the slaughterhouse, food processing companies, canteens, restaurants, can be included in the co-fermentation process. The substrate from the biogas plant is used as a fertiliser by local farmers involved in the programme.

Compared to solar, wind and hydropower, and also energy crops, biogas is reported to have the least potential. If it were to be extended artificially, as regional energy agencies are demanding, the same criticism as in the case of energy crops would apply. Furthermore, for use in public pipes and in cars, biogas would need additional purification through molecular carbon filters, activated coal, or similar, in order to eliminate compounds of hydrogen sulphide, nitrogen, and others. Local energy agencies have even requested that hydrogen be made from biogas. Given the additional processing requirements which this would involve, and given a number of less cumbersome large-scale alternatives, biogas is certainly the least suitable candidate for producing hydrogen.

Biogas may make sense within the normal boundaries of providing energy for one's own requirements in rural niches on farms and in food and wood processing factories. There, however, the real raison d'être of biogas turns out to be proper waste handling rather than energy supply. Fermenters neutralise organic wastes, particularly from intensive animal farming, which otherwise presents a big problem. This is particularly the case in large-scale intensive animal farming (i.e. more than 250 cows or 55,000 hens), where, in big reactors of 850 m^3, biogas can be produced profitably at 2–3 ct/kW. In small units which perform at about 15 ct/kWh interest in biogas is certainly not economic. But substrates from fermentation constitute high-quality fertiliser if the biomass is free of hazardous ingredients. The biogas produced is a welcome byproduct of fermentation rather than its real purpose. Biogas should therefore not be propagated in a context of energy policy, where it is at best a little sideshow, but be promoted as an appropriate way of processing organic waste that cannot otherwise be used.

Here again, though, what applies to manure and slurry may not apply to fresh leftovers from animals and food. It might be much more reasonable to do with these leftovers what has been done with them hitherto: feeding them to animals. Using leftovers with high protein content as a fuel instead of as feed looks if anything pretty dim-witted. Synthesising such molecules, if it can be done at all, is complicated and expensive. Even nature takes four units of vegetable calories (for eggs, pigs) up to ten units (for beef) to build up one unit of animal calorie in the form of protein and fat.

A telling story is the interdiction of sheep and beef meal as an animal feed in connection with the continental BSE scare in 2000 (Braungart 2002). Since then, the cadavers had simply to be burnt without being used further, in most cases not even yielding heat or electricity which represents, so to say, the minimum use of final phase-out when other options are no longer applicable. As a result, high-quality animal feed was in short supply after that interdiction, and had to be compensated for by increased imports of soya feed. In EU countries this soya mostly comes from Brazil where the additional demand is met by chopping down forested areas so as to make them into additional soya plantations.

All in all, it has to be concluded that the emphasis and support given nowadays to biofuels and energy from biomass is inappropriate and overdone. In terms of ecosystem compatibility, metabolic consistency, pollution control requirements and proper handling of wastes biofuels can even be overtly unhelpful. Among regenerative energies, biofuels are the least competitive of all, and will remain so because, in contrast to other regenerative energies, only marginal increases in productivity are to be expected, if at all. Recent innovation dynamics in this field were triggered by affirmative regulation and subsidies, pushed by biased policy and agricultural business interests. From a technological point of view, energy from biomass is immediately connective because it pertains to the same scientific and technological paradigm as carbon energy in general. This still dominates energy policy, locked up as it is within its path-dependence and happy about new demand within old territory because of its incapability to change path.

4.2.5 Ammonia and Silane: Non-obvious Alternatives to Carbon

In the search for alternatives to carbon fuels a number of possibilities have been proposed, for example ammonia (NH_3), a base chemical for producing fertilisers, gun powder and explosives. Since invention of the Haber-Bosch process ammonia can be synthesised. It can be used in certain fuel cells, and it is a suitable way of having atomic hydrogen (H) available for welding. The snag with ammonia is that its burning, i.e. oxidising, results in nitrogen oxide NO_x, an acid formant associated with a wide spectrum of environmental pollution, besides the fact that ammonia stinks, as anyone who occasionally passes by a manure reservoir knows.

A more sophisticated alternative, at least on paper, is silane (Plichta 2001). Silane can be derived from silicates Si_xO_y, a main ingredient of rock and sand, which is why silane is presented as gasoline from sand. Silanes are silicon hydrides, such as mono-silane or silane gas SiH_4 (parallel to methane or carbon gas CH_4). Silanes can also take on the form of powder or oil. Lower silanes (up to 4 Si atoms) are explosive. Higher ones, however, such as pentasilane (= 5 Si), are not. The energy density of silane is equal to gasoline, and it is easy to store and to use.

Silane not only reacts with the 20 per cent oxygen in the air, but also with the 80 per cent nitrogen. This results in a double combustion process. First, in a hot combustion chamber or furnace, silane decomposes into atoms of silicon and hydrogen. The hydrogen then burns with the 20 per cent oxygen in the air, resulting in H_2O, whereas the silicon reacts with the 80 per cent nitrogen in the air, resulting in silicon nitride Si_3N_4. Nitrogen which remains in the air can be burned up by injecting dispersed silicon powder. As a result, there are zero hazardous emissions indeed, just water H_2O and silicon nitride Si_3N_4. Silicon nitride is non-hazardous and can be processed into ammonia and water-soluble silicates. Ammonia then can fulfill its role as base chemical, and the silicates can easily be used in building materials. Without processing, silicon nitride just weathers to sand. Silane could be filled in tanks in the form of liquid silane oil. At the same time, the sandy waste silicon nitride can be taken back for recycling. Silanes can be produced everywhere. In a completely regenerative variant this could be done in typical desert countries, just like solar hydrogen; and connected to this, silane SiH_4, in contrast to pure hydrogen H_2, could be a practical store of H_2.

But silane also has its snags. One, of lesser importance, is that in an internal-combustion engine the sandy residue of silane requires different motor properties, for example ceramic motors which are more heat-resistant and less materials-sensitive (in a rocket, silane would seem to be perfect). Another snag, and crucial from an environmental point of view, is that the production of silane from silicate Si_xO_y requires a reduction agent, much in the same way as the production of iron from iron ore (= iron oxide) involves such a reduction agent, and the means of choice is: coke, in an electric arc furnace. This is why silane does not deliver on its promise of decarbonisation, and why one can remain sceptical about silane as long as there is no alternative method of silicate reduction.

4.2.6 Hydrogen: The Obvious Alternative

For the time being, the search for alternatives to carbon fuels has finally arrived where it has been heading for a long time: at hydrogen. Experts wedded to carbon fuels have over this period denied hydrogen its role as the fuel of the future. Among their counter-arguments was the fact that hydrogen may be clean-burn but is easily inflammable. Well, what else can be expected from a fuel? Hydrogen is in fact less explosive, or less damaging when exploding, than town gas, gasoline and coal dust. In World War II, when hydrogen pipelines in industrial districts were bombed, this produced long tongues of flame, but the flames died rather quickly and left the pipelines operational if the holes were not too big to be repaired.

Further counter-arguments were the high costs of producing hydrogen and lack of distributive infrastructure. Such arguments – if put forward as proof of alleged non-feasibility rather than pointing to the requirements of planning for successive steps – can only be taken seriously by people who do not know about learning curve economics (II/7.7). If there is a will among the elites to get it going, if the process is properly planned for and implemented step by step, also in stepwise combination with an adequate financial build-up and regulative standards-setting, the costs and operative infrastructure of an innovation are normally not a fundamental obstacle.

The true fundamental problem is a formative one: to have a new paradigm adopted, or at least accepted in the mind, to create knowledge, awareness and will among those who are in charge, and among those who are going to be touched by an innovation in some way. All of the rest then is a task that has to be dealt with, which certainly takes time but is no insurmountable problem. It took a quarter of a century, from the mid-1970s to the late 1990s, to have the minds of scientific, industrial and political elites seized by the clean-fuel paradigm and have hydrogen promoted as the fuel of choice. It is now definitely on its way to regular production and distribution, at first certainly on a small scale, but on ever larger scales later on. In 2002 the EU commission set up a €2.1 bn 'hydrogen vision', and in 2003 the American government followed suit with a $1.2 bn hydrogen development programme.

The properties that favour hydrogen from a technical and ecological point of view have always been obvious. Hydrogen, normally in its bi-atomic state H_2, is a dense high-energy fuel, offering a potential of more than 50.000 BTU per pound, comparing to gasoline at 16.000 BTU and coal at 11.000 BTU. Hydrogen is present in many ubiquitous compounds, among them water. Hydrogen in fact belongs to the class of renewable rather than exhaustible fuels because it either emerges from or enters the water cycle.

There may even be natural reservoirs of H_2. Certain geological findings could possibly indicate the existence of subterranean reserves of pure hydrogen. This was deduced from the presence of special bacteria. In the upper 20 km of the earth's shell, where there is hot or melted rock, conditions for hydrogen formation are considered to be favourable. There is some possibility that these reserves will not have to be tapped into, to the advantage of nature conservation, because there are methods of producing hydrogen from what is already available today, i.e. hydrocarbons in the shorter term, and, in the longer term, from what will always be there as long as there is life on earth: water.

When burned, i.e. when the hydrogen H_2 is oxidised by reacting with oxygen O, hydrogen is absolutely clean in that the exhaust consists of water H_2O in the form of droplets or steam. With regard to health, hydrogen is in no way toxic. In reality, the pureness of gaseous hydrogen will not be 100 per cent in full, so that there may also be some formation of pollutants (such as NO_X) which, however, all stay well below critical levels.

Steam certainly has an impact on weather and climate, for example in the form of mist and clouds. But exhaust steam is in practice not very relevant because even in very large volumes it will not significantly add to the natural water cycle. Moreover, exhaust steam does not accumulate in the atmosphere over a longer period of time but is completely recirculated soon, just as clouds are, by precipitation. This is why even large volumes of water droplets and steam from burning hydrogen do not contribute to climate change, in contrast to CO_2 where there is a net accumulation in the atmosphere because the carbon cycle is much smaller than the water cycle so that carbon emissions from fossils exceed the re-absorptive capacity of ecosystems.

Environmental problems related to hydrogen only arise to the extent that, and as long as, hydrogen is produced on the basis of converting hydrocarbons, as discussed in the following. Apart from that, hydrogen, on the basis of today's knowledge, is an absolutely clean end-fuel, and the more hydrogen is derived from clean regenerative sources, the more the process on the whole will become fully eco-consistent.

An additional advantage of hydrogen is that it is connective also to existing carbon-burning technologies as well as to new electrochemical technologies such as fuel cells. Hydrogen can be burned in industrial furnaces and in internal-combustion motors just like petrol or gas, but in contrast to petrol and gas hydrogen can alternatively produce electricity in fuel cells which feed electric motors in manifold applications.

For the time being, hydrogen is for the most part derived from natural gas or gasified coal. At present, the most advanced method in thermal power stations is the integrated gasification combined cycle IGCC with integrated CO_2 sequestration as mentioned above in I/4.2.3. Hydrogen produced in this way would cost about \$8–10 per gigajoule if derived from natural gas and \$10–13 if derived from coal. This is fully competitive with or even cheaper than power from nuclear and conventional coal plants at \$10–18.

Before IGCC plants will be state of the art, there are a series of more conventional ways of extracting H_2 from hydrocarbons, be these solid (coal), liquid (petrol) or gaseous (liquefied petroleum gas), and also from ammonia, methanol and ethanol from agricultural biomass fermentation or dumping site gas. In a Norsk Hydro hydrogen power plant, H_2 is obtained from decarbonisation of natural gas ($CH_4 \rightarrow C + 2H_2$). The pure carbon obtained can be used in chemical processes, for example to produce tyres. A big development programme run by Solid Energy in New Zealand is aimed at obtaining H_2 from gasification of local hard coal which is considered to be particularly suitable for extracting H_2-rich gases. Results are expected around 2005.

A different approach pursued at the Tokyo Institute of Technology is to create H_2 from rust, more precisely, from a reduction-oxidation reaction with iron and iron oxide. These materials are non-hazardous and relatively cheap. The process includes just water and iron pellets. The water is heated so as to expose the iron pellets to steam. This induces artificial rusting, amplified by a special catalyst, i.e. a reduction-oxidation reaction with iron oxide magnetite in which H_2 is released. The iron oxide can be permanently recycled. The reaction runs at 300°C. 1 kg iron produces 48 gr H_2. Whether this can really be a large-scale option remains to be seen.

Hydrogen from biomass would clearly not be a wise large-scale option, although there may be reasonable niche applications. An EU-funded development project led by a team at the university of Warwick, UK, starts from converting wet biomass into a mixture of biogas CH_4, water H_2O, carbon monoxide CO and dioxide CO_2. In a reactor the H_2 is extracted from the CH_4 and H_2O. The key element there is a semi-permeable ceramics membrane coated with palladium which only allows H_2 to pass. The purity of the hydrogen obtained is up to 95 per cent. The amount of CO and CO_2 released is equal to the amount that would be created anyway by natural decomposition of the biomass.

An electrochemical approach to producing H_2 is electrolysis, i.e. decomposition of water into H_2 and O by passing an electric current through an electrochemical cell consisting of positive and negative electrodes. Most often, electrolysis is applied in metallurgy to obtain very pure metals that need to be separated from undesired components. For energy purposes the snag with conventional electrolysis, as with the other methods mentioned, is that the electricity or the energy content of fuels required to feed the process is higher than the energy content of the hydrogen obtained. The process can be improved by additives which make the water conduct electricity better. It nevertheless takes more energy than can be recovered. In consequence, electrolysis represents a clean and reasonable option only if the electric current put into the electrochemical cell stems from clean regenerative sources other than hydrogen. There is now a major innovative race in progress to be among the first to develop such an integrated process of H_2 production which can be run on a large scale.

Adherents to 'free energy' maintain that ways of super-efficient electrolysis exist but are being ignored (Clear Tech 2002). According to Stan Meyers and Xogen Power Inc. water breaks down into hydrogen and oxygen with very little electrical input if it is hit at its own molecular resonant frequency. A researcher by the name of Freedman is said to have obtained a US patent in 1957 on a special metal alloy which splits water simply upon contact with no outer electrical impulse or additional chemical agent. If such marvels really existed, why are they neither better communicated nor considered in practice?

A recent Canadian patent is based on a process called gravitational electrolysis. A cylindrical container filled with an electrolyte solution rotates at very high speed. Centrifugal power starts to separate positive and negative ions. This creates electrical current which in turn triggers a self-sustaining process of electrolysis in which H_2 and O are separated as soon as water is fed into the rotating container. The process is reported to be highly cost-competitive. Costs are much below comparable processes such as steam reforming of methane CH_4. The system also enables smaller and more efficient systems than was hitherto possible.

A rather special concept is electrolysis on 'energy ships'. These would exploit the airflow caused by wind and a ship's speed by using large Flettner rotors on the bow which transmit wind power in travel speed, thus maximising the speed-energy differential between bow and water, and thereby driving a water turbine inside the ship. The water turbine generates electricity that can be used for producing H_2 by electrolysis, or to other ends such as running a desalination device which produces drinking water.

A more probable source of clean regenerative electricity for large-scale electrolysis of water is photovoltaics (I/4.2.6). Solar cells are the centrepiece of an energy scenario of solar hydrogen. An English firm, Hydrogen Solar Production Company, together with Professor Grätzel at EPFL in Lausanne, has developed direct photovoltaic electrolysis, an example of photolysis, i.e. electrolysis based on photovoltaics. The process makes use of a new generation of solar panels made of nanocrystalline semiconductor materials which include oxides of titanium, zinc and niobium (columbium, respectively). The molecular structure of the material represents an enormous micro- and nano-surface and is thus able to absorb much sunlight which creates more current, and as a result more H_2, than less voluminous surfaces do. The days when solar hydrogen was just a vision are gone. The development of new generations of solar cells with greatly improved performance at lower price is taking off, thus making the future of solar hydrogen almost a certainty.

A generic alternative, still shrouded in mystery-mongering despite being well documented, is Brown's gas. Yull Brown was an electrical engineer born in 1922 in Bulgaria who went to Australia in 1958 where he became a university professor and discovered a proprietary method, patented in the USA and Australia. It concerns running an electrolysis cell without membranes that creates a very special mixture of a lot of H and some H_2 as well as O and O_2 in a stochiometric mix, i.e. 2 atoms of H for 1 atom of O. Normal electrolysis creates H and O too, but immediately after creation the mono-atomic state drops to the normal bi-atomic state H_2 and O_2. Brown's gas does not, but stays in a mono-atomic state, which comes with a number of remarkable properties.

In Brown's electrolyser, 1 kWh of electricity produces 340–370 litres of gas, four times the normal yield, thus representing a positive energy balance. The method of production is highly efficient and relatively cheap compared to rival approaches of producing H and H_2. Brown's gas is non-explosive. Upon ignition with a spark on a torch tip, the gas burns by implosion, which is one of its unique properties, since normal gas explodes. Therefore, Brown's gas, in contrast to normal H_2 which immediately reacts with O_2 upon contact, is absolutely safe. Brown's gas is lighter than air, so it diffuses rapidly, thus representing another safety aspect. Normal gases are heavier than air. They collect around the lowest point on the floor and are explosive.

When normal hydrogen or oxygen is burned, the di-atomic H_2 or O_2 needs to be cracked, thereby releasing heat, i.e. losing energy. In Brown's gas, by contrast, with much H and no or just little H_2, no di-atomic state needs to be cracked. The resulting flame is a 'cool' flame which maintains an extraordinary energy potential. It has particularly interesting applications in welding, boring and cutting of metals and minerals. The 'cool' energy-rich flame of Brown's gas can even vaporise (sublimate) tungsten within half a minute, which requires a temperature near to 6,000°C, whereas normal $2H_2:O_2$ mixtures only reach 2,200–2,900°C. Brown's gas can melt bricks to a kind of volcanic material, bore and pierce bricks and ceramic tiles, steel can be welded to bricks, etc. So Brown's gas presents serious competition to already existing laser technology as well as clean-burn technologies on the basis of carbon fuels, such as porous burners and flameless oxidation (I/4.5.4).

With Brown's method, 1 unit of water yields about 1,860 units of gas, whereas normal di-atomic $H_2:O_2$ yields just 933 units. Conversely, upon the implosion of Brown's gas, the gas instantaneously collapses to 1 unit of water, leaving an almost perfect vacuum of 1.859 units. This opens up entirely new applications for anything that has to do with suction or pumping, which can be done without moving parts, a big advantage, and might also be used for propulsion of vehicles and rockets.

Brown's electrolytic cell can be produced and used on any scale, like fuel cells, small and large, stationary and mobile. They can produce the gas on demand, so that no heavy storage cylinders are required, nor their transport and cumbersome handling. If desirable, however, for example in remote areas, Brown's gas can also safely be stored. Non-continuous electricity generators such as solar cells or windmills can electrolyse the gas. Stored in some container, it can replace storage accumulators, i.e. large batteries or battery systems, not to mention conventional fossils-based generators.

Since the purity of Brown's gas is high, its use causes no problematic emissions, indeed zero hazardous emission, just water. It is safe and clean, highly effective and metabolically consistent, and its efficiency is at 90–95 per cent. Cost-efficiency is correspondingly high. The question remains as to why Brown's electrolytic cell and the stable H and H_2 yielded from it, although invented several decades ago, have not been introduced?

Whether in the form of Brown's gas, or as solar hydrogen, or from any other production approach, hydrogen presupposes a widely applicable solution to the problem of hydrogen storage which has hitherto been the main bottleneck in introducing the hydrogen economy. Here again a number of promising developments are making progress.

Conventionally, hydrogen is stored in the form of H_2 gas compressed in high-pressure tanks (200–300 bar), or as liquid H_2 in cryotanks, i.e. deep-freeze tanks at −253°C. Neither of the two methods is convenient. More practical recent developments include 'hydrogen mud'. Safe Hydrogen, a US firm, reports having developed a special 'mud' which stores hydrogen and releases it on demand at a purity of 99.999 per cent. The procedure involves mixing the 'mud' with water. The overall price would be 40 per cent below today's price of gasoline. Moreover, the process is safe and simple, non-inflammable and non-explosive.

A different approach which has reached an advanced stage of development, especially by Canadian companies, is absorbing H_2 into solid compounds, either in the form of metal hydrides or sodium hydrides. The fuel carrier obtained from the latter method is known as Borax, also known as Powerball fuel pellets. The method works at normal temperature and pressure, and the material is non-toxic and non-inflammable in the absence of water. Borax is sodium tetraborate. One unit of sodium hydride, upon contact with water, releases 100 units of hydrogen. The residual sodium can be recycled from an extra tank when sodium hydride tanks are refilled.

Meanwhile, BASF, a German chemicals corporation, has announced its nanocubes. These are composed of organic terephthal acid and inorganic zinc, representing a metal organic framework (MOF) with a highly porous structure. 2.5 grams of these nanocubes have an inner surface area the size of a football field. Hydrogen is physically taken up by the MOF at about 10 bar (the low pressure of a cigarette lighter) through a process called physisorption, which does not entail a chemical reaction. So there is no chemical bonding. Upon lowering the pressure the hydrogen is instantaneously released from the cubes. The MOFs can be filled and refilled over and over again. There is no 'memory effect' as with batteries. The advantage of nanocubes is clean, safe and convenient storage of hydrogen at ambient temperature and low pressure of 10 bar. In contrast to hydrides or 'mud', nanocubes can also be used in very small, portable applications.

Given the basic production and storage components of a hydrogen infrastructure, its distribution to the end-user is just a question of a co-ordinated effort of business management and authorities. Distribution of hydrogen could be done in a parallel system separate from existing fuel merchants and gasoline stations. More probably, however, hydrogen will be added to the product range of fuel traders and will thus be available from fuel oil suppliers and at gasoline stations. The time it takes to reach every fuel merchant and gasoline station depends on the speed at which applications of hydrogen energy are developed and diffused, for example furnaces and burners on the basis of hydrogen, and internal hydrogen combustion in motors and fuel cells.

4.2.7 Fuel Cells: An Obvious Alternative to Furnaces, Combustion Engines and Batteries

The domestication of fire led to the development of various techniques of safely maintaining campfires and open indoor fires, and of building practical stoves, ovens and kilns together with suitable chimneys. The fuel used in these energy devices was primarily wood, additionally also charcoal, and peat if locally available. The industrial revolution introduced coal, petrol and gas, and developed more powerful and more sophisticated furnaces and combustion chambers in which to burn these fuels. The process was still aimed at creating heat and pressure, which can be used to drive manifold pro-

processes, among these, since around 1900, generating electricity with the help of turbines driven by the pressure of steam and hot gas.

In the same way as fossil fuels came in tandem with steam engines, industrial furnaces and combustion chambers, hydrogen today comes in tandem with fuel cells (FCs) as a new way of providing power. In 2002, the emerging American FC industry, a network of 26 companies, has published a report *Fuel Cells and Hydrogen: The Path Forward* accompanying the start of a government programme aimed at a broad market introduction of FCs. Since FCs can be applied in any constellation which involves electricity, they are an example par excellence of what is called a key technology.

FCs are electrochemical cells, as in the case of electrolysis, with the oxidation of a fuel in an FC being simply the reverse of electrolysis. While electrolysis has an input of electrical current which decomposes water into H_2 and O_2, in a fuel cell the input is H_2 and air (oxygen), each to one of the two complementary electrodes (anode and cathode), which react through an electrolyte, such as a membrane coated by some catalyst, thereby generating an electrical current as well as heat and water as an output. Membranes and catalysts are the most sophisticated part of an FC, and the production of catalyst-coated membranes and electrodes is a new business for the chemical industry.

Not only can single FCs be bigger or smaller, but the cells can be connected to each other in modular constructions, thus piling up stacks of different size and power. There are many types and sizes of FCs. Some do not require pure H_2 but can also be run on gaseous or liquid hydrocarbons which are supplied to one electrode, for example natural gas, biogas, seam gas, and liquids such as methanol (methyl alcohol), gasoline, and even diesel. In many cases, however, the hydrocarbons need to be reformed, i.e. hydrogen must be extracted from them in the process.

The electrodes and the electrolyte in an FC are not altered by the reaction, in contrast to conventional batteries, i.e. voltaic cells, in which the materials are altered in the electrochemical process and have to be replaced if depleted. The first fuel cell was built in Great Britain as early as in the 1830s by W.R. Grove. Although the basic principles of operation have been known since then, materials science and engineering know-how were not then advanced enough to produce cells reliable and powerful enough to be able to compete with steam power.

Overall energy efficiency of FCs is at about 70–85 per cent, clearly higher than with combined-cycle thermal power plants at about 55 per cent. FCs are thus superior to traditional rival technologies. They require less primary and processed fuels, and entail less energy losses. The operation of FCs is environmentally clean. There are no or very few noxious emissions, particularly no or very few emissions of carbon and nitrogen compounds (Oertel/

Fleischer 2001, BMU 1999). FCs do not make noise. They need little maintenance because there are no moving parts. They are permanently operational
day after day round the clock.

An eco-balance of FC power has to include environmental pressure on resources and sinks caused by the production of the fuels which are put into the
cells. The ranking of carbon emissions from FCs which are applied in car
propulsion is as follows (*Technology Quarterly*, March 2001 37): the emissions related to FCs which run on H_2 from solar/wind energy are 0 units/km,
from biomass 0.2, from natural gas 0.4. FCs which run on methanol derived
from natural gas cause emissions of 0.6 units/km. Also hybrid cars have
emissions of 0.6 units/km. Today's internal-combustion engines which use
clean desulphurised gasoline cause carbon emissions of 1.7 units/km.

The cost-efficiency of FCs, however, still has some way to go. For the
moment, FCs are rather expensive. A PC25 setup of International Fuel Cells,
South Windsor, of 800 kW (in 4 units) sells for $4 million, compared to less
than $2 million for a gas turbine generator (Freedman 2002 46). The high
price of FCs comes as no surprise but is rather normal for a new development
still in the early stages of its life cycle and learning-curve economics. The
price of 1 kW from an FC today is certainly much lower than was the price of
electricity from a coal power plant around 1900. Seen like this, with a selling
price of only twice the competitive benchmark price of mature established
technologies, FCs are actually in a comfortable starting position. It is a safe
prediction that as soon as FC industries enter their take-off stage, prices of
FCs and of FC-generated kW will come down dramatically.

Does the diffusion of FCs, which generate electricity, induce an all-
electric energy future? Will vehicle propulsion systems, heating systems and
industrial heat processes all be run electrically? Probably not. Using electricity for the production of process heat and for heating houses has not always
proved to be satisfactory. A good cook prefers gas. Also with future heating
systems it might be preferable to have them based on some conventional process of burning gas and hydrogen, particularly if in combination with the
generation of electricity in a combined heat and power process. Conversely,
larger FCs directly create significant amounts of heat. Hydrogen can in any
case either be burned in a conventional way or be oxidised in an FC. The energy future will thus not correspond to an all-electric scenario, even though
the role of electricity clearly seems bound to grow.

FCs can be distinguished according to whether they are large units for stationary use in power plants, or medium-sized FCs also for stationary use in
heating systems and DHPS, then smaller FCs for mobile use in car propulsion, and portable miniature FCs in electric appliances.

Large FCs run at rather high temperatures. There are SOFCs (solid oxide
fuel cells) at about 1,000°C, and MCFCs (melted carbonate fuel cells) at

about 650°C. They are used in power plants, called hot module power plants, on the megawatt level and also in larger DHPS. The waste heat from these large high-temperature FCs is suitable for steam production. These FCs can furthermore be run on less refined fuels such as natural gas, with no reformer needed to create H_2. So there is a possible link here between a strategy of phasing in the hydrogen age while smoothly phasing out the coal age by making use of cleaner carbon (gasified coal). Most of the big electrical corporations have a stake in this new business.

Depending on size and purpose, stationary SOFCs now deliver anything from 25 to 500 kW/el. The apparatus is small enough to be transported on a truck. The electrical efficiency of SOFCs and MCFCs in combined cycles is expected to reach 80 per cent on natural gas and 65 per cent on gasified coal. For the moment, it is at about 50–70 per cent, compared to 30–40 per cent in conventional thermal power plants. Combined with heat, overall energy efficiency is at 90 to almost 100 per cent. High-temp FCs are clean, though not zero emitting. There can be low emissions of SO_x and NO_x. MCFCs also release CO_2, whereas with SOFCs the CO_2 can be sequestered or is petrified.

The typical medium-sized FC is a PAFC (phosphoric acid fuel cell). It runs on natural gas or hydrogen, at medium temperatures of about 150–250°C. Its waste heat at about 200°C is easily used. PAFCs are used especially in DHPS of about 200–300 kW electricity and more than 200 kW thermal energy. Their electrical efficiency is now at about 40 per cent, but by recycling the 200°C waste heat overall efficiency reaches 85–90 per cent. In almost all industrialised countries there are now pilot plants and test users such as hospitals, office blocks, residential units, or plants of municipal power companies, which employ PAFC-based small power stations and DHPS. Thus lead markets are developing.

This is even more dynamically the case with PEM FCs (proton-exchange membrane, or polymer-electrolyte membrane fuel cells). A PEM boom has begun which is leaving PAFCs far behind as the hitherto commercially most advanced FCs. PEM FCs are relatively small-sized, running at 80–150°C. Even at these lower temperatures waste heat can be used, on site or in the neighbourhood. Normal PEMs run only on hydrogen. If there is no H_2 to be fed in directly, it has to come from a connected reformer extracting H_2 from natural gas, methanol, gasoline or diesel. The energy efficiency of PEM FCs is comparable to those of other types of FCs. Further efficiency increases of PEMs can be expected as membranes improve and conventional electrodes are replaced with nanostructured materials. These are also reported to be cheaper than conventional ones. With further diffusion and economies of scale setting in, investment costs for power from PEM FCs will match today's benchmark of 1,000 $/kW for pulverised coal stations in a few years and then fall below it, so as to be comparable to today's industrial gas turbines.

A variant of PEM FCs are direct methanol fuel cells (DMFCs). These have special electrodes coated with a platinum catalyst that can directly consume methanol with no need for prior reforming. On contact with the electrode, methanol and water break down and CO_2 is set free. DMFCs have hitherto been most suitable in the energy range below 1kW, applied in electronic devices such as small computers, handhelds, mobile phones, traffic control, safety tech, measurement and remote control systems. There are now also bigger DMFCs that can be used in cars, for example in Daimler's prototype Necar series (Necar = new electric car). As long as hydrogen is relatively expensive and hydrocarbons are used, DMFCs may have a certain technical and economic advantage over other PEM FCs which need prior reforming. A disadvantage of DMFCs, however, is the fact that methanol (methyl alcohol) is one of about 50 volatile organic compounds (VOCs), highly toxic and carcinogenic substances such as benzene, phenol, tuluol, xylol, and isopropyl alcohol, also present in gasoline, paints, varnishes, wood preservatives, certain cosmetics and organic solvents. On environmental and health grounds all VOCs are in fact phase-out candidates to be replaced with less hazardous substances.

PEM FCs can be used in stationary as well as mobile settings. Stationary settings again include DHPS. Up-to-date pilot plants can have a capacity of 250kW. Smaller variants can serve as single house heating systems at about 1–5 kW/el and 25 kW/heat. There are various test and pilot arrangements, most of them in a context of distributed power generation, i.e. the houses are connected to the grid and can act as customers or suppliers or neutral depending on their current energy balance (I/4.2.8).

An important mobile use of PEM FCs is expected to be car propulsion. Here FCs replace internal-combustion engines by feeding an electric motor that drives the mechanical parts of the machine (I/4.9). Mass production of FC-driven cars is expected at around 2010. So electric cars may finally arrive, as well as later on FC-driven ships. The bigger impact of FCs in the near future, however, could be in home energy, DHPS and central, though not very large power stations, as well as in miniature portable applications.

Small portable PEM FCs, called miniature or mini FCs, are in the process of rapid development. They will probably be the first branch of FC technology introduced into the market on a large scale, in any portable or grid-independent application. Mini FCs are PEM FCs, in particular DMFCs. They are bound to replace conventional voltaic cells in the form of batteries and accumulators. Not only are these inefficient energy suppliers but they also represent environmentally awkward compositions of heavy metals and other highly hazardous materials. Since most batteries and accumulators are small, collection and reprocessing of used units does for the most part not work properly. More often they end up in the dustbin and then in a landfill where they contaminate soil and aquifers, or in an incinerator where they produce highly toxic airborne emissions.

Batteries and accumulators have by now approximated their limits of possible storage capacity at an energy density of 150 Wh/kg, in special cases at a maximum of 300 Wh/kg. Nanocubes, by comparison, already perform at 250 Wh/kg and are expected to reach 1,000 Wh/kg soon. Pentagon demand, a driving force once again, as in the case of microprocessors decades ago, has set a target at 1,000–3,000 Wh/kg. In comparison to batteries, portable miniature FCs have a higher energy density, are long-lived, come with a much better overall eco-balance and can be mass-produced cost-efficiently. So the future lies with portable miniature FCs. They will sell cheaper or at the same price as batteries, perform better and run many times longer.

One of those, the direct methanol FC as explained above, involves toxic methanol. The risks which this represents are on balance nevertheless lower than with batteries since there are no noxious emissions from DMFCs. Also CO_2 releases are not very relevant if DMFCs are small, and the emptied fuel cases can be reused, i.e. refilled like a cigarette lighter, or recycled as a non-hazardous material. A present day methanol micro FC measures about 5 cm by 5–13 cm, the size of a cordless phone. Most of them produce power in the range of 80–350 milliwatts. The electricity from the FC can either recharge a conventional accumulator, if electricity storage is needed, or directly feed on demand the respective device such as a calculator, mobile phone, handheld, notebook, printer or similar mobile office applications. Electrolux, a Swedish electrical corporation specialising in powered appliances, has developed the prototype of a cordless FC-fed vacuum cleaner. Powered with a DMFC, the talk time of a mobile phone is extended from today's 4 hours (best available lithium-ion battery) up to 20 hours.

In contrast to direct methanol FCs, normal mini PEM FCs run on hydrogen. Today's prototype PEM minicells produce 50 watts. So far, however, they have encountered problems with the storage of H_2. Nanocubes could be the answer. Just as with methanol, nanocubes are usable at normal room temperature and are equally easy to handle.

Batteries and accumulators, by the way, will in certain applications be replaced by miniature motors and miniature electricity generators such as free piston micro engines and miniaturised gas turbines. These devices are part of micro-electro-mechanical systems (MEMS) such as very small aircraft with a wingspan up to 15 cm, or telecom devices, minicomputers and microsensors the size of a piece of sugar or less. Most designs now use butane fuel. There are alternatives such as a nuclear MEMS generator, producing power in the range of microwatts by collecting particles produced by radioactive decay, for example nickel-63, the half-life of which is 102 years, meaning that such a device would work more than a century. The power range of these mini and micro devices is from 50 milliwatts to 10–20 watts, and they all replace batteries and accumulators with their noxious and toxic substances, thus eliminating a problem of hazardous waste. Almost all of the development work in this field is done in the USA and serves military purposes.

With FCs of any kind, although these may include high-temperature processes, a certain type of industrial high-temperature high-pressure energy technology will come to an end. Smokestack industries with highly dangerous furnaces, boilers and burners resulting in highly polluting exhaust gas banners will not exist any longer, and chemical power storage in the form of batteries and accumulators (voltaic cells) made of hazardous materials can be greatly reduced. Energy on the whole will be entirely clean, metabolically consistent and compatible with ecosystems if hydrocarbons are successively phased out in favour of pure hydrogen produced on the basis of fuelless kinetic and radiant energy.

4.2.8 Fuelless Energy: Solar, Wind, Hydro, Geothermal

Fuelless energy is radiant, thermal and kinetic energy, such as photovoltaic power, solar thermal heat, wind power, hydropower, tide and wave power, and geothermal heat. Once a playground of alternative vanguards, fuelless energy has become a frontier of sophisticated high technology.

The big environmental advantage of fuelless energy is the fact that it is fuelless, meaning zero materials-intensity as far as fuels are concerned. Necessarily, though, there are fuels and forces underlying it, particularly hydrogen as it permanently burns and fuses on the sun. The rest stems from gravitational or rotational forces as well as thermal differentials on earth. These are 'free' in the same sense as breathing air is free. But work, devices, machines and infrastructures are required to harness some of the abundant and freely available radiant, thermal and kinetic energy, for example, large areas for solar power stations covered with mirrors made of metal. This is why fuelless energy nevertheless comes with an ecological footprint. A supposedly disproportionate product materials-intensity was formerly an argument against developing 'alternative' energy. But these arguments have by now disappeared. Any conventional carbon fuels-based energy apparatus is also materials-intensive. A certain turnover of product and building materials cannot be avoided. These materials, however, for the most part earthenware, concrete, glass and metals, are particularly suitable for multiple recycling so that the environmental impact from mining and quarrying can remain relatively moderate. This is all the more true if the energy required for the recycling and reprocessing of materials is clean and eco-consistent.

The term radiant energy here refers to solar energy. There are further meanings of radiant energy, which might or might not be of interest some day, but have not yet reached a stage of research and development advanced enough to be a TEI candidate. Nikola Tesla, one of the arch-inventors of electrical engineering, demonstrated as early as in 1889 how to light a bulb with no flow of direct or alternating current, simply by what he called terrestrial stationary waves. With the help of certain devices this radiant energy can be harnessed at a few per cent of the cost of power from conventional sources (Lindemann 2001).

4.2.8.1 Photovoltaics

Solar energy can be made available either in the form of electricity from photovoltaics or in the form of heat from thermal collectors. Photovoltaics includes any technology which directly transforms sunlight into electricity. This is done by solar panels put together of photovoltaic cells. Any such cell generates low-voltage direct current on the basis of semiconductor technology. Direct current requires conversion to alternating current and to be stepped up or down to required voltages in order to be usable either in common appliances, or for storage in special solar batteries for later use, or to be fed into the power grid.

As with integrated semiconductor circuits in computing, previous generations of solar cells were made of silicon, with single-crystal, poly-crystalline and amorphous silicon being the preferred materials. These photovoltaic cells and the panels assembled from them are solid and hard like rock or glass. They can be fit on roofs, preferably at a right angle to incoming sunrays. Other types of panels can be façade-integrated as an external panelling of building walls in lieu of tiles of granite or sandstone.

Recent developments in silicon solar cells include the use of powder silicon in drop form. With this, ultra-thin layers of silicon can be produced which allows considerable materials savings resulting in up to 1/30 of previous silicon demand. Ultra-thin silicon may not, however, represent a complete breakthrough yet, because the production costs remain relatively high in spite of the materials savings. Electricity from today's 'thick' solar cells still costs 35–70 ct/kWh (comparing to 5–8 ct/kWh for conventionally generated electricity).

Efficiency of photovoltaic cells is measured as the ratio of electricity produced and solar energy received. In silicon panels this ratio is about 16 per cent, in the laboratory at maximum 24.5 per cent. This level is actually not small as is sometimes assumed. Biological energy efficiency of natural plants is just about 0.1–1.0 per cent. Higher energy yields of solar cells would of course be desirable from a technical and economic point of view, while it does not make too big a difference with regard to ecological aspects of energy conditions on earth. The sun shines on regardless. The potential of photovoltaics, which can be applied in manifold decentralised ways, is much bigger than vested interests in central thermal power would like it to be.

Researchers at the FAO have calculated that on the basis of 10 per cent energy efficiency a territory slightly greater than the size of today's peanut plantations, and a little smaller than the territory covered by cotton plantations, would be enough to satisfy the total of today's global energy demand (WDR 1992 149). On the basis of 16 per cent efficiency, land requirements for an all-solar scenario would represent a square of 420×420 kilometres, slightly greater than the size of Illinois, USA, and only a small fraction of the area of the Sahara (Witzel/Seifried 2000 6). Implanting a solar zone in a far-out corner of the Sahara is not meant here to be a serious proposal. The comparison serves to illustrate the huge potential of photovoltaics.

There are various next generation candidates whose succession struggle (II/9.4) is not settled yet. In addition to ultra-thin silicon cells there are next generation solar semiconductors made of special metals such as copper-iridium-gallium and cadmium-telluride. Yield efficiency of these cannot be increased, but the number of production steps as well as the thickness of the cells are reduced considerably. Fewer expensive manufacturing steps and fewer expensive materials both contribute to the cheaper production of special metal cells so that the price of solar power can come down. This is crucial because today's price levels of photovoltaics are still 2–3 times, in older generation variants even 10 times, the cost of conventional thermal power.

A different next generation approach is represented by solar cells made of organic materials and inorganic nanomaterials. Organic solar cells can be made of electrically conductive plastics (polymer membranes). Various prototype designs are being developed. Further research is being carried out on the so-called Grätzel solar cell which taps electricity from green leaves.

The organic plastics approach can be combined with inorganic nanomaterials, which are sometimes referred to as nano solar cells. Nanomaterials now being tested are for example tiny rods of cadmium-selenide or cadmium-telluride, the latter absorbing more sunlight than the former. These inorganic bar-shaped crystals are 7 nanometres in diameter (a billionth of a metre), 60 nanometres long, and are sandwiched between films of non-rigid polymers serving as an electrode. This combines the flexibility of plastics with the efficiency of inorganic semiconductors. The resulting material then is a nanorod-polymer composite which is about 200 nanometres thick. This no longer corresponds to a conventional understanding of what a technical cell is. Rather it consists of films and foil coatings which can be rolled out or sprayed on walls and roofs, or be ink-printed on paper and fabrics.

Solar cells can also be made of films of polymers and special ceramics. The resulting material feels like cloth. It could be used in mobile applications, for example be attached to tents and thus serve as a convenient power source to camping travellers, road shows or military units in the field.

An additional approach is based on a design of several thin layers of new materials, each layer absorbing slightly different wavelengths of light. This considerably increases electricity yield. Efficiency of solar cells could thus even reach about 40 per cent.

Solar cells made of nanorod-polymer composites will be cheap enough as to become competitive. They can thus be expected to have the potential of paving the way for the final breakthrough of photovoltaics (Fairley 2002). If there was a political will to get it done, and a co-ordinated regulative and investive effort aimed at creating lead markets and inducing the photovoltaics take-off, this could be achieved within the next ten to twenty years. Much of the resulting potential of solar power would not directly be consumed but stored, partly by feeding accumulators, although this has its technical and environmental limitations, and more importantly by running electrolytic cells which produce hydrogen.

4.2.8.2 Solar thermal power

A complementary component of a solar hydrogen future is solar thermal power. Here it does not refer to roof-top solar thermal collectors which can play a role in the energy design of buildings where they deliver warm water (I/4.8). An application of literally central interest, however, is that of concentrating solar power (CSP), i.e. power plants making use of the heat which can be created by concentrating sunrays on focal points with the help of mirrors, lenses or other reflecting parabolic surfaces. The heat is then used for generating electricity through steam turbines and, in the near future, also hot gas turbines, much as in a conventional thermal power plant. In contrast to photovoltaics, CSP generates high-voltage current. As with low-voltage current, it can be transmitted and consumed, or used for producing hydrogen.

The main forms of CSP are (a) parabolic trough plants where the heating focus is a pipe positioned along the focal axis of the trough, (b) parabolic dish reflectors where the heating points are at each paraboloid's focus, and (c) solar towers where the foci of many variable mirrors are simultaneously concentrated on a heat-absorbing medium on top of a tower.

Parabolic trough plants operate at 400°C and have a capacity of 40–80 MW. They have been in regular commercial operation for many years. Parabolic dish reflectors operate at 900°C with a capacity of just 5–25 kW since they are small (5–10m in diameter). Both trough plants and dish reflectors require a lot of land. Solar towers come with less need for land. They operate at 800–1,200°C and have at present a capacity of up to 10 MW. The tower receivers are smaller, and heat transport is easier than with dish reflectors. Solar towers and dish engines have been tested in a series of demonstration plants, some in California, some in Spain within the European SOLGATE project. The cost of electricity from trough plants and solar tower plants are by and large the same and, if run in world regions with intensive solar radiation, result in 5–13 ct/kWh which is partly fully competitive, partly near-competitive. Power from dish reflector plants, the advantage of which can be decentral use, is more expensive.

An approach to fuelless energy which is usually also mentioned in this context is that of up-current towers. In fact they represent a hybrid design combining solar heat and wind power. The solar heat is collected in greenhouse constructions surrounding the foot of the tower. Creation of hot air in the greenhouses can be amplified by solar mirrors. From the base the hot air is sucked into the tower which functions like a huge chimney. It creates and accelerates an updraft until the air escapes at the top. Electricity is generated by wind turbines installed inside the tower. Up-current towers thus are vertical wind farms. The greenhouses and the tower may not really be high technology, although the construction of very tall towers requires advanced building know-how, and the turbines employed as well as the entire controls systems are indeed high technology.

A first 50 kW prototype, 195 metres tall, was built in 1980 in the high plains of La Mancha, Spain. It worked well for eight years, until the light sheet-metal construction collapsed in a hurricane. Up-current towers with a capacity of 200 MW – the equivalent of a small nuclear plant – need to be one kilometre tall, equipped with several dozen turbines. The greenhouses at the tower's foot would measure about 7 km in diameter. EnviroMission, an Australian company, plans to build five such towers within the next 10 years in an out-of-the-way arid region in the state of Victoria. It is intended to be part of the Australian contribution to reducing emissions of CO_2.

Conservation of nature is not really an argument against huge up-current towers since for safety reasons these have to be built in uninhabited desert-like regions. Also the long distance between such places and more densely inhabited locations is not necessarily a disadvantage if hydrogen, not electricity has to be transported over long distances. But the sheer size of the construction could be questioned, as well as the low energy efficiency of only 3–4 per cent (electricity produced as a ratio of sunlight coming in). As a result, the electricity price is about 20 per cent above electricity from conventional thermal plants. This may be compensated for by durability. The planned-for lifespan would have to reach 80 years.

4.2.8.3 Wind power

State-of-the-art windmills seem to be of a more convenient size. The dominant designs of new-generation wind turbines were developed in Denmark in the 1980s. They were adapted and developed further by a number of European countries on the basis of guaranteed purchase of excess electricity at an equally guaranteed high price. Support schemes of this type triggered a wind power boom which started around the mid-1990s, particularly in Denmark, Germany, Austria, Italy and Spain. Ireland and the Netherlands have stopped such programmes.

A typical windmill today is 80–100 m tall and the rotor blades 30–35 m long, creating a rotor area half the size of a football field. Within ten years the capacity of such a wind turbine has grown from less than 50 kW to 600–800 kW, with the biggest already reaching 1,650 kW. At the beginning of the boom, most windmills were stand-alone plants, whereas later on they were concentrated locally in greater numbers in wind farms. Under favourable wind conditions costs of power generation are at a near-competitive 8–16 ct/kWh (6–8 ct/kWh conventional). Investment costs of 1,000 €/kW are already down to the benchmark of 1,000 $/kW for pulverised coal stations.

Next generation windmills could become fully competitive, for example lightweight wind turbines, as currently developed at the National Wind Technology Center, Rocky Flats, CO, or by Wind Turbine, Bellevue, WA. They have flexible hinged blades that can instantaneously readjust to changing wind conditions. There are two rather than three blades in order to reduce weight. The rotor is positioned downwind so as to prevent the flexible blades hitting the tower. The in-built flexibility allows the turbine to be 40 per cent lighter than today's standard, resulting in engines which are 20–25 per cent cheaper. The tower can also be thinner which results in less noise from 'wind shadow' (when the blade whips through the turbulent air behind the tower). Lightweight wind turbines with flexible hinged blades can finally make wind power competitive with no need for subsidisation.

The snag with wind power is that suitable locations are scarce. Not only must there be enough wind, but also enough space so that nobody feels disturbed by the wind farms. In densely populated European countries this is a problem indeed. Germany, for example, is now sown with 'power asparagus' as they are called in the local vernacular. Total capacity installed has by now surpassed 9,000 MW, the equivalent of 7 nuclear power plants of 1,300 MW. The country's biggest brown coal power station has a capacity of 2,600 MW. However, there are hardly any suitable locations remaining. Many of the wind generators were actually allowed to be erected in unsuitable locations. This has led to conflict and litigation. Lawyers were certainly among those who had unexpected windfall earnings from the subsidisation of wind power.

Conservationists have been resentful of windfarms anyway. Windmills were said to defigure landscapes and to cut into the living space of birds. Birds, it turned out, are rather adaptive if there is a certain distance between the towers (as birds have adapted to transmission lines and overhead cables which were suspected of threatening birds by the conservationists of the time). As for the landscape, it is a matter of taste whether one likes to see a battery of windmills or whether one finds them disgusting. As a matter of fact, however, there are many people who dislike them (just as they dislike transmission lines and overhead cables). Furthermore, the rotor blades can cause light flashes, which disturb people living nearby. In certain cases, the rotation can also be relatively noisy. This is partly caused by that windy whip noise, partly by older-generation gearboxes which most of the installed wind turbines need to gear the rather slow motion of the rotor up to 1,500 revs/min.

In face of these limiting factors, offshore wind farms are seen as an option. There appears to be space out there, no one would feel disturbed, and economies of scale could be fully realised. More detailed planning of the construction of offshore wind farms in the North Sea and the Baltic Sea, however, has shown these arguments to be somewhat double-edged. There is in fact not much space out there, instead there are manifold conflicts over sea use such as shipping routes, underwater cable lines, fishery, tourism and military purposes. Conservationists are opposed to the plans on principle. Last but not least, the technical problems and costs of building, operating and maintaining offshore farms are far from reliably calculable. From an engineering point of view, offshore wind farms are of course an appealing challenge. One might nevertheless ask whether it is a reasonable option, given that there are less difficult and more promising solar alternatives. Might it be that there is a new industry which has already become so path-dependent and locked into its business that it exploits any opportunity simply to carry on regardless of what this entails?

4.2.8.4 Hydropower, tidal and wave power

Hydropower represents another domain of kinetic energy. Like wind power it has a long tradition during which various mechanical principles were developed. With the arrival of electricity in the 1890s, many of the watermills were equipped with generators. Additional small and big river barrages were built in order to drive water turbines. Until the 1930s, hydroelectric power was actually in greater supply than thermoelectric power. Throughout the 20th century large river barrages (running-water power stations) and huge mountain dams (storage power stations) added to the national pride of industrial and developing countries. Beginning around the middle of the 20th century, then, the many small-scale running-water plants were increasingly bought up and shut down, or closed down by the authorities, in favour of big central power stations whose operators claimed that their monopoly was a 'natural' one. More precisely, the central power stations would not have been economic with too many co-producers, and the technical capabilities of the time were not yet advanced enough to deal flexibly with many supply sources feeding the grid rather sporadically.

The environmental problems of hydroelectric power are similar to those of wind power in that it is clean, but in conflict with conservationism and vested residential interests. There are not too many morphological and hydrological conditions where hydroelectric power is suitable. Its potential in those countries where it is an obvious option, in Canada, Norway, Switzerland and Austria, is to all intents exhausted. In Norway, in fact, all power is hydroelectric. Worldwide, hydropower represents 17 per cent of total power supply, with the biggest contributions being made in countries with large dams. Some of the huge development projects, though, had considerable problems with un-intended side-effects, for example the Assuan dam in Egypt which withheld fertile mud from the upper Nile and prevented fields downstream of the dam from being flooded and thus fertilised. Today, additional large projects face fierce resistance from conservationists.

The same, however, also occurs with small decentralised projects. Hikers and ramblers' associations fear the loss of the lovely sound of creeks burbling along, and anglers' clubs are afraid of losing their catch of wild fish. Anglers can probably be satisfied, since special measures in favour of fish migration can be taken. In addition, barrage systems are good for, actually a precondition of, productive inland-water fish farming. They in fact also contribute to the management of floods and flood prevention, as they can help to irrigate fields and meadows. On the other hand, barrages and current-water stations involve extensive building measures such as trench-lining and shoring which denude river banks and cut off natural flood reservoirs. The fact, however, that landscapes and particularly river courses have been covered in concrete during recent decades is not to be blamed on hydroelectric stations. The true reason was that farmers and settlers alike wanted to see waters flow away as quickly as possible without resulting in torrents. So it should be possible to restore the capacities of earlier small hydroelectric stations and even add one or two times as many new ones without distorting beautiful landscapes. Notwithstanding this, exaggerated shorings in other locations along rivers can and ought to be removed.

Since suitable locations on land are scarce, hydroelectric hopes, again as in the case of wind power, are turning to the sea. The tidal power plant at the estuary of the Rance into the Atlantic near Saint Malo, France, became famous, though not the model it was expected to be. A joint English-German project is exploring underwater-current turbines located in places where there are strong underwater currents near the coast. Smaller and larger units are equally possible. In a Norwegian strait the tidal current, which moves at 2.5 metres per second, drives underwater turbines, that is to say watermills, 50 metres under the sea. The blades of the watermills turn with the tide.

ABB, Switzerland, and Statoil, Norway, are expecting their watermills to be cost-competitive within ten years. An English company, Marine current Turbine, plans to install a prototype tidal watermill off the coast of Lynmouth.

A new and different approach is wave power. One design, called power buoys, tries to harness the kinetic energy of short but rather rapid up and down movements of buoys in the waves.

The designers of Limpet, short for land-installed marine-powered energy transformer, are thinking somewhat bigger. Limpet resembles a bunker on the coast. Its centrepiece is a wave-air Wells turbine which uses air flows caused by waves to drive the turbine: wave rises, air rushes in; wave falls, air rushes out; i.e. ocean waves cause the water level in a cabin to rise and to fall with each wave, thus causing an airflow in and out of a Wells turbine. It generates 500 kW, sufficient for about 400 homes. A test plant has been installed by Wavegen, a company in Inverness, Scotland.

Tidal power plants and wave power stations on the coast or near to it involve extensive building activities (shorings) on coastal sites, just as do inland barrages and dams. Since sea shores and river banks are ecologically sensitive areas, this represents a decisive disadvantage of hydropower, clean as it is apart from that, and in spite of its being operational day and night, in contrast to wind (that can be missing) and solar power (where there is sunlight only during the day). As a result, the potential of hydroelectric power can perhaps not be extended much beyond the level where it already is.

Creative ideas might still be in the air, such as the 'energy ship' as mentioned in I/4.2.5. It represents a hybrid of wind and hydropower, and would produce hydrogen and desalinate water at sea without entailing shorings, just docks for occasional mooring.

Ocean thermal energy conversion (OTEC) is a different approach to similar ends, but again it represents a plant on a coastal site. It starts from the thermal differential between warm sea surface water and colder deeper waters. Using steam turbines, it requires a medium that boils at 10–15°C such as ammonia (NH_3). Colder water can also be used for cooling systems. At the same time, condensation of warm water on carriers of cold water produces drinking water. Because of the heat differentials which are necessary for the process, the system only works satisfactorily under tropical and subtropical conditions. Efficiency is relatively low, but in principle this does not matter too much because the thermal dynamics of warm and cold water are free. One of the advantages is that such a plant is robust and that operations can be always on, day and night. OTEC represents an optimal combination of producing electricity (or hydrogen), heating/cooling and drinking water. A certain disadvantage of OTEC is the many long pipes it involves.

4.2.8.5 Geothermal heat and power

A domain in the spectrum of regenerative fuelless energy which remains to be discussed is geothermal energy. Its physical source is isotopic decay deep within the earth. Geothermal heat represents the obvious choice in Iceland where hot waters from deep down emerge naturally above ground. But geothermal heat and power can in fact have some potential in all world regions. Temperature gets warmer everywhere as one digs deeper into the earth's crust. Earth heat can be used decentrally in the energy design of individual houses and in central geothermal power stations which ideally co-generate heat and power.

A well investigated method is the hot dry rock process. Pipes are inserted in drilling holes that reach 4,000–5,000 metres into dry rock formations. Water injected at a pressure of 600 bar through ingoing pipes circulates in natural fractures or creates artificial ones, thus heating up and returning above ground through outgoing pipes at 150–200°C. Above ground the hot water and steam can be converted into power by a conventional process, or be used in heating systems. Hot dry rock plants can have a capacity of 20–50 MW. Hot water from aquifers can be used in the same way, though such deep aquifers are relatively rare.

Another method is to reuse existing boreholes of less depth. There are many such holes from prospecting for minerals, metals, fossils and gas. In this case the water may only reach a temperature below boiling point and is therefore led into a heat exchanger whose counterside is flown through by a liquid with a boiling point of 30°C (butane C_4H_{10} or pentane C_5H_{12} in a closed circuit) the vapour of which then drives a turbine.

Geothermal energy is an inexhaustible source of clean heat and power. If this path were systematically pursued, its potential is estimated at a quarter of current electricity demand. This could be achieved within a few decades. One problem, however, is those many kilometres of steel pipes which wear rather quickly and consequently have to be replaced often. While this can make the process cumbersome, it does not greatly increase its competitive price at 7–10 ct/kWh.

4.2.8.6 Outlook

The different types of regenerative fuelless energy – photovoltaics, solar thermal, wind, water currents and geothermal – represent a patchwork of different options of varying applicability and reach. Nevertheless, a clear pattern emerges from the different potentials and problems.

Solar energy has by far the largest potential and will have to come first. Solar energy also faces the least resistance and is highly connective. Where geographical conditions are favourable, large central solar thermal power stations, also large photovoltaic stations, can be installed which generate electricity or produce hydrogen to be transported and distributed over long distances. Beyond this, decentral applications of photovoltaics are of utmost importance. While still the most expensive and uncompetitive source of fuel-less energy today, new generations of solar cells (organic polymers in combination with inorganic nanomaterials) now emerging will change the picture in the future and make photovoltaic power cheaply and ubiquitously available. If developed to their full potential, including a solar-hydrogen path, photovoltaics and solar thermal together already represent levels of energy supply which would be sufficient to phase out fossil and nuclear energy within 50–70 years. In spite of the fact that current policies in too many countries (USA, Russia, China, India, OPEC) are unsupportive of decarbonisation, thus postponing the solar dawn, the symbol of the smiling sun is very likely to be on a long-term rise in the 21st century.

Solar energy can be complemented favourably by geothermal energy. But engineers and power suppliers do not seem to be particularly excited about it. As long as geothermal energy does not develop a more appealing image, it is bound to fulfill a niche role much below its potential.

Also wind and water, though representing a vast inexhaustible potential in theory, in practice face rather narrow limitations because of conflicting environmental goals. The development potential here cannot be exploited without creating social perturbations and political conflict. In countries where the wind boom is already advanced, total capacity can perhaps be doubled or tripled, though on the basis of new, more powerful turbines rather than additional plants. Hydropower can perhaps also be doubled and tripled. It can therefore be concluded that wind and water on land still have some potential which certainly should not be disregarded. In total, however, it will remain an option of lesser importance, perhaps summing up to 10–20 per cent of total power supply. Whether offshore wind power, coastal wave power, and underwater-current power have a more far-reaching potential, and how big it could be, remains to be seen.

4.2.9 Distributed Power Generation and Integrated Grid Management: Is there a Future for Central Power Stations?

Throughout most of the 20th century the overall tendency in power supply was centralisation in ever bigger power stations. These almost completely replaced smaller stations within their catchment area. Power supply became a monopoly, be this by law or as a matter of fact. The monopolists had business control of the functions of (a) power generation, (b) power transmission or physical distribution, and (c) power trading.

The tendency towards centralisation ended about twenty years ago and the pendulum seems to be swinging back towards more decentralisation. This has come about partly for political and economic reasons, partly for technical ones. Among the political reasons was the broad social movement against nuclear power and, in combination with this, a growing critique of established energy policies which were rather corporatist everywhere, and overtly statist in Europe. Industrial, ministerial and techno-scientific elites were in a position that allowed them to pursue, or indeed not to pursue, any energy strategy they pleased regardless of the costs. Meanwhile, re-regulation at national levels and within the EU have introduced increasing liberalisation of power supply, including the separation of the three functions of power generation, transmission and trading.

As for the technical reasons, a tendency towards decentralisation is taking shape through the emergence of several new developments such as

- co-generation of heat and power in local DHPS
- the boom in widely scattered wind power generators
- a re-assessment of small and medium-sized current-water plants
- and, most important of all, the introduction of small and medium-sized fuel cells in manifold stationary and mobile applications.

All of these power-generating small and medium-sized devices, in addition and connection to larger and smaller conventional power stations, fit together to form a scenario of distributed power. Distributed power does not mean a radically decentralised system with myriads of self-sufficient small units. On the contrary it involves a common grid which is as large and powerful in its macrostructure as it is widely branching in its microstructure, whether at local or global level. It is actually a worldwide power grid crossing all continents to a degree analogous to the global telecoms network. Some of the nodes in this network will have to be large and very large power stations, at least in the decades to come, while at the same time ever more of the nodes will be of a medium, small and very small size. Like today, some of these will produce base load, others will have to make sure that peak loads can be met. In contrast to the situation today, most of the units which are connected to the grid and generate the distributed power will represent supply *and* demand. Their owners will be customers *and* suppliers on their own account. Their electricity meter will alternately rotate in both directions, or there will be two meters respectively, one for incoming, one for outgoing power. A grid with distributed power has a two-way-flow of current.

To make sources of electrical power the right size, and to integrate those very many different sources in one coherent grid, is the outstanding electrotechnical project of the mid-term future. Distributed power in a fully integrated grid will bring about enormous economies of scale, thus 'negawatts', on a hitherto unknown level of overall energy efficiency (Lovins 2002).

A regulatory and economic precondition of such a scenario is a liberalised market open to new entrants on the basis of strictly separating grid management from power supply. Liberalised market competition, though, does in no way include environmental deregulation. On the contrary. Environmental standards on energy have to be as tough as possible in order to prevent price competition from leading to an environmental 'race to the bottom'. And it has to be ensured that clean fuels and fuelless energy get a chance on a more levelled playing field. This would include regulation of CO_2 sequestration which is still lacking. An environmentally fair price of electricity would then be (in the USA) at about 8–10 ct/kWh rather than today's 6–8 ct/kWh. As a consequence, clean power would be competitive or become so sooner.

Everyone then who currently generates excess electricity would have the right to feed it into the grid. For example, every stationary fuel cell in an individual home, beyond providing power and heat for own demand, can be, so to say, a small power station. The devices would belong to the house owners, or be leased from a power supply company that does the maintenance and operates the apparatus by remote control, at little or no cost to the households. In a scenario called Hydronomy (hydrogen economy), envisioned by a development task force of General Motors, decentral mobile micropower plays an important role by employing FC-powered cars when staying idle, which they do for about 90 per cent of the time. A parked FC car would help to generate cheap electricity if some connecting infrastructure existed which could be installed in private garages, or in the underground and multi-storey car parks of residential houses and office buildings.

If there were over-supply, the price would immediately drop, maybe even half- or quarter-hourly, thus causing generators to throttle back their supply, or to accept continued supplying without pay. If by contrast the grid were short of power, the price would immediately rise and cause generators to gear up or activate additional capacities. Should distributed power generators driven by solar energy, wind, water or geothermal heat not be needed for electricity supply, they could be temporarily used to electrolyse hydrogen as a suitable way of storing energy in the form of high-quality clean fuel (fuel cells would of course normally not be employed to that end). All this could in principle be done by fully automated electronic controls interaction without intervention of individuals. If such an integrated grid management could be approximated to technically, large power stations might indeed become obsolete one day.

Integrated grid management would by itself make a favourite alternative vision come true, namely to locate smaller plants closer to demand rather to transport electricity hundreds of kilometres from huge central plants. Should this be necessary, however, integrated grid management would be able to provide power from across an entire continent.

A possible new element in grid management might be flow cell batteries (Fairley 2003). They do what was hitherto not possible for lack of advanced technology: interim storage of electricity. Flow cell batteries represent a combination of a huge fuel cell arrangement and connected electrolyte tanks. The fuel cell arrangement consists of many hundreds of modules, each piled up of many hundreds of FCs. The electrolyte tanks are about 10 m tall and 20 m in diameter, made of reinforced fibre glass, and filled with concentrated salt solutions (one with sodium bromide, the other with sodium polysulphide). As in FCs for cars, the crucial part of the plant consists of special thin plastic membranes which let only positively charged ions pass through. The FCs charge and discharge the electrolyte solution on demand, preferably charging when power supply exceeds demand, for instance at night, and discharging when power demand is high. Unlike conventional batteries, FCs do not deteriorate. Such a flow cell battery can provide the power which a small town consumes in one day. Most electrical corporations have flow cell batteries under development, and a number of prototype plants are complete or under construction in the USA, France, England and Australia. It remains unclear at present, however, whether flow cell batteries will be economic; or whether their function will be that of an expensive emergency supply for grid systems haunted by frequent black-outs.

Flexible supply over short and long distances alike is also being backed by new developments in power line design. Today's power lines usually consist of steel-reinforced aluminium cable. In the USA, many lines are 30–70 years old and were not designed for the heavy loads they are carrying nowadays. A new kind of cable is made of a reinforced plastic core. The core within the core so to speak is hollow so that optical fibre can be inserted. This allows for telecoms broadband connections without the need to lay extra cables underground. The plastic core is wrapped with aluminium alloy wires. In contrast to the old steel-reinforced cables, the non-conducting plastic core of the new cables does not draw electricity from the aluminium conductor. The cable is thus prevented from heating up. As a result, power lines can carry more voltage, or can be smaller.

Electrical corporations such as Siemens, Westinghouse and others are working on grid management systems, including the sophisticated technical communications and controls which are necessary. Distributed power from smaller decentral stations could generate about 20 per cent of power by 2015. A system of distributed power generation and integrated grid management would be more stable and dependable than the current over-centralised system which is prone to breakdown upon perturbation; this is why central power stations and power lines are high-security areas. With distributed power generation and integrated grid management there would be less preoccupation with security and no black-outs.

4.3 NATURAL RESOURCES AND AGRICULTURE: TOWARDS EARTH SYSTEMS ENGINEERING

When talking about TEIs in the realm of agriculture and raw materials one deals with novel practices of production and of recovery of natural resources rather than new machines or technical infrastructures. These are certainly also part of the picture, but basic industries such as quarrying, mining, forestry, fishery and agriculture in general will first of all have to consider new regime rules of sustainable management of water, land, landscapes, ecosystems, domesticated animals, wildlife and biodiversity. Agriculture and the management of natural resources will increasingly have to become part of nationally and globally integrated regimes of earth systems engineering as mentioned in I/2.4.4.

4.3.1 Cascadic Retention Management of Water and Groundwater

Water is the paramount environmental medium for everything in the biosphere. So there is good reason to start earth systems engineering with the sustainable management of water and groundwater. Whereas the management of sewage water has much improved, and the recovery of drinking water from surface sources also has made progress, there is less action as yet with regard to deep wells and groundwater management, desalination of sea water and regional humidity management.

Beyond decarbonisation of the energy base, earth systems engineering will have to start with influencing regional water cycles related to forests, soil and surface waters. This is among the top preconditions for keeping ecosystems intact, especially conserving and fostering biodiversity and maintaining the quality of air, water, groundwater and topsoil.

Groundwater is closely linked to the protection of soil. Groundwater must no longer be contaminated and over-exploited, and former levels of groundwater deposits ought to be restored. Control of groundwater contamination requires strict regime rules in agriculture with regard to agrochemicals, manure from intensive animal farming, clean-up of brownfields and closing down of waste deposits. Sustainable utilisation of water and groundwater deposits requires detailed plans of catchment areas as well as complete regulation on cross-area long-distance imports and exports of water.

Greening of the desert may still be a far away dream. Maintaining or restoring local groundwater and surface water levels, however, need not be. With regard to long-distance water transport for irrigation and urban settlements there can be some degrees of freedom in determining the area of reference. There is certainly no need to be dogmatically localist. Water transport, even over several hundreds of kilometres from A to B, in order to help B maintain its water retention levels, is in principle no problem as long as water levels in A are much above retention and not in danger of being violated. The lake Aral experience is the typical case of an environmental crime, and illustrates what must be prevented from happening elsewhere. Tropical deforestation, however, is even much more devastating on a global scale. Forests may be quite important as a sink for CO_2. They are many times as important in their function as dynamic water reservoirs which contribute to reducing temperature and maintaining rich biodiversity.

Percolation and drainage management has been ecologically inappropriate ever since it came into existence with the building of canals and sanitation infrastructure in the 19th century. The general aim, besides making rivers navigable, was to get rid of water as quickly as possible, as if it were an undesired good. Land owners wanted to drain marshland, and settlers wanted to prevent rivers from bursting the banks. Part of the background to this was strong population growth and growing pressure on land use from the 18th to the 20th century. But water equals life. Biologically it is highly sensitive and precious, even where abundant. From an ecological point of view, pure water is clearly more precious than gold has traditionally been to economic man. So the percolation and drainage trend would have to be reversed to becoming a system of cascadic retention, i.e. ecologically benign capture and use of water. Water has to be husbanded and conserved, as cleanly as possible and in quantities which are as high as is environmentally appropriate. One would certainly not want to have again inundated cellars and flash flooding, but available water reserves can be enlarged, and water ought to be used before sending it further on its way into the ground or to the sea.

Flood prevention involves restoring at least some of the natural flooding areas along the course of rivers which have been cut off by shorings, urban and rural developments, along upper and lower reaches alike. This is neither cheap nor easy to achieve, politically comparable to the construction of fast railtracks or incineration plants because of the NIMBY syndrome (i.e. *not in my backyard*). Land owners and residents will try to stop such projects and negotiate for recompensation. It is nevertheless worthwhile since not taking such measures will result in greater damages caused by floods, storms, landslides and similar catastrophes. These are now increasing in numbers and extent because of changing climate, less moderate weather patterns and melting glaciers at the poles and in high mountains. Since the glacial limit is moving upwards and glaciers are no longer withholding water, regional management of mountain rain- and meltwater has to be tackled seriously. In the lower reaches some of the existing dams will have to be relocated further off the river bank so as to create larger flooding areas.

In all locations, no matter where, rainwater management by controlled percolation, resulting in deferred draining away of water, can contribute to improving local conditions and groundwater levels. Various new systems of rainwater percolation are being developed and tested, for example recovery and drainage systems based on basin-drainage pipes. In the surroundings of towns, artificial percolation polders can be restored or newly created. Purified effluent water would thus not feed into the draining ditches, but into the ground by irrigating the percolation polders so that the water could trickle down through the soil and find its natural aquifers. Similar ends can also be met by smaller systems of decentral rainwater management on individual sites (I/4.8).

A complement to percolation is precipitation. It depends on evaporation and cloud formation. Control of weather certainly remains out of human reach, but there are small degrees of freedom in influencing local weather conditions. It is surely worthwhile to extend this influence so that town dwellers and farmers in particularly affected world regions could experience an improvement in their living and working conditions. In the foreseeable future, rainmaking will be the centrepiece of any such strategy.

Today about 25 countries are actively engaged in rainmaking. In the USA alone there are 15 companies offering their services. China seems to have taken the lead. Purposes are fighting drought and forest fires, or driving away clouds if bright skies are the order of the day, for instance in big mass events. In Britain, according to secret dossiers recently disclosed after 40 years under wraps, the military is said to have experimented with cloud seeding as early as in 1952 in the region around Lynmouth, Devon. Allegedly they wanted to find out whether strong rainfall would be able to stop an army. 34 individuals died in a torrential cloudburst, but it remains controversial whether this was really chemically triggered or whether it was a natural catastrophe which happened by unlikely coincidence. Rainmaking is clearly a double-edged TEI, not least because of its large-scale interference into atmospheric chemistry. Promising approaches to rainmaking nevertheless re-present a major contribution to earth systems engineering.

It has long been tried to seed clouds with the help of silver compounds. If the dosage can be kept under control this may environmentally not be hazardous since silver, as used in coins, crockery and cutlery, is normally not toxic. Another approach to cloud seeding is salty sea spray. This represents a complex example of bionics, i.e. technical attempts to mimic nature. Rain mainly seems to form because of salty sea spray, continuously moving from sea onto land, and back again. Salty sea spray naturally encourages rainfall by attracting water into larger droplets, merging with each other, and resulting in rain when big enough. The rain washes pollutants out of the air. In Israel, scientists are experimenting with the artificial creation of salty sea spray in order to seed clouds and thus make it rain. If successful, it would certainly help to make arid regions on earth greener places.

Aside from precipitation and percolation, more efficient uses of water can be achieved in various ways. Similar to power generation, there are now also more decentralised possibilities of water supply. This can be provided by small water processing plants fed with surface water or captured rainwater. Traditional wells should in general not be allowed any more in order to shield groundwater aquifers. These small and technologically sophisticated water plants can be stationary and mobile. Despite their small size they include several steps of mechanical, physical and chemical treatment such as sedimentation, UV radiation, ionisation, adsorption and other methods known from larger central plants. New-generation processing units can deliver drinking water, but more often they produce process water of lower purity. Process water obtained in this way can replace unnecessary uses of high-purity expensive drinking water such as in washing machines, toilets, irrigation, construction works, fire-fighting, etc.

Drinking water can also be obtained decentrally through membrane-type filtration (reversible osmosis) of sea water or brackish water. The energy to run such smaller local units can in turn be produced by decentral regenerative energy devices (photovoltaic, wind, current-water). Fresh water obtained through desalination of sea water clearly remains an outstanding desideratum in a densely populated world. New technology in this field can be expected to include more than the sea-based energy ships and shore-based ocean thermal energy conversion plants as described in I/4.2.7.

4.3.2 Low-impact Mining

Mining industries in Europe and North America have routinely caused much environmental degradation and even overt destruction in the past. Mining industries in Russia and in developing countries continue to do so. This is why careful mining, keeping impacts on surface ecosystems and groundwater levels as low as possible, and also investing in compensatory renaturation measures during and after mining, is indeed a must. So, for the sake of conservation of sensitive ecosystems and beautiful landscapes, a basic regime rule of sustainable resource management requires that quarrying and mining shall be restrained as much as possible. Mining cannot be avoided, but it should be of low impact in that it is reduced and comes with reduced impact where it takes place. Low-impact mining entails multiple recycling of bulk minerals and all kinds of metals, and above all the phase-out of fossil fuels and uranium. Fossils as a raw production material in carbo- and petrochemistry, for instance in the production of plastics, can to a certain extent be replaced by fresh biomass (phytochemistry as discussed in I/4.4.1).

As long as fossils continue to be used, and as far as natural deposits of minerals and metal ores have to be exploited, new and environmentally less harmful or even benign ways of extracting those raw materials can be developed. TEIs in oil and gas extraction, include single shaft recovery of mineral oil, multi-directional wells and multi-phase pumping as described in I/4.2.3.

With metals, phytomining or bioleaching has opened up new horizons. Biotechnology is a rather surprising alternative in mining if one is caught up in the conventional paradigm of extraction of raw materials by hammering, blasting, crushing, pressurising and evaporating. Phytomining or bioleaching makes use of plants and bacteria, for example *Brassica juncea* for extracting gold dust (with the help of ammonium thiocyanate, though, which separates the metal from other substances) or *Thiobacillus ferrooxidans* and *Leptospirillum ferrooxidans* for extracting copper, zinc and cobalt.

Titan Resources in Perth, Australia, has developed a process called Bio-Heap, based on bacteria created by selective breeding, for refining copper ores bonded to sulphur. Billiton, South Africa, uses *Thiobacillus caldas* and *Sulfobacillus*. Also in Chile, which is the world's biggest producer of copper, the possibilities of mining biotechnology are being explored and suitable plants and microorganisms bred. In the case of extraction and basic processing of copper the advantages are obvious: efficient recovery of high-quality metal, no noxious gases, no toxic effluents, the only byproduct being iron arsenate which is environmentally stable. Operations are simple and safe since they are carried out at ambient temperature and pressure or slightly above. Operating costs of microbial heap leaching and conventional smelting of copper ores are by and large equal, but with regard to capital costs bioleaching is 10–25 per cent cheaper, depending on plant capacity (OECD 2001c 128). Permits from environmental authorities are easier to obtain for bioleaching, and measuring and reporting is less onerous. Last but not least, smaller deposits which could not so far be exploited economically, can now also be developed economically with the help of mining biotechnology.

4.3.3 Sustainable Forestry and Transition from Open Sea Fisheries to Aquaculture

The problems of area-wide clearing of rain forests and depletion of stocks of fish living in the wild are certainly known, as are the innovative practices that have to be to the core of sustainable regime rules in both of these domains. Forestry, hunting and fishery are, historically speaking, the domains where the term sustainability comes from: to manage a forest stand, or a stock of game, or fish, in a way which ensures that what is taken from the stock does not exceed natural regeneration. Do not take out more than can grow again; if you have abused, take less until stocks are replenished; also take less if stocks are to be enlarged. This is all too obvious. The difficulties here are not technical, unless we mean the new tools and technical systems which greatly enlarge the destructive potential of the stone-age technological paradigm which still prevails in rainwood clearing and open sea fisheries.

Among the reasons why it is so hard to make producer countries and individual producers comply with the obvious are

- demographic pressure on land use
- lucrative market demand which far exceeds sustainable supply
- in certain cases lack of clear property rights
- in a few cases untamed greed to achieve within a couple of years what has formerly taken decades and centuries, and
- in most cases: murky arrangements and outright corruption, i.e. a lack of sense of what is right and wrong under the rule of modern law which is based not only on human rights and civil liberties, but also on protection of the commons.

According to a release by the WWF, about 70 per cent of the wood exported from Indonesia stems from illegal logging, 30 per cent from Russia. 17 per cent of the imports in industrial countries and China are estimated to be illegal. The ecology movement and the Rio process in environmental politics have only started to create an awareness that in addition to local commons there are global commons too, namely the atmosphere and the oceans, inland water systems and continent-wide belts of forested area, fertile soils, wildlife and biodiversity. In addition to the cultural heritage there is also this natural heritage of humankind which ought to be kept intact, possibly cultivated, certainly not depleted or destroyed. Global commons can only be protected if national sovereignty feels obliged to adhere to international regimes (Litfin 1998), in this case to regimes of sustainable forestry and fishery. It looks as if it might take many more years until consciousness in developed and developing countries alike rises sufficiently to make international environmental regimes effective.

In forestry, practices of sustainable logging and forest management are well established in most of the northern countries such as the USA, Canada and Europe. In Russia there is a mixed situation. The boreal belt thus is not completely unendangered. Recent developments have added to existing traditions in forestry, particularly licensed logging and a certified wood industry, demanding that tree plantations be compatible with the location; that growth of the trees not be pushed artificially; that harvesting of the trees should not reduce the overall volume of existing stands; and finally the requirement to refrain from clear-cutting of large areas and permitting only clearings of small size, or of picking single trees out of the standing timber. The Forest Stewardship Council, based in Washington, DC, certifies loggers and tries to convince wood processing manufacturers to buy only certified wood. It was the driving force behind the setting up of the Global Forest & Trade Network, among whose members are the WWF and other conservation action groups as well as hundreds of manufacturers worldwide.

It seems however to be more difficult to influence the wood industry in subtropical dry forests and tropical rainforests in the Amazon basin, the Zaire basin and the remaining rain forests in south-east Asia, particularly Malaysia and Indonesia. Larger territories of these forests would furthermore have to be put under a regime of strict conservation. Not only do these forests have a global function of absorbing much CO_2 and keeping the water circulation higher and temperatures lower than they otherwise would be, but they also represent by far the biggest coherent reservoir of existing flora and fauna, thus also the largest gene pool on earth. Insofar as conservation of rainforests is considered to represent utility foregone to nationals, the international community would have to find ways of compensation for this loss acceptable to both sides.

With open sea fisheries the situation is comparable. Recent international agreements concerning maritime law have extended the area of national sovereignty around coastlines to stretch as far as 200 miles out to sea. So the sea within this zone has become a 'territory' under national administration, an exclusive economic zone for fisheries and underwater mining. Since 90–95 per cent of fish live in upper waters relatively close to coasts rather than far out and deep under the oceans, the 200-miles agreement has nationalised most fishing. Former quarrels over fishing rights have thus been settled, and events such as near-military confrontations between Canadian and Spanish fishermen have become an exception to the rule. Peaceful national overfishing within the zones, however, has become the rule. International efforts remain on the agenda to establish or effectively enforce fishing regimes in order to bring catch numbers down to sustainable levels. This would have to go so far as to allow for some replenishment of stocks.

Clear-cutting wood, hunting game, fishing in the wild and using open fires are among the last remnants of technological practices which date back to the prehistoric times of primitive cultures. Fishing in the wild represents a resource regime of pre-neolithic descent. It may nowadays be armed to the teeth with state of the art high-tech vessels, hypernets and military detection instruments such as sonar mapping systems and precision satellite navigation systems, but it is still primitive man who is at work there. The transition from area-wide clearings to sustainable forestry, and from large-scale open sea fishery to best-practice aquaculture pertain to the TEI trends of some historical importance.

Increasing portions of the total fish and seafood catch nowadays come from aquaculture, i.e. offshore ranching and fish farming. At present the world share of fish from aquaculture is already one third. It is of no use to engage in detailed dispute over whether shellfish or parrotfish will become extinct within 10 years, or 30 years, or sooner or later. Over-exploitation of stocks by large-scale open sea fisheries have to stop, just as unregulated hunting in the forests had to stop centuries ago. Aquaculture is indeed the solution, just as agriculture was the solution to hunting and gathering.

For sure, if not carried out properly, fish farming and offshore ranching pose environmental problems similar to intensive animal farming on land, for instance over-medication and proliferation of hazardous substances in the food chain, contamination of water and sea floor, or low quality of produce. This, however, is not necessarily so. It depends on standards and regulation, availability of capital and purchasing power, and ever growing know-how.

4.3.4 Organic Farming: Modern Near-nature Practices

Organic farming emerged about hundred years ago in the context of the social movements of conservation of nature and 'life reform' of the time. It was intended to be a counter-practice to increasing industrialisation of agriculture in the form of heavy mechanisation, agrochemicals, monoculture and intensive animal farming. There are several strands which have led to today's organic farming such as 'natural', 'biological', 'bio-dynamic' or similar. Differences depend on regional origins and degree of green fundamentalism. The general motto of organic farming is 'to work with nature, not against it'. But boundaries between reasonable ecological regime rules and sometimes not so reasonable ideological preferences for localism, traditionalism and autarky inspired by nationalism and have not always been clear.

With the upswing of the environmental movement in the 1970s–80s traditions of organic farming were revived and literally rejuvenated. A general preoccupation with health, a certain disenchantment with agroindustrial food which was becoming cheap but of ever lower quality, and a continued series of food scandals then triggered a larger boom in organic produce throughout the 1980s–90s. Today, organic farming and the labelling of organic produce is formally regulated in most countries. EU bio-directives 2092/1991 and 1804/ 1999 can be considered to represent a common denominator of what organic farming involves: integrated farming, crop rotation, no agrochemicals, only natural auxiliaries, and a ban on transgenic seeds and produce.

The idea of integrated farming is to strike a balance between animal and field farming which fits local conditions and does not exceed local resources. Animal density per hectare has thus to be much lower than in conventional agriculture, so that animals can be fed by home-grown feedstuffs, and manure and slurry is kept within the limits which can naturally be absorbed without contaminating soil and water (170 kg nitrogen per ha $\approx$ 2 cows or 12 pigs or 230 hens per ha). Externally produced feeds, in practice imports, are allowed to varying degrees ranging normally from 10 per cent with cattle to 50 per cent with poultry (i.e. 50–90 per cent of the feeds have to be home-grown). The 'preventive' use of antibiotics, hormones and anabolic steroids is strictly forbidden. Animals must not be kept in cages or on gratings, should not regularly be tethered and should have enough space in the open air.

Equally interdicted is the use of synthetic chemical fertilisers and pesticides. Auxiliaries should in general be avoided and are only permitted to the extend to which there is a proven need. Permitted aids are mineral fertilisers such as Thomas meal and Thomas lime (natural phosphate fertiliser), potassium, limestone meal and chalk. Industrial byproducts, if 'natural', are also permitted, for example limestone from sugar production, and certainly home-grown byproducts such as dung and liquid manure in those limited volumes, compost, crop residues, straw and other mulch material.

As for pesticides, there are alternative methods of plant treatment, for example authorised preparations of sulphur, inorganic copper, plant oils and soaps. Farmers are free to apply biological methods such as the use of certain insects, worms and mites (against other pest) and also microorganisms such as *Bacillus thuringiensis* and *Granulosis virus*. The most important contribution to maintaining and improving soil fertility is sophisticated crop rotation. Soil should not be deeply ploughed up. Crop seeds must be conventionally bred. Use of genetically modified seeds and the purchase of transgenic produce is rigorously excluded.

It is not easy to decide whether organic farming represents a set of innovative modern practices or rather renovates traditional practices, since in agriculture traditional practices have indeed been ecologically benign and cultivating. In any case, organic farming fully meets modern ecological requirements. It maintains and improves the land in a way which is metabolically consistent: no contamination of soil and groundwater, no depletion of groundwater, no degradation, erosion or salination of topsoil, no proneness to plant and animal disease, no toxic materials and substances, no hazardous waste. On every point exactly the ecologically benign contrary. Every percentage point of growth in organic farming comes with enriched biodiversity and ecosystems health.

Organic farmers, however, have recently also faced problems which were hitherto thought to be confined to conventional methods, particularly BSE and chemically contaminated feedstuffs. Since most organic producers cannot avoid buying some additional feedstuffs, they run a certain risk of picking up infectious or contaminated material because the supply chain is of continent-wide and actually global complexity which makes it hard to keep control of individual shipments and processed charges. Organic farmers are also concerned about the recent spread of transgenic crops in conventional farming, about possible stray crops, or transgenic crops which produce the toxin of *Bacillus thuringiensis (Bt)*. If pests developed resistance to Bt-toxin as they have done with many of the former pesticides, then organic farmers too would be affected and deprived of one of their best means of pest control.

In that organic farming tries to implement traditional near-nature practices, it may seem to be somewhat backward-looking. On the other hand, organic farming is as based on geo- and bio-sciences as any other industrial sector is science- and technology-based. Organic farmers find it in no way difficult to adopt latest generation machinery, IT, energy and building infrastructure. So organic practices could be said to complement the sophistication of traditional knowledge with some modern technology.

Productivity of organic farming is in fact higher and more competitive than defenders of conventional agroindustry would like to have it. Per unit of yield, energy efficiency and materials efficiency of organic farming is 5–10 per cent higher than with conventional agriculture. In comparison, organic farming is relatively labour-intensive, but less energy- and auxiliaries-intensive. Yield per hectare, however, is lower, for example at only 55 per cent the normal yield of potatoes, or 65 per cent that of wheat, and 83 per cent of the milk per cow. This, however, is considered to be an intended result of avoiding the ecologically devastating over-intensity of conventional agroindustrial methods.

Lower yield is compensated for by the much higher prices which the market allows for organic produce. Organic potatoes earn 3.7 times, organic wheat 3.3 times, the price of conventionally grown produce. Organic fruit sells at 50–100 per cent more than normal fruit. Organic milk, though, earns only 12 per cent more than normal milk. Furthermore, and in contrast to the small-is-beautiful image of organic farms, the size of holdings corresponds to nationally common sizes in general. As a result, and in contrast to conventional farming in industrial countries, organic farming is capable of being competitive and completely market-oriented. There are no special subsidies. Thanks to demanding consumers who are prepared to pay a premium for higher quality and taste, organic farmers have a higher profit per hectare and a higher per capita income than their conventional colleagues.

No wonder then that organic farming has of late taken off in America and Europe alike, particularly in Scandinavia, Germany, Switzerland, Austria, Italy and Greece, from next to nothing in the 1970s to about 2–9 per cent of national acreage and 2–3 per cent of the food supply at present. How far can this continue? If it is left to the market, then the niche may reach its limits at 15–20 per cent of the food supply. This is about the percentage of consumers who are spontaneously open to the offer, coming from well-off quality buyers and particularly the environmentally aware. An optimistic estimate of about 20 per cent market share for organic food in the future also corresponds to officially circulated figures.

There is no doubt that organic farming will be a kind of big business for quite some time; for how long will largely depend on the speed and extent of ecological modernisation of conventional agriculture. But there are limits determined by purchasing power, productivity, market competition and sympathy. These conditions put organic farming in agriculture in a role comparable to wind and hydropower in the realm of energy. And as with energy an important function of clean regenerative power is to challenge carbon defenders and thus to contribute to their greening (clean carbon), so organic farming in agriculture has the role of an important change agent which who exerts greening pressure on the conventional agribusiness.

4.3.5 Ecological Modernisation of Intensive Agroindustry

Intensive agroindustry will hardly turn organic, but it must undergo considerable change in order to readjust ecologically. To this end it will have to change its mindset and attitudes, develop new practices and adopt new technologies. Conventional farming has been doing so since the 1980–90s, but that learning process cannot be said to have advanced much. Most of the learning had to be enforced by regulation rather than emerging from an endogenous process. In this process, though, whether from the heart or through compliance with new regime rules, some elements of organic farming can be assimilated by conventional agroindustry.

New regime rules will especially have to make sure that animals are treated with some respect even if bred for meat. Any regime must ensure that crop variety is maintained to a certain degree, that intensity of farming becomes metabolically consistent, that irrigation does not deplete groundwater reservoirs and that hazardous agrochemicals and pharmaceuticals are replaced with more sophisticated new-generation aids. Part of the greening of the agroindustry will consist of biological and integrated pest control, and precision farming in advanced greenhouse production and closed-loop agrofactories.

To start with, conventional farmers can easily re-integrate some of the effective traditional methods of soil maintenance and amelioration, for instance covering fields during wintertime with cut green and chaffed biomass. This is an effective protection against erosion. It contributes to the health and fertility of soil and improves water retention. The use of biomass in this way is much more appropriate than burning it in not so intelligent biofuel arrangements. With biogas it is different. Dung and liquid manure must be biologically treated in biogas fermenters before the substrate is brought onto the field (I/4.2.4). Where gardening and vegetable farming involves the use of garden peat, another example, this should also be replaced with biomass and compost processed in various ways, in order to conserve peat bogs within a near-nature landscape.

Crop variety in conventional farming can be extended by adopting old and new breeds of industry crops, i.e. plants which serve as industrial feedstocks, such as natural fibres, or crops with special content such as sugar, starch, oil, amino acids and a large number of special substances for special use in the chemical and pharma industry where phytochemistry and the biotechnological paradigm are now clearly on the rise again (I/4.5). While processing capacities in the chemical industry do not seem to be a problem, with natural fibres (linen, flax, hemp, jute) there is a bottleneck in that processing infrastructures no longer exist, and new processing capacities do not yet exist. This is similar to the situation with organic food, where lack of capacity to process organic produce at present also constitutes a bottleneck.

In animal farming, new or more tightened regime rules would above all have to deal with animal density and medication. Intensive animal farming can in principle be retained, but certain criteria would have to be fulfilled, not least certain minimum ethics of how to treat animals. Cowsheds and pig barns, particularly poultry coops, have to be big enough to allow animals some space, and there has to be enough surrounding land to allow some free range. There must also be enough solid ground, as opposed to gratings, to stand upon where animals are kept indoors. Manure and slurry must be fermented in biogas plants. Furthermore, an infrastructure for long-distance transport of live animals from EU countries to North Africa and the Middle East must be implemented, otherwise these shabby and cruel conditions of transport would have to be banned.

Animal medication remains a problem. Many farmers tend not to be too particular about existing rules, and even more farmers habitually over-medicate in an attempt to prevent animal disease. In the longer run, and in combination with unfavourable conditions under which animals are kept, this in fact undermines animal health – and human health too, since many animal pharmaceuticals are proliferated through the food chain. In certain cases, expert opinion remains divided. Hormones as an additive to animal feeds are forbidden in the EU, but permitted in the US.

A typical case is the epidemic spread of resistant enterococcus in humans in the last ten years because of permanent over-medication of vancomycin in animals. Vancomycin is an antibiotic which farmers add to the feedstuffs of their animals. Target 'bugs' eventually develop resistance against the antibiotic. At the Harvard Medical School the number of persons who have caught vancomycin-resistant enterococcus has doubled within four years. 20–35 per cent of US hospitals are said to be tainted with multi-resistant bacteria. Only fifty years after introduction, most of the common antibiotics have increasingly become ineffective because of abuse in human and animal health.

There are healthier methods of safely fattening animals. Consider phytase as an enzymatic additive to the feeds of poultry or pigs. It liberates phosphate from phosphate-containing compounds (phytates) in the feed. In this way, the feed can more completely be digested and absorbed and the amount of non-absorbed phosphate which is excreted, thus the phosphate content of manure, is reduced by 30 per cent. In economic terms, farmers save on feeds, and they also save on levies on discharges. Phytase in larger volumes is produced transgenically. Animal feed is the fastest growing market for enzymes.

In principle there is nothing wrong with animal medication and feed additives. In fact these are useful (and ever more humans administer some to themselves). The question is what to rule out, and what the quality and dosage of authorised medicines and feed additives would have to be. In order to guarantee proper farm stewardship, fair competition and scare-free food consumption it is highly recommended to introduce into conventional farming what is already state-of-the-art practice in organic farming: obligatory external auditing, a complete system of demanding quality control and brand licensing, also including complete documentation of the origins and whereabouts of seeds, auxiliaries and aids, produce, animals, feeds and food.

There are new technologies of crop and animal identity verification which allow almost perfect produce tracking, from semen to sale, or seed to sale. Animals and crops can be 'fingerprinted' by various chemical and genetic methods. Information on feeds, treatments and tilling activities can also be stored electronically by swiping electronic wands against bar codes or product buttons. Produce and process verification certainly introduces another cost factor, but farmers will probably have little choice since retailers and consumers have become distrustful and want to have some proof of the pureness of produce, and clarity about where a product comes from and what it incorporates. Quality comes at a price. If conventional produce were less cheap, organic produce would still be more competitive. But to the degree to which conventional agroindustry greens, organic farming will lose its competitive advantage.

Conventional farming is slowly but steadily adopting methods of biological pest control, i.e. the substitution of low-hazard biotechnology for highly hazardous synthetic agrochemicals. The 'weapons' used can be microorganisms (viruses, bacteria) which have very special targets and are neutral to the rest, or macroorganisms such as special insects and nematodes, and the spraying of natural substances extracted from plants with selective toxicity to pests. Substances which play a role in sexual attraction are particularly effective in luring pests into traps. Biological control of this kind is part of the concept of integrated pest management. This tries to develop an optimal mix of biotechnical methods, natural substances and low-dosage agrochemicals. These methods can reduce spray volumes to a fifth or even a tenth of previous volumes. Real disadvantages are so far not known, and fears of growing parasitic insect populations seem not to be objectively based. Examples include the following:

- In hop growing, seasonal counting of mites is a bio-method for deciding whether or not to apply an acaricide. On the basis of numbers counted farmers are able to make an estimate of possible damage by mites six weeks before harvest. This reduces the application of acaricides to about 30 per cent of previous levels.
- Baculoviruses are effective against larvae of certain butterflies, beetles and similar pests. They have been used for long, but natural baculoviruses are slow in doing their job. In the meantime transgenic baculoviruses are available and permitted in Canada, the USA and the UK. They are more efficient and secrete additional toxins against scorpions, spiders and mites (Dürkop et al. 1999 112).
- Sonar devices can fight voles in fruit-growing, vineyards and private gardens by detecting body sound signals and injecting dry ice (CO_2) into the animals' earth tunnels.
- Tomatoes or pepper can be protected against soil pathogens by adding some natural soil microorganisms to the seeds. The protection is particularly efficacious against nematodes, and the plants grow bigger and stronger than they usually would.
- Mould grown on tomato juice promises to be a good herbicide, also in organic farming.
- Special extracts from moss kill fungi and deliver good protection results when applied to paprika, tomatoes and wheat, even in very low doses.
- Plants, including maize and cotton, release volatile signalling chemicals when attacked by insect pests. These chemicals attract other insects which then attack the insect pest. For example the signalling chemical which is released when the armyworm caterpillar feeds on maize attracts a parasitic wasp. The wasp lays its eggs in the caterpillar which is literally digested by the wasp larvae (OECD 1998 17).
- Kenyan agro-researchers have developed an intercropping practice which represents a combined push-and-pull approach. An important maize pest is *Chilo partellus*, a moth introduced from India. It can be responsible for losses of up to 50 per cent. The trick is to plant pull-plants immediately around the maize field, i.e. plants that are preferred by the pest, in this case *Pennisetum purpureum* (Napier grass) or *Sorghum vulgarae sudanese* (Sudan grass). 80–90 per cent of the moths will lay their eggs there. The rest is dealt with by push-plants within the maize field, i.e. plants which the pest does not like and flees from, in this case *Melinis minutiflora* (Molasse grass), or *Desmodium uncinatum* (Desmodium from South America). These plants have the additional advantageous effect of suppressing the growth of many weeds.

- Certain nematodes (for instance *Caenorhabditis elegans*) are parasites which specialise on the larvae of certain midges whom they destroy. These nematodes are as effective as they are cheap to grow and easy to use.

Integrated pest management can in turn be seen as one out of several components of precision farming. Precision farming, to circumscribe it in terms of industrial manufacturing, tries to develop lean agro-production, tilling the field only 'on demand' (real-time need) and 'just in time', with precisely targeted measures, and at the shallowest possible level as far as hazardous substances are involved.

For purposes of precision farming in the field, a number of tractor-based devices have been developed, e.g. a sensor-controlled automated rake. This reduces or even replaces chemical weedkillers. If pesticides have to be sprayed, camera-assisted weed-detection combined with an automated spot-precise pesticide sprayer help to reduce the pesticide load to only 20 per cent of previous spray loads. For driving on wet ground or bog soil, tractors can be equipped with special tyres with variable pressure. Air can automatically be let out or pumped up again on site. This helps to prevent soil compaction. Furthermore, computer-aided and satellite-based systems (global positioning) off the field, in combination with sensors on site and an onboard computer, can measure temperature, moisture, condition and composition of soil. On the basis of these data, the system automatically calculates the quantity of seeds, fertiliser, pesticides or water which have to be poured on the field, whereby this can be positioned exactly to within a square metre.

Comparable precision practices have become state of the art in intensive gardening and vegetable farming in greenhouses. At their inception, greenhouses gave much cause for environmental concern. By now, farmers, particularly in the Netherlands, have made considerable progress. In its most advanced form, greenhouse farming can today be described as an environmentally integrated closed-loop process. In addition to practices of precision farming, greenhouses can have a highly efficient energy design. Water can be recirculated and purified on site. Natural topsoil can be replaced with artificial substrates of various kind, or a thin layer of natural soil can be combined with some artificial substrate. The substrates can include recycled fibres and composted matter (the Netherlands have the highest composting rate in the world).

Greenhouses allow nearly perfect control of the conditions of plants' growth process, input of resources and emissions control. Whether the produce which leaves a greenhouse is of low or high quality does not depend on greenhouse technology but is a matter of regulation and market demand (which in turn depends on purchasing power and a sense of food quality). A greenhouse is a costly infrastructure which is only affordable on the basis of comparatively high productivity. Moreover, it in fact represents built-on

land, comparable to soil-sealed residential and industrial sites rather than traditional fields. This is why greenhouse concentrations such as in the 'glass
westland' region near Den Haag, even though environmentally clean, cannot
be implemented in too many places for reasons of conservation of nature and
landscape. On the other hand, greenhouses contribute to the conservation of
nature precisely by concentrating agroindustrial activities locally, thus leaving the rest of the countryside free for more extensive methods of farming
and other more natural ends.

What is the logical next step in the evolution of this kind of intensive-
agriculture paradigm? It is closed-loop agrofactories on several floors one on
top of the other as devised, not by chance, in the Netherlands by Jan de Wilt,
Henk van Oosten and colleagues at the National Council for Agricultural Research.

A demonstration project in the port of Rotterdam is designed as a multi-
floor plant, called Deltapark, of an unheard scale of six floors each about
1 km by 400 metres. Each floor has to fulfil a special input-output function
within a setting of industrial symbiosis (I/4.9.3). For example, floors with
little or no daylight are used for growing mushrooms or chicory. Floors
higher up have glass ceilings which allow light to penetrate, grow lettuce,
tomatoes or other vegetables and flowers. This produce is fertilised with processed manure from the ground and first floor, which is for pigs and poultry.
The animals inhale the sea breeze on the terrace and enjoy more space than
in an average farm. In the cellar, salmon and other fish swim around. In low
and dark in-between floors maggots and worms are bred, digesting some of
the excrement and building themselves up as high-protein feedstuff. Wind
turbines on the roof deliver the power to keep the process running. Numbers
are not modest: tens of thousands of fish, 300,000 pigs, 1.2 million chickens,
hundreds of millions of vegetables, hundreds of billions of small organisms
and very many gigatrillions of bacteria. It is the vision of a hyper-large-scale
integrated intensive farming system which includes fish, animals, vegetables
and flowers. One activity is designed to feed another, at zero emission, and
with an adjoint slaughterhouse the residues of which feed into an adjoint biogas plant and the in-between floors.

Given the prevailing conservative paradigm of agriculture and sentimental
images of countryside life, the Deltapark concept certainly is a shock to most
contemporaries, however accustomed they might have become to artificial
lighting and microwave fast food. In reality, intensive Deltapark farming is
not so far from what already exists in rural areas on the ground without being
noticed, and without systematically considering environmental requirements
and animal health.

4.3.6 Transgenic Crops

Genetic engineering is an outstanding example of a generic systems innovation. It is adding an important new branch to the tree of science and technology (II/2.3), it is revolutionising traditional biotechnology and has manifold innovative applications in agriculture, food, materials processing and reprocessing, chemistry, pharmacology and medicine. Modern industries can make use of genetically modified organisms (GMO) such as enzymes, viruses, bacteria, microorganisms, plants, animals and their organs – in a certain sense like sheep, horses and cows in traditional production – to produce many things, from pharmaceuticals, chemicals, raw materials, fuels, fibres, food and feeds, up to cleaning processes where dirt is broken down into separate components which are consistent with natural metabolic processes.

Genetic engineering is thus a key or defining technology par excellence. Its importance even reaches far beyond other technologies which were in their time defining such as the mechanical loom, or rail and steam boat, or the mains system, car motorisation, or computerisation, all of which propelled long waves, i.e. innovation cycles of four to six decades. Modern biotechnology is actually more complex, a techno-evolutionary field of its own, like mechanics, energy, chemistry or information and telecommunication technology. No one ever heard of the 'chemical revolution' even though it took place, roughly speaking, from the 1820s to the 1970s. But ever new domains within the chemical industry contributed to the long waves of their time, for instance dyes in the 19th century, and polymers, synthetic fibres and agrochemicals in the 20th. The 'biotech revolution' looks to be even more complex, and various new biotech developments will undoubtedly play an important role in future long waves.

As is the case in any revolution, people's sense of security is undermined. At present most people, including many experts, cannot reliably assess the innovation's potential and its limits, nor indeed the direction which things are taking. Conservative opposition and risk aversion are strong. It takes time to replace uncertain expectations with realistic experience. Not all feelings of uneasiness are unfounded, although in most cases it has turned out that in fact they were. Almost all transgenic innovations spur efficiency. But not all applications can be said to be metabolically consistent. Some are not, or not yet.

In agriculture, present developments are being moulded above all by the introduction of transgenic crops. These are crucial to the ecological modernisation of intensive farming. Beyond transgenic crops, GM microorganisms in pest control, as mentioned above, are expanding their hitherto small market share. Transgenic animals and molecular farming with animals are still in the laboratory and development stage.

Transgenic crops are also referred to as GM seeds. Calling them 'designer seeds' may be a daring future vision, but is at present a mystification which exaggerates what genetic engineering today can do. What it has done so far, is to accelerate the cross-breeding of new varieties by adding single genes, i.e. transferring genes from one kind of plant to another, also between unrelated species. It is not possible to cross-breed, say, cacti with corn, but it is possible to transfer single cactus genes that promote drought resistance into a corn plant. Transferring genes (transgenics) may or may not work in a given case, and it is intended to result in a transfer of desired properties, which in reality may or may not be the case. But as molecular and cell biology rapidly extend the knowledge base about the genomes of plants (genomics), more of the transgenic cross-breeding will lead to some useful results.

Genetic engineers can furthermore try to 'switch off' genes, for example the ones which are involved in making plants wilt. Well known are everfresh transgenic tomatoes. Though they do not stay fresh forever, they keep for longer than their natural counterparts. This is an advantage for farmers, retailers and consumers alike. Genetic engineers cannot, however, force crossings which nature does not allow, and they have not been able so far to invent new genes. They can only make use of those which occur naturally. In the laboratory, though, genetic engineers already attempt to create new genomes from scratch. This might one day indeed lead to artificial new forms of life.

As a halfway house between conventional breeding and transgenic crops, biotech methods such as genetic markers, DNA fingerprint technology and automated genetic screening can enhance the conventional crossing of crop varieties. Testing of new breeding combinations can thus be done in a short period of time. Conventional cross-breeding normally takes 8–12 years to produce a plant with improved properties. With genetic engineering methods, results can be seen within one growth period. GM is also more precise than conventional cross-breeding.

New crops that were developed by the halfway house approach include rice with a high content of iron and zinc, or soya beans, the oil of which does not need to be refined for use in cooking, has more unsaturated fatty acids and longer durability. Crossings between wild rice and commercial varieties have resulted in 10–20 per cent higher yield of good quality. In one case there was even an improved resistance to viruses. For seed companies and farmers the halfway house approach has the advantage of having improved varieties available while avoiding the possible risks of GM crops. These, however, are more political and economic (unpredictable mood swings and actions on the part of fundamentalist activists, the public, politicians and authorities) than they may have to do in special cases with health and ecology.

Transgenic crops which are now routinely grown are rice, maize, wheat, cotton, tomatoes, potatoes, soya, rape, cassava, sorghum, millet, pepper, also some beans and cabbage. Most transgenic crops are developed and grown outside the EU. A majority in most member countries tends to grotesquely cultivate fears of supposed risks of the new key technology, while it likes to ignore the risks of losing contact with realities and losing ground to the competition. The percentage of land cultivated with transgenic crop varieties is about 50–90 per cent in Latin America, Asia, Africa and the Middle East. China and India are particularly enthusiastic adopters of GMOs. But whereas China readily approves commercial GM crops, India has more political difficulties with domestic NGOs who oppose transgenics.

The poor image of transgenic crops is in large part due to the fact that the first broadly commercialised GM varieties were those with a tolerance of synthetic herbicides (chemical weedkillers) such as GM soya, maize, cotton, rape, lettuce and tomatoes. A herbicide involved is glyphosat, marketed by Monsanto as Roundup Ready (soya, wheat), or Liberty by Aventis (rape) and Pursuit by Cynamid (rape).

The case remains problematic indeed. The origin of the problem can be seen in the fact that the producers of agrochemicals and of GM seeds are often the same in that they are different divisions of the same corporation. The advantage of herbicide-resistant crops, compared to the previous practice, is that herbicides can be applied less often and in lower quantities. This saves costs and is environmentally less harmful. Less harmful is still nonetheless harmful. Herbicide-resistant crops are clearly not a metabolically consistent innovation. Eco-consistent transgenic crops are those that can do away with hazardous agrochemicals without posing comparable new risks.

A number of further weak points of herbicide-resistant crops were observed. In warm climates above 25°C, the stem of Roundup Ready soya tends to break open, and affected plants grow less and produce fewer soya beans. The cause is attributed to the 20 per cent more lignin which GM soya produces in contrast to non-GM soya. In other places, weed control by combined use of a synthetic herbicide and a herbicide-resistant crop did not translate into higher yield. Conventional rape was reported to have delivered the same yield as Roundup Ready rape, and if no herbicides were applied at all, yield was even higher; as was the income of the farmers who saved on the outlay for the herbicide.

In Denmark, glyphosat has been detected in the groundwater, in concentrations twenty times above the emission standard. Glyphosat was thought to decompose quickly and near the surface, but it did not. There is preliminary evidence of higher risk of cancer (non-Hodgkin's lymphoma) caused by intake of glyphosat. It should be noted that glyphosat is the herbicide, not the transgenic crop.

With transgenic crops tolerant of climate stress and hostile environments the picture is definitely more positive. Examples include GM crops which are more tolerant than non-modified crops to salty soil and salty water, cold and frost, drought, wetness and poor or marginal soil. The development of such crops is advancing, and widespread introduction is expected around 2010–2015. Researchers at the Indian National Research Institute for Biotechnology, Delhi, have inserted genes of a weed (*Arabidopsis thaliana*) into the genome of mustard. As a result, irrigation can be reduced by 50 per cent. A joint US/Israel research project has identified cotton genes that enable growth in a dry climate. A modified gene in Alfalfa enables this plant to grow on acid soil with high aluminium content. The effect could be transmitted to other plants as well. Another goal of researchers is to devise ways in which crops can fix nitrogen directly from the air. A new rice variety with a special maize gene absorbs 30 per cent more CO_2 from the air.

Other breeds of transgenic crops are highly resistant to pests, or highly immune to disease caused by fungi, bacteria and viruses. Researchers at the Kenya Agricultural Research Institute were successful in creating a GM sweet potato with resistance to the virus SPFMV. They have also developed a maize variety resistant to insects. Other GM crops are resistant to mites and nematodes. A GM maize from Monsanto has a resistance to rootworms. Another variety of GM maize produces the protein Avidin which makes it resistant to a number of storage pests such as moths or beetles. The substances which deter the pests are natural toxins. Practically all plants contain or secrete such substances, and normally they do not harm humans unless ingested from a poisonous mushroom or a deadly nightshade. The question arises nonetheless whether the new transgenic varieties are comparably harmless, or whether they might be more aggressive.

A much cited and controversial example are Bt-crops, especially maize, cotton and potatoes. Bt is short for *Bacillus thuringiensis.* This bacillus releases a toxic protein, the Bt-toxin, which thus is a natural insecticide and widely used in organic farming. The toxin destroys the intestinal wall of beetles, caterpillars and similar pests. In Bt-crops, the bacterial gene strand which produces the toxin has been transferred to the crops so that these produce their own insecticide and are thus pest-resistant. Bt-maize, for instance, is immune to the corn borer, the main pest in Europe and the US. In cotton, the Bt-genes create resistance to boll worms. Bt-maize is also less polluted with mycotoxins. As a result, chicken fed with Bt-maize grow significantly better than chicken on a normal diet.

Bt-crops have been attacked by GMO opponents for killing the larvae of the monarch butterfly, a non-target organism, as a consequence of eating pollen from Bt-maize. This, however, in fact barely occurs. If it happens all the same, this is unfortunate, but killing larvae and caterpillars is in principle exactly what an insecticide is supposed to do and why Bt-toxin is also applied in organic farming where it represents 80–90 per cent of the pest control.

Standard chemical insecticides such as lambda-cyhalothrin are far more destructive of monarch caterpillars. Standard agrochemicals, however, as long as they stay active, are known to be toxic also to humans and animals. One would like to know whether anti-pest toxins in Bt-crops might also have toxic effects when eaten by humans or larger animals. In principle, eating GMOs is no more of a risk than eating normal food. If, however, transgenic crops contain active toxins, this must of course be carefully checked.

A related scare now is that target pests might adapt to Bt-toxin. Studies of Bt-cotton in the US and China conducted for seven years concluded that target pests did not develop resistance. In contrast, certain populations of the corn borer have started to develop resistance to Bt-toxin. So the concerns of organic farmers about Bt-crops seem to be well-founded. Normally, Bt-toxin does not cause organisms to become resistant to it, because when sprayed as an insecticide the Bt-toxin is active only for a couple of days, whereas in Bt-crops it is active throughout the plants' life. Bt-toxin from Bt-maize was also discovered to leak into the surrounding soil, staying active there for a certain time. Therefore, refuge areas of about 15–25 per cent of the land have been proposed, where natural conditions prevail, so as to avoid one-sidedly specialised adaptation of species. The American Environmental Protection Agency demands 20 per cent of maize acres as refuge, and 50 per cent with Bt-cotton (*ISB News report*, mail out 3 February 2000).

The ultimate goal of genetic plant engineering is to create crops, vegetables and fruit which not only generate a high yield but are simultaneously of high quality and content-value with regard to substances such as starch, sugar, protein, oil, polyunsaturated fatty acids, vitamins and trace elements. Even 'nutraceuticals' could be devised which combine properties of nutrients and pharmaceuticals, such as eatable vaccines in bananas against diarrhoea (Leisinger 2001 66).

Maybe the best known GM crop of this kind is 'golden rice' which has a high content of vitamin A and iron. A special breed of GM potatoes contains about 40 times more fructose (sugar). This was achieved by inserting three bacterial genes into the DNA, each gene coding for a different enzyme involved in converting starch to fructose. As a result, the potato multiplies its own fructose content about 40 times. The product could be commercialised by 2006. Today, most fructose is made by adding vast quantities of enzymes to large chemical tanks of corn. There are now also

- transgenic potatoes with much more starch
- sugar beet with high fructose content
- more sugar in maize and strawberries resulting in better taste
- grain with glutamate-free amino acids
- transgenic sweet potatoes from South Africa which contain five times more protein than traditional ones

- fodder maize with larger content of phytate, causing in animals, especially pigs, improved resorption of phosphate
- soya beans and maize with higher content of proteins and oil, thus higher nutritional value, used as human food or animal fodder
- extra vitamins in fruits and vegetables
- carrots with extra beta-carotene content (anti-oxidant)
- tomatoes with extra lycopeen content (cancer medicine)
- coffee which is 'naturally' decaffeinated
- rice incorporating a maize gene, resulting in 35 per cent more yield
- flowers with new lively colours.

Moreover, transgenic plants can also have better processing properties, for example an improved fibre structure in GM cotton. And in addition to non-mushy tomatoes there are now also GM potatoes, vegetables, melons, strawberries, raspberries and flowers which stay fresh for longer.

A new approach in transgenics is apomixis-stimulated seeds (Charles 2003). Apomixis means asexual self-reproduction, actually self-cloning. This occurs naturally in about 400 plant species. Conventionally bred rice and wheat reproduce sexually, but the two sexes are integrated in one flower which pollinates itself (which is not far from self-cloning, and daughter generations are exact copies of parent flowers). Meanwhile, there are apomixis-stimulated cassava, potatoes, cereals, rice and corn in the laboratory stage. The hybrid crops produce seeds without sex. So there is no need for pollination. This keeps the performance of hybrid varieties which normally reproduce sexually, such as wheat, and prevents asexually reproducing plants, such as bananas, from reproducing the viruses and fungi present in their tissue.

Apomixis stimulation results both in higher yield and improved quality, reliably reproduced by the plant itself year after year. This would allow poorer and richer farmers alike to keep their own seeds of hybrid wheat, rice, corn, etc. for next season's sowing, and would thus reduce the dependence of farmers on industrial seed supply. This is why seed corporations can probably not be expected to act as promoters of apomixis, even though they fund much of the R&D which is at present done in this field. Self-cloning crops will have to be promoted by public non-profit institutions such as CIMMYT (Centro Internacional de Mejoramiento de Maíz y Trigo) in Mexico, which also helped to set up the 'green revolution' decades ago, assuming that these institutions have not become too dependent on commercial funding.

In the same way as transgenic plants can be induced to generate high content value for food, feeds and fibres, they can also be induced to generate special substances which hitherto were the business of specialty chemistry and the pharma industry. This is the approach of molecular farming with plants, and an example of phytochemistry. Transgenic plants are used as bioreactors which produce valuable substances or even products such as silk or polyester. Or they produce pharmaceuticals, for instance blood proteins in potatoes, and vaccines such as the diarrhoea vaccine produced by bananas. GM maize by Epicyte, San Diego, can mass-produce antibodies that kill human sperm and herpes viruses. Once the seeds have been harvested, the antibodies can be extracted. The capacities of the 'maize factory' can easily be reduced or expanded according to the business cycle. This in practice means another field more or less and thus avoids the costs of investing in fermenters and related fixed capital. Biotechnological production of substances by plant bioreactors is safer, better and cheaper than performing it alternatively by microorganisms in fermenters, or conventionally by chemical synthesis.

To summarise, it is obvious how many desirable things GM-enhanced agrobiotechnology can achieve, economically as much as ecologically and socially. It thus fulfils the triple goal of sustainability in a more comprehensive way than most other TEIs do. Transgenic crops can be modified to require less irrigation and to accommodate to difficult climates and soils. They improve on aspects of productivity, quality of produce and content value. These are not only welcome additions to food and feeds but are also useful for producing materials (industry crops) and introducing biotechnical phytochemistry. GM crops can reduce or even do away with mycotoxic and endotoxic contamination of conventional food and feedstuffs, a problem also present in organic produce. Finally, transgenic crops can partly or fully substitute for hazardous agrochemicals such as herbicides, pesticides, fungicides and similar agents. About 30 per cent of world yield is still destroyed by greenflies, grasshoppers, moths, beetles, worms, caterpillars, snails, fungi, viruses and bacteria, and by too much weed on the acreage, all of which can be dealt with by GM-enhanced biotechnology. Particularly if seen from the perspective of worldwide population densities and scarcity of agricultural resources it would definitely be irresponsible not to tap this huge potential.

In comparison to these advantages, fears of particular disadvantages are dwarfed, or are even untenable in face of the facts. Consumers, particularly in Europe, are afraid of GM food. But eating GM food is of no particular risk to human health. Eating conventional food and organic produce can also, and actually with more reason, be considered to be 'dangerous' because of manifold natural and industrial contaminations. Gourmets love to eat mouldy cheese and well-hung meat, accompanied by more or less concentrated alcohols, all of which are certainly more demanding for the body's detoxification organs than Bt-maize ever can be. The Royal Society, the American Medical Association and the American Association for the Advancement of Science have found it necessary to weigh in with their scientific authority in order to calm down superstitious attitudes towards GM food.

There is – among the very few scare-cases which are quoted over and over again – the case of those Brazil nut genes transferred to soya beans. Some people are allergic to nuts. Since soya beans are used in thousands of food products, that could make eating a hazardous exercise for those people with a nut allergy. But these people number very few, and they have to be very careful in nutritional matters anyway. Labelling of content is clearly what can be helpful in this case as in others. Furthermore, crop identity verification (produce tracking) is certain to develop further, which will help to keep organic produce, conventional agroindustrial crops and GM crops separate from one another.

A different fear is that GM crops would reinforce monocultures and thus be a threat to biodiversity. At first glance, it looks as if genetic engineering is concerned with creating more biodiversity, even though of an artificial kind. In the long term it could nonetheless be the case that a handful of GM crops would crystallise out as the preferred choice. This, however, is not inherent to transgenic biotechnology. Whether and how much monoculture there is depends on agricultural regime rules, guided or imposed by regulation and economic mechanisms. Neither GM crops by themselves induce monoculture, nor would a ban on GM crops rule out monoculture in conventional agroindustry.

Related to this aspect is the fear of transgenic colonialism, i.e. perturbation of the natural succession contest in flora and fauna, a preferred preoccupation of conservationists. History can record impressive examples: influenza brought from Europe to what became Latin America, in exchange for syphilis and potatoes, rabbit plague in Australia, killer bees in America, Californian 'green hell' (a particular bush) in German forests, etc. Here again it has to be said that the competitive rise and decline of populations in a living space is in principle a quite natural process. Given the fact that species migrate, mingle with or drive out others, it cannot be concluded that this is in itself bad. There is no such thing as an original natural state or an ecological standard meter. Natural dynamics are remorseless, and human intervention in natural succession has always been an integral part of human history. Unspecified requests to keep foreign or new populations out are nothing other than xenophobic anthropomorphism.

As it happens, today's GM crops have proven not to be fit for natural survival. As one decade-long test with GM rape, maize, sugar beet and potatoes revealed, all four species did badly when left untended by humans. Of 48 plots planted, 47 became extinct within four years. The message here, however, is ambivalent, because genes of GM plants are as much naturally transmitted as natural genes are. Plants pick up pollen from other plants (gene stacking). GM pollen is transmitted over many miles by wind or bees. Gene transfer seems to occur among different GM species as well as between GM species and normal species. Spreading of 'weak' genes could thus be thought to weaken natural varieties instead of simply being unsuccessful, as spreading of 'strong' genes also could strengthen natural varieties instead of just driving them out of competition. It is certainly advisable to keep a close watch on the interactions and resulting successions between conventional and GM varieties, but in principle there is no cause for extra concern. In the long run, natural selection will do its job as it always has done.

In Alberta, Canada, stray rape was found that was resistant to all herbicides in widespread use. This gave raise to fears of some new class of superweed, i.e. some wild species or agricultural breed turned wild which would incorporate GM properties of pest-resistance, herbicide-tolerance, etc. Such a superweed would be ineradicable, a farmer's nightmare and a threat to indigenous plant populations and botanic succession. Again, stray phenomena naturally occur, and have to be observed and analysed, but the superweed projection is an overshoot. Inflated populations naturally break down again. If pests develop resistance also to GM crops, and if there are some stray crops here and there, this is just confirmation that nature works as it always has done, and that also genetic engineering, an achievement though it certainly is, cannot work unheard of supernatural wonders. So far, trade and tourism have contributed more to the global spread of genes and germs. Also climate change will do more harm to existing plant populations than GM crops ever could.

4.3.7 Animals with Genetically Modified Properties

Molecular farming in animals and breeding of transgenic animal species are innovations which for the most part are still at the research stage, but will undoubtedly arrive at market sooner or later. From a medical point of view, animals can be more delicate than crops normally are, and the treatment of animals has a pronounced ethical dimension to it. For example, conventionally bred cows with 'turbo udder', or pigs with heavy bodies upon short weak legs, have been subject to medical and ethical controversy for many years. Transgenics in animals will by itself make things neither better nor worse. It is up to regulators to set rules, standards and limitations which satisfy veterinary medical criteria as well as the moral standards of the time.

It is intensive animal farming which creates an interest in genetically modified properties of animals such as rapid growth, improved digestion, consistency of meat, and resistance to disease (Wulff 1999, Dürkop et al. 1999 117). It cannot be excluded, however, that the successors to today's organic farming might also find it useful one day to have animals with such properties. At Guelph university, Ontario, Canada, a transgenic variety of pig is being bred which creates more phytase for itself, that enzyme which is now often added to the feedstuff to achieve better resorption of phosphate and thus decrease the phosphate content of manure. Success rates are still very low. There could be arguments over whether it is appropriate to baptise the new variety with the name of 'EnviroPig'.

For the moment, animal transgenics seems to work best in fish. GM fish that produce growth hormones are so far the most advanced near-market development in transgenic animals. This will give additional support to the transition to aquaculture. For example, researchers in Thailand have developed a variety of perch, a favourite dish in Thailand, with a modified gene which codes the production of growth hormone. Another example is medaka fish, kept in aquariae, in which was inserted a human growth gene. They grew faster and bigger than non-GM medaka, and also produced more fertile eggs. Bigger male medaka attract four times more females than normal ones. The same was observed in salmon. Anti-GMO protesters were not hesitant to predict that this would extinguish natural fish populations if GM medaka or salmon were released into the wild. But here the same considerations apply as in the case of GM crops which allegedly would exterminate natural plants. Also GM fish will face their limitations, the more so since for the time being GM fish have a life expectancy of only two-thirds that of natural fish (www. newscientist.com 4 Nov 1999).

The aim of molecular farming in animals is to genetically stimulate their production and secretion of special nutrients and pharmaceuticals, normally as an increased or additional content of the milk of female animals. For example, cows, sheep and goats can produce substances such as anti-thrombine III, antitrypsine and alpha lac talbumin (food additive), which can then be extracted from the milk of these animals. Since they have a metabolism similar to humans, and have the same sugar-producing processes, they can also have inserted certain human genes which make them secrete certain human proteins in their milk, for instance tissue plasminogen activator (TPA).

Besides nutrients and pharmaceuticals, special chemicals and materials are also of interest, for example transgenic spider silk. There are several varieties of spider silk with outstanding properties such as simultaneously being soft, flexible, ultra-light and ultra-strong, an ideal combination in the eyes of materials engineers. The genes of key proteins in four different types of spider silk were sequenced and cloned. Producing GM spider silk through bacteria worked well, but was not efficient enough to be applied commercially. The task has now been passed on to transgenic goats that secrete the silk proteins into their milk from which the proteins are extracted and purified. The material can commercially be applied in low-volume high-value settings such as in surgery (to produce synthetic tendons), in body armour (lighter, more flexible and stronger than Kevlar) or in industrial textiles (arrester cables for fighter jets on aircraft carriers).

Transgenic animals and molecular animal farming will need close permanent assessment of animal health and ethical questions of animal protection. Certain methods might have to be refined or changed, for example with regard to vector recombination, i.e. genetic engineering with viral genes which can cause complications or create new viruses. The advantages are nevertheless obvious again: fully biodegradable, non-toxic substances and materials of high purity with technically advanced properties; no large ecological footprints or materials rucksacks to make resources available or to process obtained materials; no pollution. This is part of a general tendency of substituting biotechnical processes under ambient conditions for physico-chemical production processes at high temperatures and high pressures, resulting in an environmentally much improved industrial metabolism.

4.4 CHEMISTRY: BENIGN SUBSTITUTION AND TRANSITION TO TRANSGENIC BIOCHEMISTRY

4.4.1 Biofeedstocks and Phytochemistry

In chemistry, as in agriculture and energy, literally everything is environmentally sensitive. But whereas energy and agriculture display a relatively clear structure, chemicals and the chemical industry is a rather complex field with branchings here and specialties there. So it is no easy task to derive a coherent picture from the many single innovations that can be identified in this realm, the more so since that which defines the chemical industry is a convention which developed over time and does not take into account the fact that a number of further industrial sectors such as gasoline and gas supply, metallurgy and other fields of materials processing are nothing other than sectors of industrial chemistry.

The chemical industry can, on one level, be described by its products: dyestuff and varnishes, soaps and cleaners, detergents and disinfectants, auxiliary chemicals such as lubricants, coolants or organic solvents, rubber, plastics and synthetic fibres, agrochemicals, pharmaceuticals, food and feed additives, to mention the most important ones. The chemical industry can also be described by characteristic production processes and by related feedstocks.

According to Grupp et al. (2002 136), the chemical industry, as far as it makes use of organic materials (not just inorganic minerals such as salts), was originally based on non-fossil biotic materials, for example plant oil, rubber latex and natural fibres. This sort of biochemistry can be referred to as phytochemistry, i.e. materials from plants (with a small portion also from animals). In the 19th century cheap coal became available, and in the first half of the 20th century mineral oil. Together with the achievements in analytical chemistry, this led to a paradigm shift from phyto- or biochemistry to carbo- and petrochemistry, and from traditional processing to synthetic chemistry. This cracks materials and molecules down to smaller parts or atoms in order to technically synthesise or re-compound them thereafter into different products and byproducts.

Today, Grupp et al. see the pendulum in organic chemistry again swinging back to phytochemistry and biotechnology. This, however, is taking place on a much higher level of development and with the biotechnology in most cases being enhanced by genetic engineering. The examples collected in the TEIs database of this book definitely support that view. Carbo- and petrochemistry, though, will certainly not be phased out but will continue to deliver the feedstocks for many base chemicals. Also biofeedstocks can provide base chemicals such as ethylene, propylene, synthesis gas and aromatic compounds. The main role of phytochemistry, however, is in specialty chemicals with complex molecules of higher order.

Petrochemistry starts, so to speak, at a great distance from end products, thus requiring many production steps of cracking and synthesising. This consumes a great deal of materials and energy, often involves hazardous chlorination, results in manifold harmful emissions, and is expensive. In biofeedstocks, natural photosynthesis has done much of the job already. Further production steps by fermentation, extraction, filtration, etc. start from a high level of incorporated energy. For chemists, traditional biochemistry had the disadvantage of being impure, whereas technical synthesis delivered products at desired degrees of purity. With today's advanced technologies it is possible to achieve desired degrees of purity also in bio-products, while there are less production steps, less energy demand, less materials losses and less or even zero problematic emissions. Accordingly, less costs are incurred, though the biofeedstocks, being agricultural products, are more expensive than coal and petrol. Phytochemical products, furthermore, have the big advantage of not of themselves representing a temporary sink of hazardous chemicals such as plastics with chlorine content. So deposing or burning of phytochemical waste is relatively harmless. The materials are biodegradable and are re-absorbed into natural cycles without causing environmental problems.

Consider vegetable oil esters. Cleaners based on these offer an alternative to organic solvent cleaners, particularly for purposes of metal cleaning which is of high environmental impact and relevant in many industries such as mechanical engineering, power plant construction and maintenance, automobile manufacturing, shipbuilding, steel mills, railways, precision mechanics and the electrical industry. Vegetable oil esters are also used in construction as a concrete release agent (formwork oils). Two important formulas are rapeseed methyl ester (rapeseed oil plus methyl alcohol) and 2-ethyl hexyl laurate (coconut oil plus 2-ethyl hexyl alcohol). These have many advantages over organic solvents. They produce the same top quality as conventional cleaners, in some applications they do even better. As a welcome side-effect they provide temporary corrosion protection. They are skin-compatible and of mild smell. Their flash point is over 100°C, so they are not dangerous and allow reduced costs of risk prevention. Their production and use involves much lower materials consumption, and they are biodegradable.

Vegetable oil esters are also economic. For example, a cable factory may need 2,400 litres of organic solvents a year, compared to just 60 litres of oil esters. The price per litre is €2.34 for organic solvents, and €5.20 for oil esters. As a result, the direct annual cost of organic solvents is €1,875, that of oil esters only €100. In addition, solvents have to be reprocessed or burned as hazardous waste at a high cost, none of which applies to esters (Handbook of Vegetable Oil Esters 2002 45).

Oleochemistry, i.e. products on the basis of oil and fat which represent half of the biofeedstocks market, is the most important segment of phytochemistry. Products are tensides (surface-active agents such as in cleaners and detergents), cosmetics, lubricants and additives to plastics. Domestic supply from rapeseed or sunflower oil is, however, not able to satisfy demand. Ninety per cent of supply comes as coconut and palm oil from overseas. Palms are grown there in agrochemicals-intensive monoculture plantations. This is another example of bad European agro-policy. This subsidises overproduction of food, and simultaneously permits (or has to permit on the basis of ill-conceived WTO trade rules) the import of crops the production of which does not meet European environmental standards.

Further important market segments of biofeedstocks are starch (25 per cent) for producing glues, packaging materials or antibiotics, cellulose (about 15 per cent) for various fibre products and plastics, and sugar (about 5 per cent), mostly as a nutrient for microorganisms. On average, the share of biofeedstocks in the chemical industry today is at about 10–15 per cent. DuPont, an American corporation, plans to generate 25 per cent of its turnover with phytochemistry in 2010, comparing to 8 per cent in 2000. For example, the production of the stretch fibre Sorona on the basis of 1,3 propandiole (PDO) is being switched from petrochemistry to phytochemistry. A pilot plant in Decatur, Illinois, is producing PDO by fermentation of cereals. In Table 4.3 a number of phytochemical products are listed, each in comparison to their like products on a petrochemical basis.

The processing of industry crops does not yet represent a well integrated manufacturing chain. Many productions still represent a stand-alone process and lack infrastructure. This is one reason for the delayed comeback of hemp and linen as natural fibres. Required is integration into a multi-connecting 4-level manufacturing chain. The first level of this chain is plant growing, i.e. farming, the second extraction of biofeedstocks in the form of oil, fat, raw starch, fibres and sugar, i.e. basic processing. At the third level base chemicals are produced such as proteins, starch acetate or polyepoxides (playing the role which olefins do in petrochemistry). At the fourth level then the raw materials are refined and beneficiated in the framework of specialty chemistry. Building up such an integrated chain would include, at least to a certain extent, a switch-over from huge central chemical plants which use homogenous raw materials (with a capacity of about 100–1,500 kt/a) to smaller decentral units. These are based on small- and medium-sized bioreactor technology which is better suited for processing smaller quantities of biofeedstocks of varied consistency. As a result, and similar to distributed power generation in energy, there will be a more 'distributed' chemical production which combines large-scale plants with medium- and small-sized capacities.

Table 4.3   Examples of petro- and phytochemical-like products

Industry segment	Petro-chemical materials	Phyto-chemical materials	Example of phyto raw material	Example of application
Reinforced fibrous materials	Carbon fibre, GF, polyamide	Plant fibres, plant resins	Hemp fibre, shellac	Toys, tool-housings, furniture
Floor coverings	PVC	Tree bark, veget. oils, fibres, resins	Cork, linseed, oil, jute	Linoleum
Textiles	Polyester	Natural plant fibres	Linen, hemp, cotton, silk, wool	Upholsters, garments, underwear
Wood glaze	Polyacrylate, glycol	Plant resins, essential oils	Dammar resin, lemon peel oil	Natural resin, oil glazes
Colours	Azo-pigments	Colouring plants	Dyer's woad, madder, etc.	Plant colours
Detergents	Linear alkyl benzene sulphonate	Vegetable oils, carbo-hydrates	Coconut fat, sugary tensides	Household detergents
Hydraulic agents	Mineral oil	Vegetable oil	Rapeseed oil, jojoba oil	High-perform. lubricants
Adhesives	Polyurethane	Natural gum, plant resin	natural latex, mastics	Carpet glue
Packaging materials	Polyethylene	Polysaccha-rides	Biopol	Shampoo bottles, shopping bags
Thermal insulation	Styrofoam, polystyrene	Pulp, straw, lignin	Sunflower pulp and straw	Building materials

Source: Research Forum Environment. Austrian Federal Ministry of Science.

Such a scenario and the examples given in Table 4.3 do not necessarily include genetic engineering. They can also be conceived of as natural phytochemistry, in analogy and connection to organic farming. Among the best known examples of such natural product lines are natural dyes and varnishes, and clothes made of natural fibres only and produced without hazardous chemicals and dyestuffs. But as in the case with organic farming, natural phytochemical products represent a niche market with a potential of perhaps 10–20 per cent, satisfying the demand for alternative health and lifestyle. This has shown itself not to be contagious to the majority of users and consumers. Natural phytochemical products may in the future even face difficulties as other new-generation chemicals also emerge with ever more properties of green chemistry, thus shedding the image associated with hazardous products and processes.

This is why biofeedstocks and phytochemistry will for the most part come hand in hand with transgenic crops and GM-enhanced biomaterials processing. A case in point is transgenic fibres, for example conifers as pulp trees. Conifer wood requires more intensive pulping to remove lignin than hardwood. But the quality of conifer cellulose fibres is superior. Transgenic conifers are modified to produce less lignin, an even better fibre quality and higher yield. Some cellulose fibres (cotton, linen, hemp, lyocell, rayon) and proteinaceous fibres (wool, silk) are also genetically modified in order to obtain improved or novel fibre features.

4.4.2 Chemical Biotechnology

In the same way as agricultural biotechnology is referred to as 'green' transgenics and medical biotech as 'red', chemical biotechnology is called 'white', whereas in materials (re)processing and waste treatment it is called 'grey' or 'brown'. In general, all biotechnological processes in chemistry, as represented by the examples to be given subsequently, have a number of environmental advantages:

- Production processes consume much less energy and
- are much safer. This results from the fact that biotechnological processes run at normal ambient temperatures and pressures. High-temperature high-pressure processes which are both dangerous and pollution-intensive are no longer involved.
- Feedstocks can be less refined.
- Processes result in a high yield of target products because of highly specific reactions. Hence there are far fewer byproducts and less waste.
- Separation of intermediate products is in most cases not necessary.
- Catalysts based on toxic heavy metals are replaced by enzymes.

- Organic solvents are replaced by water.
- There is neither much hazardous waste to be treated nor large quantities of (often hazardous) auxiliary chemicals to be recycled.
- There are no or only a few noxious emissions.
- Emitted substances are biologically degradable.

Chemical biotechnology is for the most part process technology. Broadly speaking, there are two classes of bio-processes, the one being microbial bio-synthesis, the other biocatalytic processes (GM-enhanced fermentation, enzymatic catalysis).

Microbial biosynthesis refers to the production of substances by microorganisms, most often bacteria, or yeast or some other microbes. These can produce, for example, polyester, a synthetic fibre which in this case is bio-synthetic rather than chemosynthetic. The microbes are especially bred and in many cases genetically modified to do their job more effectively and efficiently. Transgenic microorganisms can be 10–100 times more efficient than their naturally occurring brethren.

Corynebacterium glutamicum (GM) synthesises L-serin, an amino acid used widely in the pharmaceutical and food industry. This replaces the production of L-serin by chemical hydrolysis which entails high energy demand, low yield and high amounts of waste salt.

Escheria coli (GM) can synthesise pyruvate, a specialty chemical in the production of pharmaceuticals, cosmetics, agrochemicals and nutritional additives. The method has a yield of nearly 100 per cent without problematic environmental impacts, whereas the traditional production method requires much energy for the pyrolysis of grape acid, and involves organic solvents and heavy metals.

Saccharomyces cerevisiae (GM), which is a yeast, can produce L-G3P (L-Glycerol-3-phosphate), a specialty chemical used for enzymatic synthesis of carbonhydrate-based pharmaceuticals. The metabolism of the yeast cells is reprogrammed (metabolic engineering) as to increase anabolism and decrease catabolism of L-G3P, resulting in an optimised, economically competitive yield. The product is of very high enantiomer purity, which adds to its advantages over environmentally highly problematic chemical synthesis of L-G3P.

If the keratin in poultry feathers can be cracked, they are a source of valuable polypeptides used in animal feeds or for producing amino acids. Animal feeds is a particularly big market. Hitherto, the keratin cracking and recovery of polypeptides was done by acid or alkaline hydrolysis, i.e. concentrated hydrochloric acid or caustic soda, at a temperature of 150°C . But most feathers have to be burned in incineration plants at a high cost. Now the job can be done by extremophilic bacteria, i.e. bacteria working under conditions somewhat above or below normal ambient temperatures and pH values. In this case, the cracking of the keratin in poultry feathers is done by *Thermoan-aerobacter keratinophilus (GM)* at about 70°C and neutral pH value.

Also production of hydrogen H_2 through bacterial photosynthesis might one day be an alternative. Photolytic production of hydrogen can be achieved by bacterial chloroplasts, working as photosynthesisers, in water under sunshine. Whether it can in fact be done is beyond question. The open question is whether any of the microbes experimented with could one day do it efficiently enough. Research with purple bacteria has been carried out for many years without progressing beyond the laboratory stage. In any case it is a good example of ecologically fully adapted bionics.

Particular hopes are pinned on GM enzymes. They are the natural catalytic tool in any biological metabolism. Since they consist of special proteins, the field is also referred to as proteomics. In the human body about 1,000 metabolic reactions are catalysed by enzymes. The number of existing enzymes is estimated at about 7,000. Only 80–100 of these are used in industrial processes. In principle, there is no chemical transformation which could not be performed by a suitable enzyme. Any physico-chemical production process could theoretically be replaced by a biotechnological one. Enzymes, however, are difficult to handle so that proteomics requires some sophistication. There is no all-purpose biocatalyst. Most enzymes are active only in water solutions within a rather narrow band of temperature and pH value. They only react with one or a few substrates, and they seldomly react with xenobiotics (non-natural substances). They are not very stable either. In spite of a number of biocatalytic state-of-the-art processes, biocatalytic research and development in general still has a long way to go.

Among the 23 topsellers of bulk chemicals are 11 that can now be produced biocatalytically, i.e. by fermentation through GM enzymes: Ethylene, ethylenoxide, dichlorethane, propylene, formaldehyde, propylenoxide, glycol, adipin acid, acetone, methanol and isopropanol (Dürkop et al. 1999 26). So far, however, there are not many large-scale industrial plants such as those large-scale fermenters which serve the production of ethanol.

In the fermentation of specialty chemicals biocatalysts play a comparatively more important role already. Proteolytic enzymes, glycosidases, lipases and others are used in the production of a number of pharmaceuticals and fine chemicals, and in food, cleaning agents, textiles and leather goods (OECD 1998 30, 51). Also the synthesis of ammonium acrylate can be based on bacterial enzymes. Ammonium acrylate is a key intermediate in the manufacture of acrylic polymers. The conventional chemical process is energy-intensive and gives rise to byproducts which are difficult to remove. Bacterial enzymes directly synthesise ammonium acrylate of the same quality and are much less energy-demanding. The process has now been successfully operational for many years.

L-carnitine is an essential co-factor in the transport of long-chain fatty acids. It plays a role in intensive animal breeding. Lonza, a company that produces L-carnitine transgenically, has published an eco-balance comparing it to conventional chemical production: Waste for incineration per tonne of L-carnitine is down from 4.5 to 0.5 t. Salts are down from 3.3 to 0.8 t. Total organic carbon in wastewater is down from 750 to 360 kg per tonne. Wastewater is down from 220 to 40 m^3 (OECD 1998 51).

Special enzymes also synthesise certain chiral compounds, so-called enantiomeres. Pure enantiomeres are used for producing medicines against arteriosclerosis. The enzyme D-amino acid oxidase (either produced in the process by the yeast *Trigonopsis variabilis* or in an isolated form) catalyses the production of pure L-amino acids, which are important in the food and pharma industry, for example in the production of penicillin. In this case, as in most of the other cases, good results are only achievable if microorganisms or enzymes are genetically modified. Materials and energy demand as well as costs then only represent 25 per cent of the previous chemical process. Emissions of zinc compounds per tonne of L-amino acids are down from 1.8 t to zero emission. Waste for incineration is down from 31 to 0.3 t. Effluent COD, however, is up from 0.1 to 1.7 t. Emissions of organic matter are down from 7.5 to 1 kg. In the chemical process, environmental costs for waste incineration and sewage water purification represent 21 per cent of production costs. In the transgenic enzymatic process they represent just 1 per cent.

Extremophilic bacteria and microorganisms produce extremophilic enzymes. An enzyme from *Anaerobranca gottschalkii (GM)* reacts with an unusually broad spectrum of substrates also at higher temperatures and under alkaline conditions. They transform starch into valuable carbohydrates such as dextrin and saccharin. These have many applications in pharmaceuticals and food processing.

High-quality amino acids can be produced by proteases (protein-decomposing enzymes) from extremophilic microorganisms which work at about 80°C. Amino acids are needed in big volumes in food, feedstocks, pharmaceuticals and infusion solutions. The biocatalytic process replaces conventional chemical hydrolysis with low yield, high input of feedstocks and energy, and high output of waste.

Certain enzymes are used as an aid to other enzymes, for example cross-linked enzyme crystals (CLECs). CLECs are zeolite-like lattice structures which impart stability to any catalytic site under a wide range of reaction conditions. Operational stability is an important requirement for industrial catalysts. CLECs play a role in the production of both bulk and specialty chemicals.

Beyond production processes, enzymes can also be a useful content of products. Proteases, amylases, lipases and cellulases are used in cleaning agents for washing machines, dish-washers and for special tube and pipe cleaners. In tube cleaners also microorganisms can be used (instead of enzymes). The washing temperature can be low, which saves energy. Chlorine bleaching agents are eliminated. Dish-washers are less aggressive because of a higher alkaline milieu. The creation of AOX is reduced considerably.

An analysis of 21 case studies of biotechnology in chemical processes, most of them representing enzyme-catalysed and microbial syntheses as described above, from various countries on behalf of the OECD concluded 'that the application of biotechnology has invariably led to a process more environmentally friendly than the one it replaces'. This holds true despite the fact 'that the role of environmental effects tends to be secondary to economic and product quality factors when companies consider adopting a new process' (OECD 2001c 35).

4.4.3 Biosensors (DNA Chip, Enzymatic, Microbial, Optical)

GM-enhanced microbial biosynthesis and biocatalytic processes would not be possible without analytical tools of genetic engineering such as DNA-chip technology, optical, enzymatic and cell biosensors. These tools make use of specific reactions of biological molecules upon contact with other substances (biomolecular recognition). They have many applications, beyond chemistry also in medicine and in the quality control of bio-products of any kind, from food processing to pollution control and waste treatment.

DNA-chip technology, also known as DNA fingerprinting, consists of DNA micro-arrays coated onto a silicon chip. This allows quick checking for the presence of certain substances in a substrate, and identification of organic substances of any kind. The bits of DNA are used as biomarkers, for example Celegans Toxchip, from *Caenorhabditis elegans* (a nematode), delivering immediate results about the presence and activity of certain toxic substances. The method saves cost- and time-intensive laboratory work. Results obtained are of higher precision and reliability than previous chemical analyses.

In breeding, DNA chips were developed for the quick analysis of a genome in hybridisation, for example of biotechnologically promising bacteria such as *Escheria coli*. There are oligo-nucleotid chips containing short single-strand DNA molecules (15–60 DNA elements) and fragment chips containing longer double-strand DNA molecules (100–4,000 base pairs). The technology allows tens of thousands of analyses to be carried out automatically in parallel, increasing the efficiency of materials throughput by a factor of 1,000. In the processing of food, feeds, materials and waste, the potential of quality control is significantly enlarged.

In enzymatic and microbial biosensors, enzymes and yeasts fulfil the function of DNA bits in chip technology. For example, a combined enzyme system with D-serin dehydratase and laktatdehydrogenase can be used for real-time in-process detection and recovery of serin, an amino acid, from molasses. Molasses contain a number of valuable amino acids and oligosaccharides. With the help of this combined biosensor, a 60 per cent increase in serin concentration is achieved, while energy consumption is 30 per cent less, water thousands of m^3 less and chemicals hundreds of tonnes less. Such results are of course also highly profitable.

A multi-analytical hand-held device on the basis of PCS-enzymes measures sugars such as glucose, saccharose, lactate, glutamate, ascorbate and hydrogen peroxide. This has applications in the control of biotechnical production processes, food processing and clinical diagnostics. The device is easy to handle, quick (checks on 50 parameters within 2–3 minutes) and cheap.

In industrial sewage water from galvanics or similar processes, detection of toxic heavy metal ions from copper, zinc and cadmium is of particular importance. This can be done by combining a microbial biosensor in the form of immobilised transgenic yeast cells with a flow-injection analytical system.

Quality control of wine is now also based on enzymatic biosensors which replace chemical analyses based on heavy metals. Ethanol content during brewing of beer can be determined by an in-process, online flow-injection analysis with special enzymes. This saves large volumes of hitherto wasted water and beer. A 20-minute test to detect mould (fusarium) in cereals helps to prevent gushing of beer in breweries, and accordingly saves resources and money. The 'virtual cellarmaster' is a biosensor-aided software which optimises the biochemical milieu of acetic acid bacteria in vinegar production. This results in a double-concentration yield while reducing time, energy and resource input. The elimination of problematic chemicals with the help of biotechnology is of course particularly advantageous to anything that has to do with food and beverages.

Optical biosensors make use of one or several optically tagged receptor molecules at the surface of cell membranes. If the receptors receive a substance they are specialised at, cell signals are transmitted and amplified by way of fluorescence resonant energy transfer (FRET). The technology works at the picomolar level. Technically, bio-fluid layers are coated onto biocompatible (or bio-indifferent, inert) materials such as glass or silicon or teflon. There are different methods of measuring the light waves involved. Optical biosensors can detect DNA, other molecules, and especially organismic antibodies. They do their job within a few minutes, are simple to handle (no additional reagents) and are insensitive to variations in temperature.

Biosensors based on reflectometric interference spectroscopy allow the measurement of the concentration of the target product in a fermentation medium. This is an important factor of process control in the biotechnological production of chemical, pharmaceutical and nutritional substances.

An optical biosensor on the basis of the surface plasmone resonance method can measure the concentration of a medication (for instance Cyclosporin A) during and after transplantation surgery. This replaces high-performance liquid chromatography (HPLC) which involves considerable quantities of highly pure, thus expensive, and hazardous organic solvents.

An optical cell biosensor – although still in its early research stages – uses laboratory-grown human cells to detect toxic chemicals such as lead nitrate, acetaldehyde or sodium arsenate. The cells are put in a tube with water and some low-density lipoprotein (compound of fat and protein). A cell's uptake of lipoprotein is slowed down by toxic chemicals in the water, resulting in dimmer fluorescence that can be detected even in very small traces. It takes two hours to do this, compared with the two days which conventional laboratory tests require.

The advantages of biosensors are environmental as well as economic. Conventional chemical processes are replaced which usually involve problematic substances and auxiliaries such as heavy metals and organic solvents. The fact that biosensors deliver instant in-process results does away with expensive, cumbersome, time- and manpower-consuming off-site analyses. This results in high cost savings. Improved precision of the analyses enables faster and more specific intervention, thus eliminating perturbations, making product consistency more stable and homogenous, and resulting in a much higher yield of target substances. In consequence, feedstocks, energy, water and auxiliary chemicals are saved, and emissions are reduced, all of which again results in corresponding cost benefits.

4.4.4 Benign Substitution and Elimination of Hazardous Substances

Benign substitution refers to the substitution of harmless or low-impact substances for hazardous high-impact substances such as heavy metals, halogenated compounds and organic solvents. Benign substitution is aimed at a reduction or complete elimination of pollutants and hazardous emissions from the production, use and disposal of products. Some scholars define the principle of substitution of hazardous substances as the 'identification and replacement of hazardous substances by less risky, inherently safer substances or processes' and consider this to be the main element of the greening of chemistry (SubChem 2002 14).

Benign substitution in this sense has by now reached the status of an almost classical approach to ecological modernisation of the chemical and materials processing industry. Previous passages have in fact included examples of benign substitution, with the phased-in items being of a biological and biotechnical nature. The chemistry involved is modern biochemistry, and the chemical industry is in the process of converting into a biochemical or biotechnological industry (as it is in the process of becoming the industry of micro- and nanomaterials, both organic and inorganic). Chemistry will none-theless remain chemistry in a more conventional sense.

Benign substitution has already inspired previous concepts of clean chem-istry. According to a paper by Herold and Roland (1996) of Henkel Automo-tive Chemicals, the goal is to achieve 'elimination of hazardous ingredients like heavy metals, nitrite, amines, etc., and design of chemistry to be easily analysed and controlled, recycled, biodegradable and eliminated from waste water by neutralisation, coagulation and precipitation'. Jackson (1993 153) has a similar definition under the term of non-hazardous chemical products and processes. Benign substitution, positively speaking, is about developing 'inherently safe' products and processes, whereby 'safe' is short for all of the criteria concerning the environment (metabolic consistency), health and op-erational safety (Ökopol 2001).

Among the successful substitutions of the recent past are zeolites for phos-phate in washing agents. Zeolites are alumo-silicate materials with a special molecular structure which enables them to take in and release water (depending on temperature), absorb organic particles, and act as an ion ex-changer (which softens hard water). No phosphate in washing agents means no nutrients in the effluent, and correspondingly less algae flowering and less eutrophication in surface waters. (The remaining phosphate problem can be traced back to intensive animal farming).

The Montreal Protocol of 1986 for the protection of the ozone layer was successful in phasing out CFCs, at least in industrial countries. CFCs are xenobiotic compounds previously used as spraying and foaming agents and as coolants. They were identified as the main culprit behind stratospheric ozone depletion. The CFC case was the first to shed light on the fact that sub-stitution is not automatically benign. In certain cases it is simply problem substitution rather than problem solution. Substitutes may do away with cer-tain disadvantages, while retaining or introducing others. HCFC-22 (chloro-difluoromethane), one of the halocarbon replacements for CFCs in refrigera-tors, is less damaging to the ozone layer, but is still a greenhouse gas. The global warming potential of SF_6, another such replacement which is used for electrical insulation in the grid system, is 20,000 times that of CO_2. On eco-balance, however, 1 kg of SF_6 saves 25,000 kg of CO_2 by a 21 per cent reduc-tion of the overall global warming effect related to electricity distribution. European producers and users have decided to apply SF_6 in closed casings only, no longer in open applications (similar to new practices concerning or-ganic solvents).

In certain industrial applications, CFCs as coolants, or other chemicals such as FHCs or ammonia, can be replaced simply with water. Water-ice mixtures, where applicable, are highly efficient coolants, also for cold transport in mining applications. Energy requirements, and costs, are 10–30 per cent less than with like systems which use chemicals.

Accompanying the CFC debate there were calls for a general ban on xenobiotic compounds. The real bone of contention, however, is not the CFCs being xenobiotic but the extraordinary persistence of CFCs in the environment. Their half-life period is 30–40 years, and complete natural degradation takes as long as up to 160 years. Suitable chemical alternatives, propane and butane (higher valence synthetic gases), are completely decomposed within just one week. A general ban on xenobiotics would actually represent something like 'dechemicalisation'. This does not make sense. It rather mirrors an unreflected ideology of naturalness.

In most cases, benign substitution typically deals with heavy metals and organic solvents, but also with non-biocompatible mineral fibres, formaldehyde and other substances.

The first successful case of phasing-out of a heavy metal was the elimination of lead from gasoline. Lead emissions from car traffic are down today to about 5 per cent of what they were in the 1980s. Among the replacements is benzene, which is highly toxic when inhaled. At the gasoline station the benzene can largely be withheld through a suction pipe. On balance, small quantities of benzene pose a much lower risk than large volumes of lead. The even better alternative, of course, would be the ultimate substitution of hydrogen for gasoline. This would not involve problem substitution.

In the 1990s, Ciba Polymer, a division of a Swiss chemicals corporation (which has since been merged into Novartis), introduced an agricultural foil without nickel additives. The biggest market for such foils is Spain, but the Spanish farmers did not adopt the new product. It lacked the typical shade of green they were used to, and thus they associated it with some old-generation variant which they remembered was unsatisfactory.

With its dyestuffs, Ciba was more successful. Most conventional colour pigments consist of or include heavy metals. A heavy metal-free alternative was developed, called DPP pigments (Diketo-diarylo-pyrrolo-pyrrol). This has been in use now for many years and seems to be an example of truly benign substitution. An additional goal of new formula dyes is to increase the effectiveness of dyestuff transfer into fibres, or effectiveness of dyestuff fixation on surfaces. Conventional paints and coatings achieve about 65 per cent dyes transfer. New ones achieve 95 per cent. This too results in materials savings, less waste and correspondingly lower costs.

Among the alternative pigments which are currently being tested in the laboratory or in practice are iron oxide (yellow, reddish-brown, black), titandioxide (white) and spinel-derived pigments (blue, green, brown, black). Spinel is $MgAl_2O_4$. Either the magnesium or the aluminium in it is replaced by manganese, cobalt, chromium, copper, nickel or iron. It remains to be seen whether all of this is benign substitution, or whether it may also include cases of problem substitution.

In the automobile industry, screws and special small parts are coated with cadmium which provides protection against corrosion. These metal surfaces are now alternatively being coated with Deltaseal. This provides the same anti-corrosion protection, but optical appearance is not stable. So far, nobody has seriously complained, which saves many tonnes of cadmium a year. In a similar way, chromium and nickel as anti-corrosion protectants of the car body are being phased out.

Heavy metals conventionally also serve as additives to plastics, for example cadmium, lead and barium as polymer stabilisers (different from softening agents such as diethylhexyl phthalate which are equally hazardous). New formula polymers try to reduce or even do away with these additives.

With mineral fibres, the point in case was asbestos fibres, once a universal protection against fire and heat. After evidence of their being carcinogenic, asbestos fibres had to be eliminated from building materials, brakes, etc. Some replacements represent typical examples of problem substitution, for example bromine, a halogen such as chlorine, as a protection against fire.

Mineral fibres are generally considered to be carcinogenic, but they are indispensable for purposes of insulation and sealing. The typical materials of choice are basalt fibres and glass wool, in a few cases also synthetic resins. These materials remain alien to living organisms. They cannot be metabolically absorbed and thus accumulate in the tissue. So industry has successfully looked for ways to create mineral wool and mineral foams which are bio-soluble. Fibres are modified in such a way that they can be absorbed and metabolically transformed by living organisms. An additional development of late are silicon fibres substituting for ceramic fibres. And of course there are 'natural' alternatives such as paper flakes, wool or expanded clay (Sub-Chem 2002 94).

A number of particularly noxious chemicals such as formaldehyde, PCBs and PCPs have preoccupied environmental and health protectionists for long. They are being eliminated (PCBs have been interdicted in many countries) or significantly reduced, or restricted to applications in closed casings which can be returned to the producer and reprocessed in a closed-loop process. Formaldehyde, used as disinfectant and protectant in textiles, or as binder in chipboard, has been eliminated or reduced. In Europe formaldehyde in shirts is now below 300 ppm, down from 1,500 ppm in the 1970s. In underwear there are now about 50 ppm, down from 1,000 ppm.

Another field of particular concern are organic solvents, both in production processes and in the application of chemical products. Within industrial closed-loop processes which can be made 99.x per cent exchange-proof to the outside, chemists prefer to keep organic solvents as a process auxiliary because available alternatives are in many cases less effective and more complicated. The solvents can for a certain time be recycled and reused. In the end they have to be burned as hazardous waste.

With regard to products, i.e. open applications of organic solvents such as in dyes and varnishes, there is a general consensus on their substitution being desirable. The portion of organic solvents in a dyes formula can be as high as 75 per cent. Solvents are, for example, tuluol in gravure printing, mineral oil in offset printing, and alcohols and esters in flexo printing. They all serve as a thinner necessary to coat dyestuffs onto a surface. Further organic solvents are benzene and citrus terpene.

Dichlormethane is used as a solvent pickling (paint remover). A good substitute for it is dialkaline ester as pioneered by DuPont. Dialkaline ester can be derived from byproducts of nylon. It is non-hazardous to humans and the environment. While dichlormethane is being phased out on the European continent, its use is increasing in Britain (SubChem 2002 86).

In printing, placing text or colour on plastics surfaces, for instance on car fittings, CDs and boxes, is particularly solvents-intensive. A new hermetic-system printing technology, which replaces the older tampon system, comes with a 90 per cent reduction of organic solvents.

Another benign substitute for organic solvents is supercritical carbon dioxide ($SCCO_2$). One of its applications is in the production of decaf-coffee where it removes caffeine from coffee beans. Further applications are in dry cleaning of textiles, metals, silicon chips, or strip-off of old photoresist, where $SCCO_2$ also replaces perchlorethylene.

Most research in the field was aimed at replacing organic solvent-based systems by water-based dyes and varnishes. In the beginning, water-based paints still had certain disadvantages such as less colour brightness, or substantially longer drying time, or they required more dyes or caused more overspray per unit. These problems have by now been overcome or at least significantly reduced. One example is Simaco Inkjet dyes. Colour quality is reported to be satisfactory. Organic solvents are completely eliminated, in the production as well as in the application of the dyes. Ink containers can easily be cleaned and reused. Water and residual dyes can also be recycled. There is no problem of hazardous waste. In consequence, the system is much cheaper, even if the dyestuff as such has a higher price.

AURO Phytochemistry, Braunschweig, has developed a new generation of water-soluble paints and coatings that can be used in almost any applications of dispersions, varnishes, priming and finishing coatings, on walls, doors and floors. The binders of the natural paints are based on linseed oil, sunflower seed oil and rapeseed oil. The quality of the natural paints is high, meeting Euronorm EN 927.

4.4.5 Low-impact Chemistry

Is there a common denominator of TEI trends in the realm of chemistry and chemicals? In the 1980s one would perhaps have said there is a shift from 'hard' to 'soft' chemistry, though this may sound a bit too anthropomorphic. Also the discourse on 'clean chemistry' may appear to be a bit too clean to be close to the realities of the chemical industry. 'Sustainable chemistry' is in no way specific. Since, however, there is a need for some distinction, I prefer to speak of hazardous high-impact chemistry versus less risky low-impact chemistry. 'Impact' refers to negative effects regarding the environment, health and operational safety. High-impact chemicals would include those which have been discussed in the previous chapter and which need to be eliminated if possible, particularly

- explosive mixtures or compounds
- toxic heavy metals
- halogenated compounds
- POPs in general (persistent organic pollutants such as PCBs)
- organic solvents
- highly concentrated acids and alkaline solutions.

Low-impact chemicals would include most of the rest, with particular attention given to water as the preferred medium of solution. It remains to be seen whether in the end there will be a real broad transition from high-impact to low-impact chemistry. In any case it can be said that there is an intentional shift of emphasis from higher to lower impact. The components of this shift, and maybe real transition, from high-impact to low-impact chemistry are the shift from coal and petrol (fossil biomass) to biofeedstocks (phytochemistry), from synthetic chemistry to transgenic biotechnology, and more conventional innovation processes of benign substitution.

Scheringer (2002) discusses the main elements of low-impact chemistry as 'chemistry of short range'. This refers to low mobility of chemicals in the environment, and natural degradation within a short half-life period, i.e. no long-term persistence and bio-accumulation in the environment and organisms. It is assumed that whether chemicals are toxic, carcinogenic or mutagenic, depends on their mobilisation and persistence (non-biodegradability) in the environment or in organisms.

Another attempt to describe what low-impact chemistry involves can be found in *Green Chemistry* by Anastas/Warner (1998) and Anastas/Williamson (1998). The model of green chemistry includes aspects of add-on technology, such as treatment of waste after it is formed, and features of the efficiency revolution such as energy saving or waste reduction. More importantly, it also includes aspects aimed at changing metabolic consistency:

- Synthetic methods, wherever practicable, should be designed to use and generate substances that possess little or no toxicity.
- The use of auxiliary substances (solvents, separation agents, etc.) should be made unnecessary wherever possible and innocuous when used.
- Raw materials and chemical feedstocks should be renewable rather than depleting wherever practicable.
- Derivatives (blocking groups, de/protection, temporary modification) should be minimised or avoided because such steps require additional agents.
- Catalytic reactions are superior to stochiometric ones.
- Chemical products, after use, should not persist in the environment and break down into innocuous degradation products.
- New analytical methods should be applied that allow for real-time, in-process control prior to the formation of hazardous substances.
- Substances used in chemical processes should have little potential for chemical accidents (releases, explosions, fires).

4.5　MATERIALS AND MATERIALS PROCESSING

In a certain sense all technologies make use of materials on the basis of some processing of materials: food, fossil fuels, biotic working materials such as wood and fibres, and inorganic working materials such as metals, sand, stone, salt and other minerals. These materials come nowadays in natural and artificial varieties. Insofar as the processing of materials involves their transformation, this represents a chemical process, and most of the industrial sectors related to materials processing represent in fact domains of chemical engineering: metallurgy, cement making, pulp and paper industry, textiles, leather tanning, wood industry and food processing.

4.5.1 New Materials

Beyond innovations in plastics and synthetic fibres the recent past has seen a veritable boom in the development of new materials. There are new metal alloys with extreme properties, new ceramics resistant to heat and aggressive chemicals, secondary materials from reprocessed waste materials, or biotic compound materials made of various biofeedstocks and fibres. For example, plant fibres and soya protein yield compound materials that can fulfil the same functions as plastics now fulfil in the interiors of casings, cars, trains, ships or in the furnishings of homes and offices. The most suitable fibres seem to be reed, flax, cocoa and sisal. They are cheap, technically as malleable, robust and light as plastics, harmless and biodegradable.

There are also new composite carbon fibre-reinforced plastics, or carbon fibres reinforced by synthetic resin, as a substitute for aluminium and steel in aeroplanes, ground vehicles and building. Density of carbon fibre-reinforced synthetic resin is 1.5 gr/ccm compared to 2.7 in aluminium and 8.0 in steel, translating into high energy and cost savings.

New metal-like materials, such as organic metals, are hitherto unknown hybrids. These are compounds of C, H, O, N and sulphur with properties very similar to those of metals. One such organic metal is Polyanilin. This is a kind of electricity-conducting plastic. It is non-melting, non-soluble and non-corrosive which also makes it a good coating material. Aerogels and metallic glass, or transparent glass-like metals, are simultaneously ultra-strong, ultra-hard and ultra-light. Such properties make them an almost ideal working material for bridges, high-rise buildings and vehicles.

Some of the new materials enable new technologies, for example polytronics, which is short for polymers and electronics. Semiconducting plastics consists of some organic polymer which is chemically doped with metal atoms. Polytronic material can also be made of very thin layers of silicon semiconductors printed on a polymer sheet (organic thin-film transistors). This might become another key innovation in that it has the potential of widely spread applications from chip manufacturing and computers to smart cards, labels (physically flexible memories), sheet batteries, flexible screens, semiconducting coating of solar cells, or in LEPs (light emitting polymers), formerly also known as LEDs (light emitting diodes). They are very cheap compared to silicon wafers. They can be printed onto almost everything, in any application where electronic components can be or need to be flat and flexible. The environmental burden of polytronics is much less than in the production of silicon wafers or conventional solar cells. In particular there is much less toxic waste, and organic polymers are easily disposable.

Superconductors have attracted particular attention for several years. These are special electricity-conducting materials which cause no or very little loss of voltage because their electrical resistance is practically zero. With superconductors, power can be transported over very long distances at almost no loss, including solar power from remote areas. The snag is that for the time being all superconductors are only operational at very low or very high temperatures. Constructing transmission lines with such properties is economically unfeasible. So superconductors can unfortunately not be included in a realistic account of TEIs.

Another frontier of high-tech materials is crystal breeding. Artificial crystals bred according to specific requirements with regard to molecular structure and physico-chemical properties are essential for many applications in electronics, optics, laser technology, opto-electronics, ultra-sound and energy technology. Further applications are in chemistry, pharmaceuticals and cosmetics. Breed-crystals replace chemicals or metals, and obviate their ecological footprint. Available for consumers, for example, are crystal deodorant sticks consisting of ammonia alum as an alternative to deodorants containing alcohols or chlorinated compounds. The crystals are non-hazardous and do not represent a waste problem.

In connection with these new materials, certain substitutive tendencies can be observed:

- In the manufacture of clothes, upholstery and industrial textiles there seems to be a situation of 'anything goes': Synthetics replace natural fibres, just as natural fabric replaces synthetics, and both synthetic and natural fabrics may be combined.
- Biotic compound materials replace plastics and wood in the interiors and furnishings of buildings, vehicles, casings and furniture.
- Near-nature solid materials with properties of clay, wood, straw, or similar, replace steel and concrete building materials in certain applications.
- Parts made of plastics replace parts made of metal in machines of all kinds, for instance chain drives made of polyacetal. The factor of eco-efficiency involved in this is 10–20 due to very low weight which results in less energy demand in use, low noise and low wear and tear even under extreme physico-chemical conditions, thus much longer durability.
- Composite materials are being substituted for metals and reinforced concrete in vehicles and buildings above and below ground.

Given the diversity of new materials it is not possible to make one general statement on their environmental importance and metabolic consistency. In fact this largely remains to be assessed. But the general tendency is obvious: The new materials, or their special applications respectively, help to boost energy efficiency. They also help to reduce materials throughput, and they often substitute for hazardous chemicals or metals.

Whereas materials such as biotic compounds, organic polymers, ceramics and breed-crystals do not entail a particular waste problem, this may be different with hybrid compound materials. Too little is known about their natural degradation and environmental effects. So these new compound materials might have to be kept within a closed take-back regime in order to properly reprocess them. Since these materials are used in buildings and vehicles, recovery and reprocessing should not cause particular difficulties.

4.5.2 Nanomaterials and Nanotechnology

A nanometre is a billionth of a metre (10^{-9}), a micrometre a millionth of a metre (10^{-6}). Nanotechnology refers to materials of the order of several nanometres to several hundreds of nanometres. A nanotransistor is a hundred times smaller than today's silicon chip transistors. The dividing line between nano- and micromaterials is blurred.

Nanomaterials are now most often nanotubes consisting of carbon. Carbon nanotubes (CNTs) are basically soot, in the same sense as diamonds are compressed coal. Carbon atoms in a nanotube are positioned in a chicken-wire lattice that wraps into a hollow pipe. The resulting molecular structure gives the nanotubes a number of useful properties such as excellent conductivity. CNTs are extremely thin (0.4–100 nm) but can be relatively long (1 mm). Multiwall nanotubes are stuffed one within another (like Russian dolls). Ultimately, the 'walls' of such structures are only one single atom thick.

Carbon nanomaterial was first discovered in 1985 at Rice University, Houston, in the form of soccer ball-shaped molecules. These were called fullerenes or buckyballs, in memory of Buckminster Fuller. Carbon nanomaterials are now widely researched and developed in industrial applications also in Europe and Japan. Infineon, a German electronics corporation, has developed a patent on how to grow CNTs in an industrial standard process derived from the technique of producing silicon wafers.

Applications can include many things. In medicine CNTs are being experimented with as nanocontainers of pharmaceuticals. In electronics electricity-conducting CNTs can result in cheaper flat-panel full-colour TVs, new X-ray devices, brighter lighting or superfast computers. These will have nanoprocessors made of CNTs suspended between two silicon wafers (connection on = a one; off = a zero). Such nanoprocessors are 100 times faster than today's microprocessors, and memories based on this technology have 1,000 times the data density of conventional memories.

Superfast computers could possibly represent the next generation but one. Also next generation computer memories, MRAMs (magnetic random-access memory, also called magneto-resistive memories) and FeRAMs (ferroelectric random-access memories), will already be important mass-applications of nanotechnology. They will replace present DRAMs (dynamic random-access memories). Magnetic random-access memories have an ultrathin conductive layer between two magnetic plates with a thickness of only 10 layers of atoms. Data are stored in the spin of electrons inside these tiny magnetic sandwiches. They do not lose magnetism, i.e. their memory, even if the power goes out. Such a memory is retained for at least 10 years. MRAMs store as much data as DRAMs, but write faster, at almost instant access, at the same time consuming less energy. MRAMs and FeRAMs represent a textbook example of increasing eco-efficiency in the succession course of new product generations. It is an American development the funds for which partly came from the US Defense Advanced Research Projects Agency.

Another field of interest are nanosensors. These are nanotubes with gas molecules, or some organic molecules, attached to their surface. Nanosensors can be used as detectors, foe example to detect gas. The cost of a nano gas detector is ten times less than that of a conventional detector. Similarly, detectors for other substances can be designed which are of particular interest in chemistry, medicine and environmental measuring and control. Designing new nanomaterials is a task as complex and demanding as genetic engineering. Computer modelling on the basis of quantum physics seems to be particularly good at this task.

Nanofluids are nanomaterials added to fluids. For example, nano-size copper particles (which conduct heat, or cold respectively) connected to CNTs are added to radiator fluids in cooling systems. The resulting liquid-solid mixture has more than double the cooling capacity of conventional coolants such as ethylene glycol. The development of such mixtures has not so far been possible because the heat-conducting particles have been so big that they settle out of the fluid or otherwise cause some damage to the engine or cooling system. Nanoparticles are small and light enough to stay integrated in the fluid. This makes smaller and lighter cooling systems possible, particularly in motor vehicles and power plants (also in combination with fuel cells). Biofluids thus enable smaller systems, i.e. less materials requirements, better energy performance and phase-out of conventional chemicals.

As is the case with other new materials, nanomaterials boost overall productivity in that they save much larger quantities of natural resources, materials and energy than they themselves represent. Or they enable much higher technical performances without increasing resource and energy demand.

Nanotechnology also enables thermoelectric materials. Thermoelectric refers to the direct conversion of heat into electricity, or vice versa. A suitable application for this, for example, would be converting the heat in car exhaust pipes or jet engines into enough electricity for radio, air conditioning, lighting and controls. Further applications might be microprocessor cooling, or household appliances (fridges), and ultimately even power generation.

The working principle of thermoelectric conversion is that heat differences in some materials trigger a flow of electrons towards the colder end. The principle has been known for a long time, but no suitable materials, efficient and cheap enough, were available. Now there are new semiconducting nanomaterials that block the heat flow while enhancing the electrical flow. This is caused by tiny superlattice structures, i.e. ultra-thin layers less than five nanometres thick. The entire nanofilm consists of several alternating layers. They block the propagation of atomic vibrations, thus the heat, but still let the electrons flow and thus generate current. A Japanese company manufactures prototype thermoelectric wristwatches powered by the body heat of the wearer. When fully developed, thermoelectric materials will save electricity in power plants, and increase the energy efficiency of combustion engines and fuel cells. However, they seem to require precious metals such as gold and platinum, i.e. metals with a considerable mining rucksack.

Silicon and other mineral components are also candidates for the creation of inorganic micromaterials. A recent development still in the early research stage is silica micro components. Silica structures are naturally produced by *Radiolaria*, marine microorganisms that form their siliceous skeletons from silica in the sea water. The siliceous structures are of interest as industrial microcomponents. By mimicking *Radiolaria*, a great variety of tiny solid shapes and structures can be produced from silicic acid (silica in salt water).

Zeolites are an example of mineral nanomaterials which have been state of the art for 10–20 years (for example in replacing phosphates in washing powder). Zeolites are porous minerals. The diameter of the pores is around 1 nm. Zeolites thus have an immense inner surface which enables them to serve as an ion exchanger, or filter, or separator, or carrier of catalyst materials, including microbes performing biocatalysis.

An intriguing perspective of nanotechnology is the possibility of creating 'self-healing' materials. For example, nanotubes can be filled with proteins which form adhesives. When the nanotubes break open, the proteins are released. Incorporated in materials under particular stress, such as aeroplane wings, small micro cracks which otherwise would propagate over time could thus in time be repaired.

An 'Action Group on Erosion, Technology and Concentration' located in Winnipeg, Canada, the self-defined mission of which is to provide radical counter-expertise, has requested a moratorium on nano research and the use of nanotubes until there is more evidence of alleged health hazards. The reason is that inhaling nanotubes in large quantities might be detrimental to health, as is inhaling soot, for example from Diesel engines. In contrast to airborne soot particles from diesel exhausts, which are indeed inhaled, carbon nanotubes are embedded in or compounded to some carrier material, or are encased in a container (such as the nanocubes which store H_2). So there is no cause for alarm concerning the risks of inhaling nanotubes. By extension though, it might be reasonable from a health and environment point of view to urge a moratorium on Diesel engines because these disseminate vast amounts of minute soot particles most of which are too tiny to cling to filters. The risk situation might be different with medical nanotubes that are injected into animal and human bodies. Although this is certainly not done in vast amounts, effects have to be observed thoroughly as a matter of course.

Beyond inorganic nanomaterials, there is nanobiotechnology. GM-enhanced microbial biosynthesis and biocatalytic processes as discussed in 4.4.2–3 are nothing other than organic micro- and nanotechnology. Cell walls, for example, can be regarded as microtubules made of nanocomponents. The ribosomes in a cell plasma, with a diameter of about 10–20 nm, are the nanotool for the synthesis of proteins. Enzymes are another kind of nanotool, itself made of protein, for ripping molecules apart or joining them together. Molecular, genetic and cell engineering represent biotechnology at the nano and micro level.

4.5.3 Surface Technology

Surface technology is one of the domains whose environmental impact tends to be underestimated. It encompasses everything from galvanisation of metal surfaces, to coating of materials, to the cleaning and disinfection of surfaces. Most of conventional industrial surface working represents hazardous chemistry, and this too is being revolutionised by nanotechnology as well as by plasma technology.

Nanobiotechnology can, for example, do the metal finishing by enzymatic degreasing on the basis of biodegradable tensides and microorganisms in a closed rinse water system. This replaces the alkaline process in the electroplating industry. Hydroxide sludge is down from 30 t in the alkaline process to 15 t in the biotech process. 20 per cent sulphuric acid is replaced by 8 per cent sulphuric and hydrochloric acid. Water consumption is down from 8,000 to 800 m^3 (OECD 1998 52).

Nanobiotechnology can produce self-cleaning and self-disinfecting plastics. Enzymes are attached to CNTs which are in turn incorporated into a polymer. Depending on the enzymes, the resulting plastic material kills microbes, or degrades fatty matter and oil sludge. The method can also be applied in medicine to the surface of artificial organs or organ parts in surgery.

In the self-cleaning and self-protection of surfaces, the lotus effect has won fame. There are no chemical or biochemical reactions, just micromechanics: a micro-rough surface is created, i.e. a bristly surface of micro to nano size. The leaves of lotus flowers have such micro-rough surfaces. The real surface of such a material consists only of the tips of the micro-bristles which makes the surface much smaller than its spatial size is (complementary principle to inner enlargement of surfaces by porous structure). Compared to normal surfaces, the contact between external materials and the surface is 90 per cent reduced. As a consequence, water and dirt do not have enough contact to adhere to the surface. This is why water drops do not remain on the surface, and sweep away dirt particles which would otherwise remain with the water. Also wind can blow particles off the surface more easily. Microscopic polymer bristles acting like spikes can destroy the membrane of bacteria. A nanostructured surface can thus stay sterile.

All of these approaches eliminate 'chemical warfare' by detergents and disinfectants. In certain applications they can even eliminate dyes and paint and yet bring colour to things. This can be achieved by nanostructured light-refracting surfaces. This represents a bionic blue butterfly effect by mimicking the colours of certain butterfly wings. These are actually colourless (i.e. they contain no pigments), but have nanostructured surfaces which refract incident light. Structures with the shape of a nano-sized 'fir tree' generate marvellous shades of blue.

Thin-film chemistry can produce ultra-thin layers in the nano range. This is known as colloidal (or electrostatic) self-assembly. Some material is repeatedly first washed in purified water, which creates a negative charge on the surface, then immersed in a solution of charged particles, which makes the positively charged ions adhere to the material in a thin sandwiched layer. Such coatings display perfect purity. Any surface of an object can be coated with such thin films at room temperature. Thin films of metals deposited by electrostatic self-assembly are as conductive as bulk metals, whereas thin films created by other methods are not. Applications are in magnetic materials as used in phones and computers, or in the scratch-resistant coating of windows, glasses and solar cells. Colloidal self-assembly of coatings saves much energy and materials, and eliminates hazardous chemicals.

Thin-layer coating can also be achieved by a kind of thermal spraying as, for instance, in the Pyrosil process with silane. Hydrosilicon gases are added to the gas of a burner above the treated surface. This creates a very thin and very hard layer of silicate, thus safely protecting the material. A further layer of silicates is made less dense/hard and is thus a good basis for primers or immediate coating with paints or glues. These nano-layers are of the order of 1,000 times thinner than conventional layers of some micrometres created with the help of chromium or phosphate in a metal sheet bath. In contrast to the latter, there is neither sewage water nor hazardous waste, nor are there secondary cleaning or rinsing processes. And much less material is required for creating the layers.

Old coatings on metals can be removed by inductive decoating. An induction coil creates electrical current underneath the coating, evaporating some of it so that the rest begins to lose hold and flakes off. The rest can be removed mechanically by brushing. The old coatings can be recycled or burned. The method is quick, saves energy, replaces hazardous pickling and organic solvents, and does not create hazardous process waste.

Plasma treatment of surfaces is another TEI approach to materials working. A plasma is sometimes described by physicists as the fourth state of aggregation beyond the gaseous state. A plasma is usually created in a vacuum by floating gas through the electric field between two electrodes (a light arc or a similar device). This creates some 'vaporisation' of the gas, a plasma, i.e. a cloud of charged particles (ions) of any kind: electrons, neutral gas particles, photons, molecules. Such a plasma looks like fire but its temperature is only about 70°C.

To obtain a carbon coating, a methane plasma is used. The resulting carbon coating has properties between those of diamond and graphite. Surfaces of any kind – including plastics and textiles – can thus be made hydrophilic or hydrophobe, or -philic or -phobe of other substances, or be sterilised, or be made ultra-strong. In this way, textiles can be made water-resistant without chemical impregnation. In medicine and chemistry, plasma treatment can turn surfaces into special membranes. Selective effectiveness of plasma-created membranes is at 95 per cent, compared to 60 per cent in normal membranes. In optics, unbreakable, scratchproof spectacles can be made. Plasma treatment is of growing importance also in the finishing of automobile parts, in electronics (screens) and glass (building and vehicle windows).

Polyolefins, i.e. polyethylene and polypropylene, are the most widespread plastics in everyday use. They are difficult to coat and thus need surface treatment by some thermal or wet-chemical process. Plasma treatment of polyolefins is now a clean alternative. The plastic is plasma-treated in its form as powder, before moulding. This modifies the material in such a way that its surface can afterwards easily be coated with anything. The plastic powder itself can be processed as usual in any of the known processes such as powder coating, slush moulding or rotation sintering.

With the help of a plasma of ionised helium and oxygen, materials can also be disinfected or decontaminated. The oxygen ions in the plasma are highly reactive and thus neutralise germs/pathogens, for instance anthrax. Surfaces can be cleaned without the damage involved in using aggressive chemicals, or chlorine dioxide gas which is corrosive.

Vacuum-sputtering is a plasma treatment which for example improves the performance of solar-thermal absorbers. Ionised particles of the coating material are 'sputtered' onto the copper sheets of absorbers. The coating keeps 25 years at minimum. Solar absorption with such a coating is 95 per cent and thermal emission 5 per cent (at 100°C). In comparison to a conventional coating on the basis of black chromium, there is no toxic exhaust gas and no sewage water and waste. Energy requirements are reduced from 10 to 1 kWh per m^2 copper sheet.

There are also more conventional alternatives in surface working. One of the state-of-the-art approaches is dry powder coating, for example of dyes and varnish. This was triggered by tougher environmental standards in industrial countries (VOC directives), and has already eliminated vast amounts of organic solvents and reduced demand for water. The use of organic solvents in dry powder processes is only about 25 per cent of what it used to be. (More advanced nano and plasma technologies, though, completely eliminate water and organic solvents from the process.)

As in other applications, organic solvents and water in coating processes can also be replaced by supercritical CO_2 ($SCCO_2$), for example by spin coating which is a $SCCO_2$-based precise control of thin-film creation for photoresists and other polymers. In photolithography, spin coating replaces water-based developer solution. $SCCO_2$ is also used as an alternative solvent in high-tech grinding and polishing in chip making.

Finally, some of those previously discussed new materials can make for good coatings, especially organic metals. For example, a rust inhibitor on the basis of polyphenylenamin (an organic metal) is directly dispersed in varnish on the surface. The coating is absolutely insoluble and does not melt. No prior surface treatment is needed (such as chromatisation or phosphorisation formerly). Materials treated in this way have considerably prolonged maintenance cycles because of less corrosion. They also require less varnishing. Coatings of organic metals do not represent hazardous waste when removed. This results in cost savings in each respect.

4.5.4 Materials Processing

4.5.4.1 Clean-burn technology
Steel, aluminium, glass, cement and similar industries operate with large blast furnaces or melting furnaces. Their environmental problems are the same as in conventional power plants. Since around 1990 furnaces can commercially be equipped with regenerative combustion systems. The working principle of regenerative combustion is heat exchange between flue gas and combustion air in a regenerator, for example a ceramic honeycomb generator. This pre-heats incoming air to a temperature close to furnace gases, resulting in impressive fuel savings of 50–60 per cent. Many users, however, remain reluctant because a typical disadvantage of 1G regenerative combustion systems are excessive NO_x emissions, in addition to high investment and maintenance costs. There are now 2G regenerative combustion systems available which tackle the NO_x problem by exhaust gas recirculation, i.e. the exhaust is rechanneled into the combustion chamber, or products of combustion inside the furnace are redirected into the root of the burner flame. This creates still more complete burn-up rates and reduces NO_x and CO emissions to very low levels. CO_2 of course remains proportional to the carbon fuel input.

Beyond large-scale furnaces, clean-burn technologies are now becoming available. These are required to produce process heat, in order to heat up working materials, or to dry humid working materials, or to provide steam not only in laundries, hospitals and steam turbines, but also in private households and in cars and caravans, for air heating and warm water systems. Conventional burners and combustion chambers, even if advanced in their way, have rather low energy efficiency, are relatively large, and run on fuel oil or gas, thus causing the typical airborne emissions of fossil fuels.

Clean burning, strictly speaking, means creating the very hot blue flame that appears when the metabolic rate of the burning process comes close to 100 per cent so that almost no problematic emissions build up. This can best be achieved by using pure oxygen instead of ambient air. With special types of porous burners any liquid fuel is vaporised on the surface of some porous burner material where the emerging gas can promote blue-flame combustion. Porous structures of ceramics seem to be particularly appropriate for achieving this. The snag is, it takes bulky pressurised tanks and pumps to push the liquid fast enough through the pores as to make them vaporise. There is now a new microtech approach to it (by Vapore, Richmond, CA) which makes use of capillary forces. These inherent forces can pump a liquid through fine micrometre-scale pores. Heat applied to the surface of the ceramic vaporises the liquid and the resultant gas escapes through an orifice from where it can be burned. The principle can also be applied to fuel cells. By making gas vapor on-site in this way, more complicated reformer devices can be replaced.

An alternative to clean hot flame burning is flameless oxidation of gas or hydrogen. Flameless oxidation is not truly flameless, but the flame which can be seen at the outlet of the burner is no normal flame because oxidation occurs inside a porous body. Such porous burners typically consist of silicon or metal foams or aluminium oxide fibres. The critical point is pore size. If pores are too small, the flame at the outlet is quenched; if pores exceed the critical width, flames may propagate inside the porous body. This type of porous burner has a number of advantages over conventional open-flame burning. The burners are very energy-efficient and cause very low emissions on gas and no emissions on hydrogen. They have a variable dynamic power range of 1:20 (compared to 1:3 in conventional processes). And they have a high power density, which makes small devices feasible requiring only a tenth the volume of conventional burners.

4.5.4.2 Advanced membrane technology
Membrane technology plays an ever more important role in industrial processes where substances have to be extracted from some substrate, or have to be separated from one another. High-tech membranes are made of polymers, or ceramics, or fibre materials in combination with metals.

As described in previous chapters, catalyst-coated membranes are the centrepiece of fuel cells. They can also be applied to the extraction of hydrogen. Drinking water can be extracted by membrane-type filtration (reversible osmosis). Zeolites are nothing other than porous nanofilters. As will be described in I/4.10, pollutants are eliminated from sewage water through ceramic membranes, cellulose-stainless steel membranes and ultrafiltration (i.e. low pressure on one side of a semi-permeable diaphragm).

In the production of chlorine, there was a typical transition between the 1970s and the 1990s from the amalgam technology (dating back to the late 19th century) via an interim diaphragm method to semi-permeable membrane technology (Erdmann 2001). Chlorine is obtained from the electrolysis of salt in water, resulting in chlorine as the target product at the anode, and caustic soda and hydrogen as byproducts at the cathode. The task is to keep the chlorine separate from the byproducts in order to prevent a new reaction. The amalgam method uses mercury as the cathode. This results in mercury-containing effluent, besides being energy-intensive. With the diaphragm method, anode and cathode space are separated by a diaphragm which allows current to pass, not however the products of electrolysis. This is less energy-intensive, but results in asbestos waste since the diaphragm is coated with a layer of asbestos fibres. The semi-permeable membrane technology requires still less energy and causes no problematic emissions. It was, however, only in the 1980s that new materials which fulfilled the demanding requirements of a high-performance low-pollution membrane became available.

In Erdmann this example illustrates how regulation can either foster or hinder the introduction of TEIs. In the mid-1970s, the German regulator forced immediate and very expensive add-on measures aimed at removing mercury on to the chlorine producing industry. This caused a financial lock-in on the outdated amalgam path for another 20 years. The Japanese regulator, by contrast, simply put a ban on the amalgam technology in 1975, becoming effective in 1985, thus allowing for an appropriate transition time of ten years. As a consequence, Japanese producers were free to make the best choice between the diaphragm and membrane methods so that around 1990 almost all had chosen the membrane technology.

4.5.4.3 Sonic devices and lasers

Mechanical techniques such as mixing, milling, crushing, cutting, drilling and assembling parts represent technological paths which are rather mature. In spite of manifold incremental improvements, these techniques tend to involve more or less heavy machines with many moving parts. These are energy- and materials-intensive, and not always particularly efficient in that they entail suboptimal yield rates and correspondingly much waste. By and large this situation seems bound to remain unchanged. In certain cases, though, laser technology and new sonic devices can deliver better results.

Lasers can carry out tasks such as cutting, boring, drilling or welding with high precision in the range of micrometres. Since this is well known and not exactly an innovation any longer, there is no need to go further into lasers here. Meanwhile there are 3G and 4G high-performance lasers in various fields of application.

Still relatively new is the use of laser-based optical sensors in food quality control and environmental monitoring. For example, laser in-situ analysis (LISA) is used for controlling beer quality during different steps in the brewing process. LISA measures diffusion, reflection and absorption of light. This replaces a number of cumbersome ex-situ controls which involve hazardous chemicals. Similarly, optical sensors monitor biofouling in bottling and canning plants, resulting in a 50 per cent reduction of chemical detergents and disinfectants, and 20 per cent energy reduction.

Laser is short for Light Amplification by Stimulated Emission of Radiation. This indicates the working principle of lasers: Photons, which tend to travel in all directions by nature and thus create diffuse light, are forced in a unidirectional line and thus focus and amplify the power of light waves. Something similar can be achieved with sound waves too. Standing in front of a bass reflex super woofer clearly demonstrates what sound resonance can do inside a solid-liquid body. It is also known that high-frequency sound can smash glass. So high-frequency sonic systems can be applied to the processing of comparable materials.

Low-frequency sonic systems can be used for doing what was formerly done by mechanical mixing, stirring or shaking. Montec Research in the Butte, MT, USA, has developed a sonar mixing system below 100 hertz, for applications in mining, petroleum processing or waste treatment. This takes 60 per cent less mixing time than with conventional mechanics and also less energy. Low-frequency sound waves represent a 'soft' approach which does less or no damage to the materials that are being mixed. High-frequency sonic systems, by contrast, can damage more delicate materials.

4.5.4.4 Powder metallurgy
Powder metallurgy has successfully been established as a state-of-the-art TEI since the 1980s. The principle of powder-metallurgical injection moulding and sintering, similar to the extrusion of plastics, is that metals, in the form of small-sized granules, are pressed instead of being melted. This results in much higher resource productivity, does not create swarf waste and requires little or no further processing.

Foamed metals are a special branch of powder metallurgy, useful for weight reduction and energy efficiency.

Dry metal-working replaces cooling lubricants, resulting in much less waste and no hazardous waste. In wire drawing, mechanical descaling can replace chemical descaling done by pickling with highly concentrated acids. The metal scales which have to be removed are a very hard oxide layer on the pre-processed material which comes from the hot-rolling plant. The dry mechanical approach consists of a cascade of special brushes, with final put-on of a lubricant for further processing.

4.5.4.5 Sulphur-free and chlorine-free pulping

Pulping and paper making is an old technology, in existence since about 2,000 ago years in China, from where it diffused westwards through the Middle East and the Mediterranean arriving 600–700 years ago in Italy and southern Germany. Basically, the technology remains unchanged, though it has of course been mechanised and chemicalised. The big environmental problem, besides the sheer amount of wood fibres consumed, has been sulphur and chlorine in the pulping process. Only since around 1990 can a number of TEIs, which can be retrofitted to existing plants, deal with the problem.

Among these were modified delignification, chlorine-free bleaching, reuse of condensates and purification of all air- and water-borne emissions on the release side. If an average paper mill had to adopt all of the best available technologies from scratch, it would have to invest about €100 million (Calleja et al. 2002 18).

Modified cooking is aimed at delignification of wood pulp in sulphur-based pulping. Methods are extended and continuous cooking. Sulphur is partly reduced, and for the rest recycled. Modified cooking also contributes to reducing chlorine in subsequent pulp bleaching (AOX emissions). An extension to extended cooking is delignification with oxygen, which puts an oxygen reactor between pulping and bleaching. This was introduced particularly in Japan where oxygen is cheaply available. An alternative approach is the Organocell method which does away with sulphur by using methanol or ethanol and caustic soda.

Most of the TEIs in pulping were developed by actors located in the main pulping nations, such as Kamyr/Kvaerner or Sunds Defibrator in Sweden, and Beloit or Union Camp in the USA. A driving force, however, was environmental action groups such as Greenpeace in the main paper-consuming nations. Industry was cooperative not only because of its newly developed environmental awareness but also because of reduced operating costs of the pulping TEIs which more than compensated for the high investments.

Bleaching of pulp fibres was conventionally done with chlorine. A first step for phasing out chlorine was the substitution of chlorine dioxide for elementary chlorine, resulting in ECF-pulp (i.e. elementary chlorine-free). ECF-pulp is cheap and easy to retrofit, but corrosive, and does not in itself satisfy the criterion of zero emission. Applying additional measures then results in TCF-pulp (totally chlorine-free). One of these measures is ozone delignification, where the ozone acts as a bleaching agent during cooking.

The ozone has to be produced on-site, which increases costs. The process on the whole, however, is cheaper than previous methods, because it results in very low emissions and a high degree of recyclability of the effluent. All of these new pulping technologies diffused worldwide during the 1990s, having at present reached market shares of 85–100 per cent (Beise et al. 2003).

4.5.4.6 Biotechnical processing in pulping, tanning, textiles and food

TEIs based on biotechnological approaches can be found in all industries which process biotic materials: food and wood processing, textiles manufacturing, leather tanning and also pulping and paper making. Everything which has been said in previous chapters about biofeedstocks and their GM-enhanced biotechnological processing also applies in these sectors.

Biotechnological pulping and paper making includes GM conifers as biofeedstocks (I/4.4.1). The conifers should of course be grown within a regime of sustainable forestry. Particularly in developing countries this is for the most part not yet the case. Sulphur-based thermo-mechanical pulping can furthermore be improved by cracking lignin with the help of cellulase and hemi-cellulase enzymes. The same end can also be achieved by specially bred white rot fungi. Other enzymes can be targeted on fibre bleaching and brightening or on the removal of undesired substances such as resin (OECD 2001c 99–125). In paper making, GM starch is already used on a regular basis for smoothing papermass. In paper recycling, enzymes and microbes can perform the de-inking.

In textile manufacturing, state-of-the-art biotechnology includes enzymatic separation of fibres, the use of catalases for removal of residues of H_2O_2 in bleaching water, enzymatic stone-washing (artificial ageing), creation of smooth surfaces (bio-polishing) in cotton fabric, and regeneration of cellulose fibres.

Plant residuals in wool and cotton can be removed by enzymatic degradation. This replaces the methods of 'carbonisation' which use sulphuric acid, and 'burning' of the material at about 100°C which is unfavourable both for the fibres and the environment. It also reduces peroxide bleaching of the material which involves large amounts of alkaline solution and saline effluent load.

In wool processing, extremophilic proteases (protein-decomposing) and lipases (fat-decomposing) replace chlorine oxidants and softeners. This results in less pollution of effluent water, particularly with AOX, and less solid waste. Enzymes isolated from *Thermoanaerobacter keratinophilus* reduce proneness of wool to becoming felted. In both cases wool properties are improved with regard to dyeing and wearing comfort.

Conventional anti-shrinking treatment of underwear involves various chemicals and results in contaminated sewage water (polyvinylacetate, polyvinylchloride, magnesiumchloride, zinc nitrate). This can be replaced with a sophisticated process which uses hot water steam, natural auxiliaries and mechanical processing, resulting in zero pollution.

In leather processing, enzymes can help with the removal of hairs, fatty matter and globulines (proteins from animal tissue). Natural tanning agents on phytobasis can replace chemical bate agents.

Food processing is biotechnological by nature. Traditional processes of breaking down biotic materials in order to obtain content substances, as well as processes of fermentation and conservation, are being superseded by modern biotechnology in ever growing numbers and applications. Among the transgenic enzymes which are now used on a regular basis in food processing are amylases (liquefaction of starch, baking), maltases (baking, fruit processing), pektinesterases (juice and wine), proteases (meat processing, baking) and lipases (fat and oil processing). Furthermore, there are a number of transgenically produced food additives such as fruit acid, amino acids, vitamins, taste and aroma agents, hydrocolloides (for keeping moisture content) and further enzymes such as cellulases and glukosidases.

Examples for the use of biosensors in food and beverage processing have been given in I/4.4.3.

The use of GM yeast is widespread, particularly in brewing. In the UK a GM yeast for bread baking has been licensed which converts glucose and maltose simultaneously (not the former after the latter).

Biotechnology also contributes to transforming food processing, at least partly, into a kind of biochemical design food manufacturing in which many feedstocks can serve many purposes. One never knows what the meal one is served is really made of. Consumers have become used to liquors and beers made on the basis of any crop you like. A new one is beer brewed from whey (90 per cent of which is normally poured away). It has just 3° alcohol, a low calorific content, and is rich in trace elements and vitamins.

According to Dürkop et al. (1999 32) the main advantages of GM-enhanced biotechnology in food processing are

- food safety, i.e. reduction of pathogenic germs, avoidance of toxins, reduction of undesired substances (such as nitrates), and development of hypoallergic food
- food quality, i.e. reliable and possibly improved physiological usefulness as well as flavour enhancing, though many flavour-enhanced products definitely do not achieve the premium quality required to satisfy more cultured tastes
- avoidance of waste, i.e. improved storability, less oxidation of fatty matter, reduction of microbial degradation
- lower costs, i.e. higher yield of processes, lower cost of raw materials, optimised use of production plants.

4.6 BUILDING

4.6.1 Regime Rules of Green Town Planning

In 1933 the functionalist Charter of Athens was adopted at one of the annual meetings of CIAM, i.e. Congrès International d'Architecture Moderne, with Le Corbusier as one of its co-founders. The Charter of Athens proclaimed strict separation of the functions of vocational life (production sites), housing (residential areas), recreation (leisure and sporting grounds) and mobility (travel and transport areas). The buildings envisaged were large, uniform, and built of reinforced concrete and glass. To conservationists this was the fall from ecological grace.

For completeness sake it has to be said that the fall already started in the life-reform movement of the 1880s–90s with the 'green' concept of garden towns, i.e. suburbias surrounding towns and consisting of small stand-alone houses with an adjoint garden, and later on a garage. Middle-class town dwellers were no longer satisfied with bringing some bits of nature into town, for example municipal parks, street trees and roadside-planting, but wanted their own private living spot be transplanted into a natural surrounding. Migration to the green belt, urban sprawl over large agglomeration areas and traffic congestion are among the results. The effects of this in terms of conservation of nature have been disastrous. Whether this could have been prevented remains an open question. Population growth, industrialisation, growing living standards, and mobilisation and migration (from countryside to towns, from international peripheries to centres) forced traditional self-contained towns to open (many of which still had medieval walls around the centre, and some adjacent and relatively small middle-class town extensions from the 19th century).

Most town planners now agree on the ecological and functional preferability of concentration and density to widely scattered populations and buildings, not least because this is the only way to use public transport and supply infrastructures to full enough capacity and thus cost-efficiently. Urban concentration and density certainly create environmental 'hot spots', but the rest is left intact. So town planners would like to reverse the centrifugal trend of town growth, and also the trend to functional separation. Operative functions such as housing, shopping, working or schooling ought to be spatially reintegrated according to the vision of the 'short distance town'. Re-gentrification of town centres here and there seems to suggest such a turn-around. This, however, might just be the one swallow that does not a summer make, besides that it does not make a significant ecological difference whether it is the middle or the lower classes who live outside or inside a town centre.

Nonetheless, the regime rules as such can be put forward: A building ought to be located as much as possible in proximity to other activities connected to the function of the building, or to its inhabitants or users (short distances, less traffic over longer distances). There ought to be a preference for, and a financial premium on, the renovation and reuse of old building substance instead of constructing new buildings. Equally, regulation on land use has to make sure that redevelopment of urban and industrial brownfields is preferable to the development of greenfields.

Since the transsecular cycle of S-shaped population growth has come to its end in Europe, and has passed its inflection point elsewhere too, thus slowly approaching its retentive stage, the growth of agglomeration areas will almost certainly also come to an end. But urban agglomeration areas are bound to remain. They will not shrink back to earlier structures of traditional town-countryside relations. Shrinking towns are a limited phenomenon, a subordinate counter-fluctuation related to transitions from old-industrial districts to dynamic new high-tech regions. So the picture on the whole exhibits retention, and some fluctuation around the retentive level. Indeed this holds true for urbanisation as much as for the levels of consumption and turnover volumes of resources.

This will best be indicated by the level of population and real living standards. The present European trend to declining numbers of native population (i.e. a fertility rate of less than about 2.2 children per woman) will sooner or later bounce back to some reproductive level. The same can be said for today's most prolific countries where fertility rates are already in train of falling to reproductive and temporarily sub-reproductive levels; the result being a retentive stage of population numbers and urban agglomeration areas.

4.6.2 Architectural Principles of Environment-oriented Building

As far as the type of buildings is concerned, town planners and architects can help each other. From an ecological point of view, the critical issues in building are land use and the energy design of buildings. In both respects what people actually prefer most is ecologically the least adapted, namely standalone single-family homes. In comparison to multi-floor houses and high-rise buildings they require much more land and much more energy to heat or to cool. Town planning can make sure that multi-floor houses in enclosed blocks come first. If a development plan allows open street carées, high-rise buildings are preferable to smaller ones. Regulative building standards, architects and owners would have to make sure that living in multi-floor buildings is worthwhile. It is well known that quite often this is not the case, which in turn adds to urban sprawl.

With regard to other aspects, architects can to a certain degree compensate for the ecological deficits of urban agglomerations which town planners cannot do away with:

- If applicable, architects nowadays take into account the north-south axis and make the ground plan of a building fit with the angle of sun incidence. The principle is to expose living rooms or working rooms, equipped with relatively big windows or glass doors, to the southern direction, and align sleeping rooms, kitchens, bath and store rooms, with smaller windows, to the north (in the southern hemisphere it is the other way round). This helps make optimum use of solar radiation. Daylighting may also profit from north-south alignment. Daylighting, though, i.e. making use as much as possible of natural light rather than artificial lighting, is a health and environment feature of its own, particularly relevant in larger buildings and broad office floors. Light wells, suitable location of windows, and new varieties of glass can all contribute to a friendlier room atmosphere by daylighting.

- Roof-top turf, where applicable, gathers dew and rainwater. This contributes to improving local micro-climate.

- Minimisation of soil sealing has become an environmental state-of-the-art practice. This also helps to retain water.

- A new challenge for architects and the construction industry is design for re-usability of construction elements and recyclability of materials. This is of particular relevance in functional buildings. For example, construction elements made of reinforced concrete, as used in halls, could easily be recovered and reused if this was planned for. The elements require only little re-processing. Materials savings of this practice are 80 per cent, cost savings 20 per cent.

- It is furthermore up to the architects to integrate into the plan of a building an environmentally intelligent energy and water design. An appropriate choice of building materials is one aspect of the response to this challenge.

4.6.3 New Building Materials

In terms of sheer volume, building is the most materials-intensive realm, and building rubble represents by far the largest faction of solid waste. It does not represent the most urgent waste problem though, because most of these materials, such as steel, glass, concrete, bricks and timber, are normally non-hazardous, do not contain active organic matter and have a long life expectancy. Hazardous building materials, most often substances added in materials processing, are successively being eliminated by benign substitution. Ever more building materials are recycled, as aggregate secondary material in new products, or as complete cycleware with little downcycling involved. In road construction, old pavements can by now be fully recycled. New concrete pavements are made with nearly 100 per cent of granules of old concrete, which is recycled on-site and bound with fresh cement.

For construction elements which are under the permanent strain of pressure, vibration, drag, etc., such as in high-rise buildings or bridges, there are new materials available, for example polymer matrix composites (PMC). These are carbon fibres or glass fibres encased in plastic. Tube-formed PMCs replace steel and reinforced concrete. They are lightweight, and so translate into less energy demand and faster construction, and they have a longer durability than steel and concrete, yielding cost and materials savings.

Materials for walls, floors and ceilings, and the coatings and paints applied to them, are of particular relevance with regard to insulation and health. Indoor pollution is another health and environment problem the impact of which tends to be underestimated.

A number of alternatives to asbestos and other mineral fibres have been mentioned already in I/4.4.4. Mineral wool made of biodegradable fibres definitely does not pose a health hazard to workers and occupants any longer.

For purposes of house wrap a great variety of thermo-active building materials is now commercially available. Among these is 20fold expanded perlit, a volcanic rock converted into a kind of foamed mineral. The insulation effect is enabled through the presence of air within the structure. The material is light, effective and cost-efficient. A similar effect is achieved with sandwich walls made of two layers of slit-bricks or wood. The body-cavity, i.e. the hollow space in between, is about 25–30 cm thick and filled with insulating material such as recycled styropor or biosoluble mineral fibres. Such sandwich walls provide high insulation, low noise and high fire resistance. Good insulation is also guaranteed by perforated bricks made of wood shavings, cement and water. The physical properties of these bricks are similar to those of clay bricks, but they are much lighter.

Windows which consist of special glass, or have been given a special coating or surface structure, also including holographic-optical elements, provide transparent heat insulation. Another principle of transparent heat insulation, in parallel to sandwich walls, is sandwich glass with some transparent insulating material in between, for example polymer foils, comb-like cavities or microporous material. Conventionally insulated windows lose 2.6 W/m^2, whereas the value for heat-insulating windows is 1.6 W/m^2.

A competing approach to insulation materials is thermo paints. These are reflecting wall coatings which prevent heat emission across the wall. Thermo paints come with energy savings of 10–20 per cent.

Wooden wall and ceiling materials can be made of cycleware and yet have high quality, in contrast to conventional chipboard which often contains problematic substances. One such approach uses lean wood from thinning of domestic forests. The wood is processed according to an innovative method which combines cross-laying, gluing (with non-hazardous glues) and low-pressure pressing. The wooden material obtained fulfils any criteria from fire safety via stability and longevity to insulation. It contributes to improved room climate and living quality. When it becomes rubble, it does not represent a particular waste problem but can favourably be recycled for some use. A material with similar properties which can fulfil the same functions are board, panels or sheets made of wood, straw, sawdust or similar residues, which are soaked in water glass and hardened under hot gas. Fibre materials made of used paper, straw, waste pulp, or similar, are penetrating ever more segments of indoor-applications, furnishings and furniture. Fibre materials almost always create a good room climate and good acoustics. They are cheap, and easily recyclable or biodegradable after use.

4.6.4 Energy Design of Buildings

In northern latitudes about one third of the entire energy demand is caused by room heating. In private households the percentage is even about 60–80 per cent. Older, poorly insulated buildings need about 22–27 litres of fuel oil per m^2 annually. Contemporary standards for new buildings require 15–20 litres, stricter ones 10–12 litres. Low-energy houses perform at 3–7 litres.

This is why house wrap is an important issue in ecological terms too, and why most of the new building materials focus on heat insulation in the first instance. An older, poorly insulated 150 m^2 single-family home consumes 4,000 litres fuel oil a year which emits 20 kg NO_x, 34 kg sulphur dioxide and 10,800 kg CO_2. If the same house is retrofitted with insulation materials on outer walls, under the cellar ceiling and the roof, and equipped with insulation windows, fuel consumption and emissions fall more than half to 1,800 litres fuel oil, 9 kg NO_x, 15 kg sulphur dioxide and 4,860 kg CO_2. Newly built houses with insulating materials already integrated in the construction body do better still. Besides recycling, the energy design of houses is the field where the 'efficiency revolution' can make significant contributions indeed, and where it has done so to a certain extent already.

As a partial alternative to insulation, standards can also be fulfilled by floor heating in combination with a heat pump. In the energy design of buildings there is a general competition between construction measures (new materials, architectural and construction principles) and heating/cooling systems (new furnaces, boilers, radiators). In most cases there is some reasonable combination of construction measures and heating technology.

Whether construction solutions driven to the extremes of a zero-energy edifice are still reasonable can be doubted. A hermetically insulated indoor climate may not be the last word in buildings that are worth living and working in. People might wish to leave doors and windows open and have cosy indoor climates facilitated by breathable novel building materials with the functional properties of wood and clay. Introduction of clean fuels and improved heating/cooling technology which allow more air circulation and exchange between inside and outside the house would seem to be preferable. But low- or zero-energy buildings certainly have the merit of having pioneered a number of characteristic components of an environment-oriented energy design. Low-energy buildings usually include:

- North-south alignment, daylighting and a winter garden
- House wrap, insulating walls and windows, relatively small windows
- Photovoltaic panels roof-top or façade-integrated
- Solar thermal collectors roof-top, for warm water and heating rooms
- Local heat and power system
- Energy-efficient appliances within the house, from low-energy refrigerators to long-lived low-watt light bulbs.

A low energy design may also include some of the following components:

- A solar air and ventilation system, i.e. solar air conditioning and solar assisted ventilation with heat recovery
- Heat recirculation in pipes beneath the floor
- Heat pumps, as formerly applied in combination with electric heating. In Switzerland, 40 per cent of all new buildings are now being equipped with heat pumps.
- Lighting control by automatic switch-on-off or dimming on the basis of light and movement sensors.

Solar thermal collectors and calorific boilers can be combined. Instead of having two different systems with two different hot water containers, both sources of heat – the roof-top solar collector and the calorific boiler – feed the same hot water container from which the central heating system and the warm water circulation are fed. The combination produces a reduction in investment costs, heat loss, emissions, space requirements, and enables easier system control.

Some architects and energy designers have a special preference for burning local wood as a fuel in decentral heating systems, from single small houses up to district heat and power systems. This can be fresh firewood, or waste wood, or lean wood, in pieces or chaff-cut or in pressed pellets. The advantages, it is argued, are savings of mineral oil, and greenhouse neutrality because of a closed carbon cycle. As discussed in I/4.2.3, the advantages of burning wood are questionable. It is not really a step towards improved metabolic consistency. Burning organic macromolecules, if at all, ought to be the very last step in a cascade of prior uses and reuses of wooden material. In general, wooden material ought to be used for purposes of soil melioration (which might involve simply leaving it where it is) or as a biofeedstock in materials processing industries.

By contrast, geothermal heating deserves a positive appraisal, though not many architects have been won over to it so far. Below frost level, earth temperature is constantly at about 10°C. In wintertime, amplified by heat pumps, this can be used for heating; in summertime it can directly be used for cooling. Particularly low-energy houses tend to be heat traps because they are so well insulated that heat cannot escape. At 30–35°C outdoor temperature, geothermal cooling provides a comfortable 24–26°C indoor. Probes, i.e. pipes arranged in loops and filled with water, are inserted into the earth, and a water pump creates a constant water flow downwards and upwards. Almost all of today's systems tap into heat at a depth of about 100 m. This range is also applicable to single-family houses. Investment costs for geothermal heating for a newly built 150 m² home are at €17,400, compared to €12,500 for conventional heating. This is competitive since there are no fuel costs, just the costs for powering the heat pump.

The ecological advantage of geothermal heating is partly negated as long as the electricity for heat pumps comes from conventional power stations. Geothermal heat would make use of all its advantages by tapping into deep heat below 400 m. Water heated from this depth is hot enough to immediately drive modern turbines, including microturbines, for generating electricity (thus also contributing to distributed power generation as outlined in I/4.2.8). The waste heat of this process can further be used for district-heating of houses and glasshouses.

4.6.5 Water Design of Buildings

Despite an extended discourse on alternative water design, most architects in industrial countries do not really consider elements thereof in their plans. Maybe water is too abundant and cheap to care about. A different water design also includes additional efforts and makes the infrastructure of a house more complicated. So it may be that environment-oriented water design will have more relevance for countries which are troubled by scarcity of water in general and clean freshwater in particular. In Israel, water recycling has become a regular practice.

Outdoor water design, suitable in northern countries where rainwater can be led off, should include water permeable pavements on footways, roadways or parking lots. If a pavement can be made water permeable both man and nature profit. Such pavements can for example be implemented with water permeable polymer fabric covered with sand and a solid porous surface, laid over a gravel base which keeps pollutants in shallow groundwater where microbes break them down.

Systems for the capture of rainwater for irrigation, and for percolation of rainwater on site, have already been discussed in I/4.3.1 in connection with the approach of cascadic retention management of water and groundwater.

As for indoor uses, water design is aimed at saving and recycling water, which includes purification of used water, or rainwater, on site. Filtered rainwater meets the requirements of washing machines and certainly of toilets. Used water which is purified in-house with the help of special devices can in principle be used for any purpose, except for drinking and preparing meals, but including taking a shower. One such device, AquaCycle by Pontos Ltd., has now been launched at the European market. The small purification plant consists of a pre-filter, followed by 2-step biological purification and final UV disinfection. Space requirements are small, and it is cost-competitive to tap water. It can deliver 2,400 litres a day, enough for about 5–6 families. In somewhat larger calibration, it is suitable for residential blocks, hotels, hospitals and factories of similar size.

Installation of water-saving fittings, such as variable toilet flush, or a run-through-sieve at the water faucet, are state-of-the-art. A new inliner-system in warm-water pipes (principle 'pipe in pipe') avoids the cold-water losses of single pipe systems as well as the energy losses of circular warm water systems. It also reduces the formation of germs (legionelles).

A good facility management can have at least a certain number of these TEIs implemented by contracting. Contracting means that a house owner gives a service company the job of looking after the water supply, or electricity supply, or house/office heating. Contracting companies specialise in searching for and realising eco-efficiency gains. In most cases this involves removing old systems and introducing TEIs. The contracted services pay for themselves through cost savings obtained.

4.7   VEHICLES AND TRAFFIC

4.7.1 Traffic Avoidance, No. Optimisation of Modal Split, Yes

What has been said on the spread of large urban agglomerations, can be said in much the same way with regard to traffic. Urbanisation and traffic are indeed two sides of the same coin. Traffic density is bound to remain with us, no matter whether the traffic medium is car, bus, tram, train, underground, bike, foot or ship. This also includes high-speed long-distance traffic by aircraft, fast rail and high-speed magnetic levitation trains.

Traffic, i.e. transport of persons and goods, together with telecommunications, is an operation which links operations, a cooperative necessity of labour division and functional divisions, an expression of the vastly enlarged capacities of modern society, and of greater degrees of freedom for the people. Any attempt to actually bring traffic down would cause severe dysfunctions. Minimising traffic means minimising human operations, which would in practice translate into a considerable loss of living options and living standards. People who urge traffic avoidance apparently do not understand what they are talking about.

Desirable, certainly, and occurring anyway, is traffic optimisation. Clever logistics cannot avoid traffic, but it can avoid idle journeys of freight cars and wagons, or suboptimal loading of transport vehicles. In densely populated large towns high-frequency public transport, especially underground and rapid city rail, can substitute for passenger-car city rides. This is not possible, at least not at reasonable costs, in less densely populated outskirts and peripheries. Seeking to strike optimal proportions of the modal split in traffic is a general top priority of traffic planning. Traffic or mobility optimisation, however, does not lead to less traffic, but rather to a continued growth of vehicle use and transport infrastructures on the stabilising basis of increased efficiency (II/7.7.1). This includes increased eco-efficiency, measured as environmental pressure (resources consumed and emissions) caused per freight mile or passenger mile.

Any mode of transport – ship, train, car, aircraft – entails a large ecological footprint, though there are relative differences. Vehicles require special traffic infrastructures which occupy a lot of land (harbours and canals, rail and stations, motor roads, airports). Proximity to traffic there is unpleasant because of the high noise levels which impair living quality near such traffic zones. Traffic can be dangerous, and disconnects living spaces for humans and animals. In cities, dense traffic and parked cars in the streets considerably limit the usability of public space. Instant individual mobility must be of very high value that most people are prepared to pay this price.

Similar to building, vehicles and traffic infrastructures represent vast amounts of materials, including the eco-rucksacks of resource recovery and processing, and the final waste problem when old vehicles are phased out. Finally, and also in parallel to building, the energy design of vehicles is the outstanding environmental issue. The energy design also incorporates onboard power supply in cars since there are ever more onboard systems which consume electricity. Given the range of diverse environmental problems related to traffic in general, and car traffic in particular, there is no ideal way to the greening of traffic. Nevertheless, for the time being it can be said that consumption of mineral oil and resulting exhaust gases are by far the most urgent environmental problem, particularly with regard to car and air traffic, though railway and shipping traffic in principle have the same problem.

In advanced industrial countries the peak (inflection point) of traffic growth rates and of environmental problems related to traffic has been passed already. Developing countries will follow suit. As the growth of population and urban agglomeration areas on the whole peter out, so will growth of traffic, in connection to growth of production and real income. But this will not happen in the near future. The transsecular transition from traditional to modern society, as far as industrial production and technology is concerned, has taken about 150–200 years so far. If the endogenous trajectory of the underlying path is not deflected by some external shock or internal breakdown, it will be another 100–150 years before the entire industrial or technological transition will be achieved globally (II/7.8.2). The slowdown of dynamics of growth and structural change, though, will be perceptible earlier. Traffic frequencies in advanced countries are slowly but surely in the process of approximating a retentive stage.

4.7.2 Concerned Car Use (Soft Driving, Car-sharing)

In the context of recent research and policies on the 'sustainable household' or 'sustainable consumption' the question of whether and how to use cars was among the favourite subjects, together with buying organic produce and being an ecologically aware tourist. Commuters were requested to join up in commuter groups, using one car for all instead of one car for each. Car-sharing communities have been set up. These are cooperative rent-a-car companies based on shared ownership and mutuality. There may well have been more research projects on car-sharing in the 1990s than there are car-sharing associations altogether.

The appeal to those concerned includes a strong social element: overcome individualism, join a community, share green ideas and everyday routines with like-minded people. This is not the place to discuss how far this is socially sound and progressive or whether it might represent a regressive impulse. The fact, however, that practices such as commuter groups and car-sharing involve a special value base is an important pointer when assessing the potential of such practices for being adopted and widely diffused.

Any sober empirical sociologist could make the correct prediction: There is little potential for such 'social' practices to spread beyond a relatively small niche because the overwhelming majority of people holds firmly to the habits of consumerism and individualism. Many people prefer a traffic jam to riding a crowded train or bus because inside their car they feel undisturbed within a mobile sphere of privacy. In addition, most people's lives, whether private and working, are so diverse as to make organising the shared use of cars rather complicated. Commuter groups are feasible only for people who still live within an old-industrial and therefore fairly rigid time schedule. Given the inconvenience of not having a car instantly available, the mere fact that car-sharing saves some money may not convince even the thrifty.

Technically speaking, collective commuting and car-sharing are measures aimed at increasing eco-efficiency. Almost all cars represent underused capacity. They lie idle 90 per cent of the time, and when they move around, it is most often just the driver who is transported instead of the four or five who could comfortably be seated in a normal passenger car. From this one could conclude the following: Put as many passengers as possible in a moving cabin; instead of always having a car waiting around, call one only on actual demand, in the same way as one calls a taxi, or rents a car on an hourly basis. On paper, this would increase efficiency by a factor of 4–10, and as an abstract technical principle it may be impeccable. The flaws, however, are social. It seems strange how 'social' people propagate such an extreme engineering point of view regardless of the social realities of car driving. In the same vein one could call for pot- and pan-sharing. Most of them lie idle most of the time. Similarly, one could also call for a ban on detached family homes and expect everybody to move in tenement blocks.

A different approach to concerned car driving, and in fact not a new one, is soft driving, i.e. no unnecessary acceleration, not driving at high revs/min, avoiding sharp braking. This can save 15–20 per cent of gasoline consumption and is certainly a reasonable practice everybody can follow, apart from hotheads and show-offs.

It should be noted that any method of concerned car use results in some efficiency gain within the context of established technologies. It does not include (certainly not preclude either) structural change towards improved metabolic consistency of cars.

4.7.3 Incremental Improvements of Internal-combustion Engines

Internal-combustion engines explode a fuel inside a cylinder. The ensuing pressure drives a piston which is at the root of mechanical motion. At a certain period of time an alternative to 'explosion motors' was seen in external combustion, as in a Stirling motor, where the burning takes place outside the cylinder. A Stirling motor is a hot gas or air expansion machine. There are two cylinders in direct correspondence with each other, a working cylinder and a compression cylinder. An external burner, propagating a normal flame such as in a conventional heating system, heats up the gas in the working cylinder (most often helium). The expanding gas pushes the piston which drives the crankshaft. The piston in the compression cylinder is directly linked to this, so that it pushes cooler gas from the compression cylinder into the working cylinder. The process does not come to a standstill because the residual energy in the crank acts as a flywheel that pushes the working piston back into the cylinder and pulls the compression piston out of its cylinder, thus soaking in the gas.

Green-minded people have put some hope in the 'soft', 'unexplosive' design of the Stirling motor. José Lutzenberger in Brazil once introduced it to me as the magnificent green motor of the future (dating back to 1818, similar to the fuel cell which was invented in the late 1830s). The Stirling motor, however, is suitable for continuous processes without much acceleration or slowing-down, as in stationary applications such as DHPS. The motor is unsuitable for driving a car, especially given prevailing driving habits (quick start, rapid acceleration, abrupt braking).

So car propulsion seemed restrained to improvements of the internal-combustion design; or so it seemed to conventional motor engineers who dared not try a radical change of technological path. In this they are similar to path-dependent thermal power plant engineers who have spent their vocational life with how to efficiently burn fossil fuels in large furnaces.

The series of innovations in internal-combustion engines which were triggered by environmental concerns started around 1980 with the three-way exhaust catalyst. It was criticised for being an add-on-measure which was materials-consumptive, and which replaced the evil of hazardous exhaust with hazardous waste. In addition it increased specific fuel consumption because catalysts need a slightly fatter gas mixture, i.e. more dispersed gasoline relative to air in the combustion chamber. Whereas working near open catalysts is prohibited in factories for health reasons, the same open catalysts have been introduced in the streets.

An alternative at the time was the lean-burn engine. A lean mixture in internal-combustion motors tries to maximise the portion of air in the combustion chamber, while minimising the portion of dispersed fuel, resulting in less fuel consumption and less emissions from burning. The competition between the two approaches represented a classical constellation of innovation contest (II/9.4.4). The integrated lean-burn concept is clearly superior to catalysing fat-burn exhausts end-of-pipe. But it was the three-way catalyst which won over the lean-burn concept. The reason, very probably, was a typical lock-in mishap which started in the 1970s in the US where the automobile industry, under political pressure, had introduced catalysts first.

Another vain hope in the greening of the automobile is the Diesel engine. Compared to Otto motors Diesels consume less, and also less refined fuel which is sold for less money than normal gasoline. Green-minded people advocated Diesel engines despite their sooty smoke plume which everybody could see and smell, besides 6–8 times higher emissions of NO_x in Diesels compared to Otto motors. Diesel exhaust cannot be treated by three-way catalytic converters because of the high oxygen content of the exhaust gas which destroys the catalyst. Otto motor cars had to be equipped with exhaust catalysts whereas diesel cars were allowed for many years to continue sooting around before they had to be equipped at least with soot filters. This was particularly inadequate in face of the growing fleet of delivery vehicles and heavy trucks which have the biggest motors and longest mileages of all cars.

There were also some motor-integrated improvements. In the 1980s, Rinaldo Rinolfi at FIAT invented the common rail injection. It feeds the fuel into a manifold (or rail) which creates a steady injection pressure regardless of the motor's speed. In consequence, the fuel is burnt at a constant high temperature resulting in lower engine noise and much less noxious emissions.

Diesel engines, though, cannot do away with the soot. The filters even seem to have aggravated the problem. The reason is that only relatively large grains of soot are captured by today's filters, whereas small and ultra-small soot particles in the range below 10 micrometres or even below 2.5 micrometres (PM 10, PM 2.5) can pass unimpeded. These microparticles, however, are particularly hazardous. Being ultra-light, they stay in the air and form aerosols which are inhaled by everybody. This is comparable to passive smoking. Carcinogenity of diesel exhaust is ten times that of normal gasoline. This has been known for over ten years. Adequate political reaction is still lacking, except in Dakha, Bangladesh, were all diesel vehicles have now been rigorously prohibited. Elsewhere, meanwhile, the market share of diesel cars is on the rise. In Germany all commercial vehicles and 40–50 per cent of new passenger cars now have Diesel engines. The first 3-litre car by Volkswagen was a Lupo diesel, the demand for which, though, is not overwhelming. According to the growth-efficiency principle people prefer more spacious cars with bigger motors.

New findings published by Mark Z. Jacobson in the *Journal of Geophysical Research* indicate that diesel soot particles have to be put on the list of greenhouse substances. One gram of soot has a greenhouse effect many thousand times that of one gram of CO_2. If these findings are not to be dismissed completely, diesel soot could turn out to be the worst of all greenhouse substances. Efficiency increases in Diesel motors thus seem to represent a case of ill-directed progress in the wrong place.

Hopes to bail out diesel cars now rest upon next-generation soot filters that catch particles of about 100 nm or, alternatively, plasma exhaust cleaning as pursued by DaimlerChrysler, GM, Ford and Peugeot Citröen in cooperation with American special technology firms. In a plasma device the oxygen molecules of the exhaust can be ionised, thus making the exhaust gas accessible to subsequent catalytic treatment as usual. In the plasma, soot particles are also dissolved. If this were confirmed, diesel cars would be rid of their soot emissions and 90 per cent of NO_x – and thus may last longer than they otherwise would. Normal diesel cars cannot meet the new US diesel emission standards taking effect in 2007. Plasma devices would however be relatively expensive.

Elsbeth motors are another development with a certain undeserved green appeal, because this type of motor is designed for running on plant oils. Enthusiasm for Elsbeth motors probably has the same origin as the questionable appeal of burning biomass. An Elsbeth motor is a modification of a Diesel engine rather than being a different kind of motor. It consumes any kind of cheap plant oil quite indiscriminately and can be switched back to diesel any time. Envisaged market segments are lorries and buses as well as ship motors and stationary generators. Financial break-even in trucks is reported to be reached after six months. Cost savings are said to be about €10,000 p.a. and truck. But it must be remembered that Elsbeth motors involve bad use of land and biomass, and are in fact heavily polluting.

Otto-motor engineers too have been busy achieving efficiency gains by incremental innovations. Among these, also partly applicable to diesel- and hydrogen-fuelled cars, are the fast starter-generator, the pressure wave turbocharger, direct fuel injection, the cranked connecting rod, variable valve systems, the camless engine, cartronics and mechatronics. This list is just exemplary and certainly far from being complete.

A fast starter-generator shuts down the engine whenever a car is not moving, and immediately restarts it upon a move of the gas pedal. A starter-generator with relatively simple electronics reduces gasoline consumption by 20 per cent. This certainly makes a difference.

Turbochargers have so far been applied in cars with larger motors. Turbochargers accelerate the rods by using the force of exhaust gases to spin a turbine and compress the air which flows into the combustion chambers. With a pressure wave supercharger as developed by the Swiss Federal Institute of Technology, Zurich, this principle can also be applied to smaller motors. The developers claim that the device can enable a small car to run 28 km on 1 litre gasoline, i.e. make it a 3-litre-car. Pressure wave superchargers are expected to reach the market around 2005–07.

With direct fuel injection there is no carburettor. The fuel is directly injected into the cylinders. This is state of the art in Diesel motors. Now there are different ways of doing this in Otto motors too. In spite of direct fuel injection requiring higher pressure at about 250 bar (compared normally to 120 bar at maximum) it can accomplish fuel savings of about 15 per cent.

A Swabian by the name of Siegfried Meyer, a man who is neither an industry engineer nor has academic credentials, thus a typical outsider to the competition (II/9.4.5), has invented a cranked connecting rod. The German engineers national weekly devoted a long article to it, and the first wave of letters to the editor were remarkably hostile, accusing the inventor of being a swindler and charlatan. The controversy arises from the dominant motor engineers' paradigm which has it that connecting rods can only be straight.

The cranked rod has, however, been tested in everyday routine operation. It simply requires small changes in ignition and injection to be retrofitted. It saves 25 per cent of fuel in a large truck, in a bus even 40 per cent. This is much more than the 10 per cent fuel savings from BMW's Valvetronic that was presented in 2000. The cranked rod can also be used in passenger cars, or in ship motors and propeller aircraft. Astonishingly, there are even less emissions than the reduction in fuel consumption would suggest, especially reductions of CO and VOCs. The established know-alls, however, are not interested. To them, the cranked contraption was 'not invented here'. The inventor is reported to have found partners in Taiwan.

Variable valve systems open valves only partway when little power is needed. They still have camshafts. In a camless engine, the camshaft is replaced by electromechanical actuators. This increases fuel economy by 10–20 per cent and strengthens engine torque at low speed by 15–20 per cent (i.e. faster acceleration).

The examples reflect the typical pattern of incremental change. Sophisticated as they may be, they do not bring about structural change towards improved metabolic consistency. They just contribute to a structure-conservationist type of 'efficiency revolution'. Increased efficiency, however, will not only translate into rebound growth of cars and traffic, but is at 20–40 per cent not very revolutionary either. This is due to the advanced structuration stage of maturity and to approximating marginal utility in internal-combustion engines (II/7.4). After all, this is a technology which has intensively been researched and developed for 130 years.

There are nevertheless innovative elements and components which may be useful also beyond Otto and Diesel motors. Cartronics, for example, is an electronic controls approach aimed at interconnecting and fine-tuning various components of a car and its engine, such as brakes and downshifting, gear-switching and transmission controls, engine temperature as well as generation and consumption of electricity. Sensors and software mediate between the driver's impulses and the engine. Sensors also can detect combustion misfires from the composition of exhaust gases, and feed this information back to an actuator which reconfigures the engine's operations. A sensor-and-software battery manager, a box of the size of a deck of cards, controls battery charge. If this threatens to fall below a certain threshold level, the engine idle rate is automatically activated to speed recharging of the battery, or functions of less priority such as seat-heating are briefly interrupted. Such cartronics systems are now available in luxury car models and will generally be available by 2005. Mechatronics goes beyond cartronics in that it also includes the substitution of electromechanical actuators for conventional mechanics such as hydraulic cylinders. In the same way wires replace brake fluid lines.

Certain development efforts have been geared towards hybrid cars. These are cars with a combination of two propulsion systems in one car, most often a Diesel engine or an Otto motor combined with an electric motor.

Mild hybridisation is an approach developed by Ricardo, an English engine consultancy, and Valeo, a French car-parts maker. The approach uses a number of the components listed above such as common rail fuel injection, variable valves and pressure wave turbo charging together with an electric motor that compensates for lack of pulling power at low engine revs. Although this raises mass and weight, it results in increased fuel efficiency, for example in an Opel Astra up to 20 per cent (4 litres per 100 km) and 30–35 per cent less emissions of CO_2.

4.7.4 New Propulsion Systems: Hydrogen and Fuel Cell Cars

Due to the ecological disadvantages of internal combustion of carbon fuels, the automotive industry has already been at the crossroads for 10–20 years. They face the choice of whether to reluctantly remain on the familiar path and lose ground, or whether to take the plunge and switch over to a generic systems innovation in car propulsion. Switches are set towards a change of path: towards clean hydrogen-fuelled cars or fuel cell-powered e-cars.

A hybrid approach which is exactly at the crossroads is an electric motor combined with a gas turbine. This was developed in the US by the Rosen brothers Ben and Harold (one of whom launched Compaq Computers). The concept consists of an e-car the electric motor of which is fed by a generator coupled to a gas turbine. A special flywheel allows the turbine to run at constant speed (while the car does not). The design is much cleaner than internal-combustion cars, yet not as good as e-cars powered by FCs.

A hydrogen motor car is still driven by an internal-combustion motor. The motor, however, runs on hydrogen instead of gasoline. This is the approach which has in recent years been favoured by BMW.

Ford has presented a new Model U (an allusion to the famous Model T) on the occasion of its 100th anniversary. In contrast to a pure hydrogen motor car, the Model U is a hybrid approach which combines hydrogen combustion in a conventional motor with an electric motor. Fuel efficiency, according to Ford, is 25 per cent higher than in most advanced conventional cars, and emissions are next to zero.

The next step then is to drop internal combustion and replace it with electrochemical fuel cells (I/4.2.6). An FC-powered car is an electric car, i.e. a car driven by an electric motor with the electricity generated onboard by an FC. FCs are fed either directly with hydrogen or by a reformer which extracts the hydrogen onboard from natural gas or methanol. The FC-powered car has been pioneered by DaimlerChrysler with its Necar series since 1994 (Necar =

new electric car). Almost all of the global car manufacturers are now in the process of developing prototypes, on their own or in some joint venture. Meanwhile, the bandwagon effect has fully set in so that development finish and market launch are set for take off. Japanese manufacturers (Nissan, Toyota, Honda) have prepared the public for a commercial FC-car launch within the immediate future. This car, however, will be for 1G lead users. Veritable mass production of FC-cars, according to American and European manufacturers, will probably not take off before 2010.

The FC in a car tends to be a PEM-FC, though there are also prototypes of alkaline fuel cells (AFCs) which were originally designed for use in spacecraft. Zevco, an Anglo-Belgian company (Zevco = Zero emissions vehicle company), has started a prototype test in London cabs. AFCs have superior power-to-weight ratios. They are relatively simple (fewer pumps and compressors than PEM-FCs). However, AFCs do not run on normal air. They need bottled, therefore expensive, pure oxygen. This could be the reason why the general trend may not include AFCs.

Ecological performance of FC-cars is impressive while technical and economic performance are satisfactory. FC-powered e-cars need 25–50 per cent less energy than comparable internal-combustion motor cars. An FC-powered car which runs on regenerative hydrogen needs about 20 kWh/100km, on hydrogen from steam reforming 32, on methanol 39. Hydrogen combustion takes 43 kWh/100km when regenerative hydrogen is used, almost 70 when steam-reformed hydrogen is used, compared to 45 in a conventional Diesel engine, and 52 in an Otto motor. E-cars equipped with conventional batteries need 60 kWh/100km.

The bottleneck is hydrogen stations. Complete supply systems, it is said, cannot be expected before 2010–15. Car manufacturers and producers of industrial gases are currently running local pilot tests with hydrogen stations. Equally unclear at present remains the best way of producing hydrogen, and how the transition from conventional carbon-based sources to regenerative hydrogen will be managed. Production of hydrogen in central IGCC plants on the basis of coal with CO_2 sequestration would seem to be a solution for a certain period of time during which a solar hydrogen infrastructure could successively be developed.

Also the innovation contest for the dominant design of hydrogen storage has not yet been settled. Nanocubes might be a suitable technology. A favourite approach at present is sodium hydrides such as borax Powerballs (I/4.2.5). DaimlerChrysler has developed a Sodium Van on the basis of a hydrogen-on-demand concept using Borax technology. This has performed well in a test on a 50 km course in Death Valley carried out by the US magazine *Car and Driver*. The producer of hydrogen-on-demand is Millennium Cell Inc., New Jersey. The system uses a liquid made of 75 per cent water and 25 per cent sodium boron hydride, a washing powder-like substance. The resulting liquid can be handled just in the same way as gasoline, with the advantage that it is neither inflammable nor explosive. The producer of the sodium boron hydride is US Borax. The range of the Sodium Van is 480 km. Maximum speed is 130 km/h. Wheel-to-wheel efficiency comes close to 20 per cent, compared to 14 per cent in conventional cars.

4.7.5 Materials and Car Recycling

The second biggest environmental challenge for cars and other vehicles after propulsion, is materials demand and solid waste. After long years of delaying tactics, the automobile industry has finally given way to adopting the principle of product stewardship, and to acceptance of EU regulations on obligatory take-back and recycling of cars.

The car industry, sales agencies, retailers, garages and reprocessing industries are in the process of setting up complex systems of car disassembly, reuse of car parts (which has a long-standing tradition) and recycling of recovered materials both within and outside car manufacturing.

These endeavours have in turn fostered the adoption of principles of design for disassembly and recyclability in car construction and manufacturing. Optimum recovery and recycling requires that materials be as pure as possible. A case in point is aluminium. Ever more vehicles have an aluminium chassis and body. This is strong but relatively lightweight, thus contributing to improved energy economies, and it is suitable for permanent recycling. Design for recyclability also requires materials to be clearly marked in order to be identifiable without special examination. This is of particular relevance with regard to plastics as well as textiles and filling materials made of synthetic and natural fibres.

Part of the present design trend is to use cycleware. This is a common practice now with bigger parts made of plastics such as bumpers, panel dashboards or inside door panels. This is being complemented by biotic materials such as pulp fibre materials made from used paper or straw for use inside the passenger cabin. Such biotic materials have good plasticity and contribute to comfortable indoor climate and acoustics. They are cheap, and after use they are easily recyclable or biologically degradable. Resource productivity of biotic cycleware is 7–12 times higher than with plastics.

Design for recyclability is certainly more easily postulated than followed through. Actual trends in materials requirements, in-detail applications and manufacturer and customer preferences quite often run counter to the principles of design for recyclability. Consider composite materials such as carbon fibres reinforced by synthetic resin, or new materials such as glassy metals, the nature of which is impure. They are used because they are strong and reliable, but also light, thus saving energy, and long-lived, thus saving materials. They may, however, represent a problem in terms of recovering as a pure material that can be reused.

A similar problem arises from mixing materials which occurs almost unavoidably. How can a small controls box containing dozens of different small-scale metals and plastic parts be recycled? Metals most often represent alloys, just as plastics may contain various additives. Each round of recycling creates further mixing of ingredients and deterioration of materials properties. This is known as the problem of downcycling. What can be done is to be as strict as possible with regard to the pureness of first-use materials so that later mixing of materials can be reduced as much as possible. Moreover, technologies of disassembly, recovery and reprocessing can be expected to progress in line with growing demand for such activities from the side of manufacturers on the basis of tighter materials standards.

4.7.6 Alternative Propulsion in Ships and Aircraft

Motor ships today are equipped with large Diesel engines. In principle, the alternatives in car propulsion can be transferred to ships, notably fuel cells. In ships these would have to be of much larger size, comparable to the PAFCs and SOFCs from smaller and larger power plants.

The military has ordered 1G FC-powered submarines already. These can remain submerged for several weeks at a time. They cause no noise and no exhaust gas, so they cannot be detected in this way from the surface. The motivation here is purely technical rather than ecological. But if novel military technology is shown to work, it will be adopted in civil applications.

With aircraft it is different. Clean propulsion in aeroplanes is a missing feature. FCs would appear to be applicable in aircraft in combination with propeller propulsion, which is to say in smaller planes, or helicopters or propeller-driven airships. In spite of some development work, no prototype of a fuel cell-powered propeller machine seems to exist so far.

Jet planes which use hot gas turbines could fly on hydrogen rather than kerosene. It remains unclear, though, whether vapour trails from hydrogen-fuelled jet planes flying in large numbers at high altitudes would contribute to climate change. Below tropopause (9–13 km height) condensation trails and exhaust gases are particularly persistent. But there does not seem to be a serious development of hydrogen-fuelled jet planes anyway.

There are certainly next generation conventional jet engines under development. One goal is to reduce emissions of NO_x by 50–60 per cent which is important with regard to ozone levels. But given the rising importance of air traffic, cleaner carbon will not be sufficient. Carbon-free clean aircraft propulsion is a desideratum of outstanding importance.

For the moment engineers are preoccupied with increasing the fuel efficiency of old-type aircraft by redesign of construction components and by introducing new materials. Forty aircraft companies across the world are cooperating in an R&D project called Power Optimised Aircraft. It is funded to the tune of €100 million and aimed at slimming the body and equipment of an aircraft. A special target is the many heavyweight pipes of the hydraulic system, including the fluids therein. The two basic approaches are, first, to introduce low-weight new materials, and second, similar to mechatronics in cars, to substitute electromechanical actuators for conventional mechanics, in an attempt to make functions aboard electric as much as possible. The time horizon of this joint R&D is 2010. Results will be of some advantage to all kinds of aircraft regardless of their particular propulsion system.

A historical alternative to internal combustion in aircraft was the airship. The accident at Lakehurst in 1937, where the Hindenburg went up in flames and hundreds of passengers died, does not really explain the discontinuation of the airship's trajectory of development. Other traffic carriers too have had their catastrophes. Cars have accidents, trains are derailed, ships sink, aeroplanes crash. Why should airships be infallible?

A disadvantage of airships is their vulnerability to stormy weather. From an environmental point of view, though, airships have only advantages. They are almost silent. They can be of any size, particularly of a very large size unattainable to aeroplanes. They are highly fuel-efficient and cause few emissions. If equipped with FCs, they would be zero emission vehicles. Airships, however, are not very fast. As compensation they can start and land in any place. This would make them an ideal vehicle for door-to-door transport, particularly of heavy freight on long- and short-distance deliveries to remote destinations difficult to reach by normal ways. Aeronautics enthusiasts have never given up the airship. In the 1990s, Cargolifter, an airship renaissance company so-to-speak, tried a fresh start. During the start-up boom which lasted until 2000, the new company did not face too many difficulties in raising capital. Since then, however, it was having serious difficulties getting off the ground and had to file for bankruptcy in 2002.

The latest frontiers in novel aircraft propulsion are mini-engines and micro-generators (MEMS) as discussed in I/4.2.6, which drive small, unmanned planes. Since these are very small, it remains unclear at present what their ecological relevance might be.

4.7.7 Fast Rail versus High-speed Magnetic Levitation (Maglev)

Magnetic levitation trains (maglevs) run on an elevated guideway and have no wheels. They glide over the guideway with no contact to it, similar to a hovercraft gliding over water on an air cushion. In contrast to a hovercraft, which makes an infernal noise, a maglev is silent, and very fast, quickly attaining 400–500 km/h. This makes it the ideal vehicle for continental high-speed long-distance travel from one city centre to another, substituting for continental flights in the range of 200–800 kilometres.

The maglev principle was patented in 1934. Prototype systems were developed in the 1970s–80s in Germany and Japan. In the Japanese design, levitation is based on electromagnetic repulsion. This requires permanent magnets or superconductors, which is technically rather complicated and involves a lot of mass. The German approach, by contrast, is based on the attractive forces of electromagnetic induction. Levitation is caused by support magnets at the upper side of inwardly cranked runners or skids. These, from below the guideway, are attracted upwards to ferromagnetic stator packs mounted to the underside of the guideway. Attraction between the two pulls the vehicle up. Propulsion then is caused by a synchronous longstator linear motor which creates a magnetic travelling field. Braking is achieved by reversing this field at which point the motor becomes a generator whose current is re-fed into the on-board supply batteries of the system.

According to electrical engineers, electromagnetic levitation and propulsion is a relatively simple and robust system. It consumes less energy than is required for air conditioning inside the passenger cabin. Overall energy demand is a third less than in fast trains, and three to five times less than in car highway traffic and air traffic. In contrast to conventional trains, a maglev can climb rather steep gradients and follow tight curves. Maglevs are zero emission vehicles. At a speed of 200 km/h a maglev is completely silent. At 300–400 km/h the noise of the wind gets gradually louder, but never reaches levels known from conventional trains, not to mention aeroplanes. The area below the elevated guideway can be used for any purpose, as these guideways can be built in parallel aside already existing rail or road routes. The magnetic field inside the passenger cabin is at 100 μTesla. This is only double the natural magnetic field of the earth at 50 μTesla, compared to the 500 of a colour TV, or the 1,000 μTesla of hair dryers and electric stoves.

Both pioneer countries, though, have little space left, and already have at their disposal a large traffic infrastructure, including fast rail. They thus have had difficulties with constructing a long-distance maglev reference line within their own country, the more so since there are two dogged defenders of the status quo: on the one hand radical conservationists, and on the other established railway monopolies together with national airlines.

In 1975 Ernest Callenbach wrote *Ecotopia*, a fantasy on the California dream turned all ecological. The hero of the novel was transported to the scene of the action by letting him glide on a magnificently fast and silent maglev, an obvious element in an ecological vision of a future modern society. Green backbenchers and deep-green militants have soon forgotten about

Ecotopia. Many of them have chosen an uninventive sufficiency approach, even resisting efficiency progress in old technologies for many years. They oppose traffic infrastructures on principle, be these motor roads, or runways or new rail routes – except bike and bus lanes. Maglevs, which are an additional new system, requiring new infrastructure and representing high-speed long-distance traffic bypassing 'locals' into the bargain, must be the devil's own invention to them. They fiercely mobilise against any attempt to have a regular maglev route built.

With railway companies and national airlines it seems at first glance to be different in that they have declared themselves the natural joint operators of high-speed maglev traffic. Looking into this in a more realistic way, the claim proves to be an illusion, to put it mildly. The truth is more the opposite: magnetic levitation is disruptive both to fast rail and to domestic or continental flights. Introducing maglev implies shutting down fast railway lines to the same destination, which incurs large sunk costs to the railway companies. Maybe politicians and railway CEOs were not aware enough of this, but in the deeper layers of their corporate gut this was the way it was treated, possibly through sheer routine, in path-dependence and mental lock-in. Instead of treating maglev as a strategic investment, it was handed over to unimaginative bookkeepers. It was a big mistake to have entrusted the railway monopolists and the national airlines in place with the task of setting the maglev on track. One could equally have asked a heathen to baptise a Christian.

Plans aimed at introducing high-speed maglev around 1990 were in conflict with plans to implement fast rail since the beginning of the 1980s. Presumably, the French TGV (train à grande vitesse) was tipping the scales in favour of the conservative fix which was ready-at-hand, thus probably postponing the maglev for decades. The French were able to see that magnetic levitation was coming but did not have the new technology available. After the French had launched the TGV, the Germans and Japanese were forced to recognise it as the benchmark of the time. Thus was the short-term logic, even though fast trains have obvious disadvantages compared to magnetic levitation. At a speed of 200–250 km/h fast trains are faster than conventional trains, but much slower than maglevs at 400–500 km/h. Fast trains nevertheless also require completely new rail tracks and perform on eco-balance much worse than maglevs. One cannot fail to remember that 'being earlier is the rival of being better'.

The innovation contest between fast rail and high-speed maglev is comparable to the succession struggle between cleaner carbon and hydrogen/fuelless energy, or the succession struggle between efficiency-trimming of old car propulsion designs and the introduction of FC-powered cars – the one representing the last impulses of a mature system, the other the first attempts of a still-to-be clarified new one.

In the late 1990s, the government of the USA, a continent by itself, decided to examine the adoption of maglev traffic. China, also a large continent and now rapidly industrialising, has implemented a reference route in Shanghai. Whereas Europe has bet on the wrong horse, which will possibly lock it in to the fast railway track for decades, the maglev does not yet seem to be a lost cause elsewhere on the globe.

4.8 UTILITY GOODS IN USE

When considering final materials and products in segments such as clothes, furniture, lamps, office and household appliances, photography, etc., one can draw a long list of many hundreds of products with novel features of environmental intention or effect. In principle, however, and in analogy to buildings and vehicles, the significance of these innovations usually comes down either to the energy design involved or to the choice of materials involved, or both in combination. Social management innovations of the car-sharing type could also be included.

4.8.1 Shared Ownership, Non-ownership, Modular Design

Consumer protection agencies here and there have launched campaigns aimed at motivating people to team up. The rationale is the same as in the case of collective commuting: one washing machine for three families instead of one each; one deep-freeze for all instead of one each. In Swiss blocks of rented flats the old custom of having a common laundry room in the cellar or on the loft has widely survived. It could be revived elsewhere too, and some architect has tried to implement such a collective infrastructure in new buildings. For the most part, however, it does not work well. Most people simply have no real motivation to give up the degrees of individualism and privacy they are used to.

A similar element is to have people rent or lease goods instead of owning things (Stahel/Giarini 1993). The reasons behind this are twofold; on one hand there is the issue of underused capacity, on the other the issue of take-back in the context of product stewardship and recycling. Why have expensive bulky skiing equipment if it can easily be rented on site? Why own a TV set if it can be leased at less cost? Such leasing agreements could include regular delivery and installation of the latest TV model and disposal of the old unit. The leasing approach, however, is ambivalent with regard to the goal of increasing both efficiency and sufficiency. Not only are retailers nowadays prepared to take back an old appliance if one buys a new one, more importantly, the leasing business drives up turnover of new models. In contrast to the idea of longevity and extended use of products, leasing fosters fast throughput. It could be that an important motive behind the notion was in fact the pursuit of radical social reform, namely the introduction of non-ownership, rather than sufficiency and eco-efficiency. No wonder then that concepts of non-ownership or of teaming up in shared ownership are rather confused and deliver poor results.

As an answer, adherents of the non-ownership approach have propagated modular design, aimed at breaking down complex products into a number of components which represent relatively coherent units within the whole. If such a unit or module is broken or becomes obsolete, it can be replaced with a single new module without having to replace the entire appliance. One does not throw away a car if a spare part solves the problem. Here again, a proven principle is implemented in areas where in real practice it is not applicable. Modular design was developed in order to ease the assembly and repair of complex products. As a rule, modules incorporate more elements and materials than would be ideal from an ecological sufficiency point of view. If the controls of, say, a dishwasher do not work for some reason, repair is not carried out by identifying the problem and replacing a switch or a wire, but by replacing the entire controls module.

Furthermore, modular innovation does not work either. New generation products come as a modified system. The modules within the system tend each to be modified because they have to be specifically integrated within the entire product. One cannot fit latest generation seats into the passenger cabin of a 10-year-old car. One cannot exchange the drum of a penultimate generation washing machine against the drum of the latest model. Certain coats for spring and autumn have detachable inside lining, or sports jackets have detachable sleeves, just as there are casual-wear trousers with detachable legs. This was immediately mistaken as an inspired eco-discovery. If the sleeves of a jacket or the knees of a pair of trousers were worn out, one would not need to buy new clothes. Clothes would be designed in modules which are zipped together so that one could zip new sleeves or new trouser legs on to the still usable rest of it. Technically, this is perfectly feasible. In practice it might simply be the cause of general merriment.

4.8.2 Power-efficient Office and Household Appliances

As far as electric appliances or electronic devices are concerned, a main feature of the past industry offensive aimed at increasing eco-efficiency was to reduce the energy demand of such apparatus. In the past 15 years, electricity consumption of freezers, washing machines, TV and hi-fi sets and similar appliances was reduced by 20–40 per cent, an achievement indeed. Green-peace has helped a small manufacturer of refrigerators to launch an 'eco-fridge'. This also consumes significantly less energy in comparison to previous models, partly due to better insulation. In addition, it uses an alternative non-greenhouse-sensitive cooling agent. The cooling agent becomes environmentally relevant after use of a refrigerator when it is disposed of or reprocessed. Established manufacturers, who first had denied the feasibility of alternative cooling agents, followed suit.

New approaches to refrigeration are not only power-efficient but they also represent examples of benign substitution (I/4.4.4). In closed-cycle air refrigeration air as a coolant replaces ammonia or CFCs (which heat up, thus releasing heat, upon compression, and cool down upon expansion, thereby also cooling down surrounding space). To use air as a coolant has formerly not been possible because this requires special compressors which have only recently become available, with a pump shaft that can rotate at 30,000 revs/min. Also sound waves, which are pressure waves, can be used to heat up or cool down substances. This is the idea behind thermo-acoustic refrigeration in which sound-pressurised helium is used to conduct heat away from natural gas, as a step of liquefaction of gas. Finally, there is magnetic refrigeration. Its starting point is the effect that certain substances (strange ones such as gadolinium alloys) heat up when exposed to the magnetic field of strong permanent magnets. Whether these new refrigeration technologies will really be implemented in household refrigerators or air-conditioning systems remains to be seen. Their application in larger industrial plants and processes, however, is likely.

Another innovative product line is low-energy light bulbs. They are sold at a higher price than conventional bulbs, but are several times more long-lived and consume much less energy. The emitted light is just as agreeable as that of other bulbs. NASA has obtained a new patent on an energy-saving non-flickering light tube which produces 96 per cent of the spectrum of sunlight. In many hotels, hospitals, school buildings, offices, etc., energy-saving light bulbs have become state of the art. In private households, by contrast, only a minority make use of these. This may be because in private households running costs are normally not calculated too precisely.

An additional example of energy-saving lighting are LEPs (light-emitting polymers, or light-emitting diodes LEDs) as discussed in I/4.5.1. These are not bulbs, but rather thin and flexible sheets of glowing plastics. Beyond room lighting, they can be used for flexible signs, billboards or computer pull-out screens. Brightness and colour depend on the materials used. There are organic and inorganic LEDs. New organic LEDs (OLEDs) survive up to 80°C for 10,000 hours. They also can produce soft white light suited for indoor lighting. The lighting industry thinks LEDs will replace light bulbs and also fluorescent tubes in the not too distant future.

LEDs are non-flickering, shock-resistant, durable and glow brightly for tens of thousands of hours. Since there is no particular heat-up as in bulbs, LEDs are much more energy-efficient. They also have low weight. Brightness of light is measured in lumen (lm), i.e. sensitivity of the human eye. For the time being the best LEDs emit 25 lm/W, which is twice as efficient as light bulbs are, and a third more efficient than fluorescent tubes. Performance is still improving. LEDs are expected to reach 80–150 lm/W. This is envisaged by around 2010.

Power-efficient appliances are certainly worth the endeavour. One must nonetheless keep proportions in mind. In most developed countries lighting represents just 5–11 per cent of electricity use (15 per cent in Japan) and only 1–2 per cent of total final energy demand. It is clearly more important to have energy-efficient vehicles and heating/cooling systems in buildings.

4.8.3 Product Design and Choice of Materials

Even if the power consumption of utility goods is important, and can indeed be given first priority in the present carbon fuel era, metabolic consistency of materials might in the long run be considered to be of even greater importance. The main task is to eliminate pollutants and hazardous substances in end products. In this respect, many things have already been achieved, although much remains to be done; this is all the more so because with globalisation a lot of low-quality materials are swamping the markets.

Choice of functionally appropriate and aesthetically agreeable materials is in any case an important part of product design. The principles of environment-friendly design tend to give highest priority to the reduction of energy demand. But all of the remaining issues then centre around the question of how to deal with materials and recycling (Tischner et al. 2000 43). My own compilation of ecodesign principles reads as follows:

- Try to be energy-efficient, without however making concessions as regards user functions, performance, reliability and comfort.
- Use materials of high purity and quality which are recyclable, unproblematic if disposed of, and not particularly toxic if they catch fire.
- Reduce materials diversity if possible.
- Long-lived parts or modules should be reusable after some refurbishment.
- Parts should be marked in order to easily identify the material they consist of when disassembled and recycled.
- Develop a design that facilitates disassembly. For example, avoid or reduce glued and screwed joints.
- Avoid reliance on user behaviour in attaining EHS goals as discussed in I/3.3.1, i.e. that products ought to be environmentally adapted, non detrimental to health and safe in operation.

Some readers may miss product longevity as another principle of design for environment. Product longevity has become a widely accepted idea (Stahel/ Giarini 1993). Commonly, long-lived end products are thought to be of less environmental impact than short-lived ones. However, whether long-lived products such as houses, furniture or vehicles actually cause less environmental impact than short-lived products such as food and clothes, is not at all obvious. In general, short-lived goods have a much higher frequency of turn-over, and thus high environmental impact. But long-lived goods, that tend simultaneously to be complex goods, come with a much higher ecological footprint. In the end, the environmental performance largely depends on the product design and the kind of materials involved, for example, whether one is dealing with easily replaceable minerals or more difficult to provide metals; or whether a product has an energy design of high or low impact; or whether the materials can easily be recycled or be harmlessly dispersed in the environment. In the latter case one must additionally ask whether appropriate measures have been planned for, for example by ensuring the pureness and non-toxicity of materials, the organisation of recovery, etc.

The actual meaning of product longevity comes down to a moral call upon consumers to show virtue by being less demanding, or by demanding things less often. This is an ethical principle which represents moderation of consumption and careful use of goods rather than representing an engineering and design principle. The supporters of product longevity have thus been criticised for not being accessible to the innovative aspects of recurrent production cycles. In terms of life cycle analysis, production proceeds in a learning curve. Recurrent production tends to become ever more productive, involving increased eco-efficiency. From time to time an older generation of a product will also be replaced by a new generation variant or even by a new kind of product fulfilling the same function. New generation products or new kinds of products most often tend to be technically improved compared to the previous like product, not only including increased energy efficiency of machines, propulsion or heating systems, but also including improved product quality and improved features of metabolic consistency. For example, running outdated thermal power plants on dirty coal up to 50 or even 70 years, instead of having them replaced by new generation technology after 30 years, contributed much to the devastation of the environment in Eastern European countries until their industrial breakdown towards the end of the 1980s.

Another design aspect that should be stressed in this context regards the choice of materials. Some people tend to think that ecodesign would automatically involve natural materials such as wood or wool. A preference for such natural materials, however, is of an aesthetic rather than eco-functional nature. Artificial materials such as plastics and synthetic fibres, seen from an ecological point of view, do in principle not deserve the bad reputation they have with certain people. Synthetic materials, if properly designed and of high quality, are in many cases ecologically preferable to natural materials, quite apart from the fact that they are more functional and perform better both technically and economically. Even if it were possible to make all utility goods of natural materials, this would certainly not support the conservation of nature, given the vast volumes of materials turnover caused by the sheer number of consumers multiplied by their level of affluence.

Irrespective of whether natural or synthetic, end products or utility goods should as much as possible fulfil all of the EHS requirements. Expressed more generally, they should be on the whole inherently safe. In practice, EHS risks of some sort cannot be avoided completely. This also depends on a given state of knowledge. The question then arises as to whether those risks can reasonably be expected to be controlled.

A case in point is the use of PVC (polyvinylchloride), a chlorinated polymer used for floor coverings, windows, doors, tubes, pipes and interiors. In principle this is a strong, resistant, long-lived, efficacious material. It is also non-toxic under normal conditions, but highly toxic when it catches fire, and slightly toxic when it decays. The toxicity is the result of the release of dioxins. So one could conclude that PVC can be used quite safely in a number of open-air and underground applications, but should not be used indoors. Moreover, products made of PVC must as much as possible be kept in a closed recycling loop and not be allowed to decay uncontrolled in the open.

How far EHS risks related to utility goods are considered to be acceptable will always remain a controversial matter. Scientific findings and personal attitudes towards risk do not always converge. For example, people do not refrain from driving a car, despite the fact that the risk of an accident is comparatively high. The owners of public buildings had to restructure entire complexes of buildings in order to remove asbestos materials. In contrast we are free to smoke and drink hard liquors in spite of the fact that their regular intake is more than a thousand times as unhealthy as life-long exposure to 1,000 asbestos fibres per m^3 of air inhaled. Even the probability of being the victim of a lethal football accident in American schools is ten times higher than that of developing a mesotheliom from exposition to asbestos fibres. This is not to argue in favour of an asbestos relaunch. But risks related to the everyday use of utility goods should be seen in due proportion and treated adequately. To have non-contaminated high-quality food has certainly much higher priority than to keep a distance from switched-on halogen headlamps, or to be brief when using a mobile phone out of fear that their pulsed microwaves might do some damage to the brain (in the same way as they conveniently heat up meals in microwave ovens).

To be sure, if EHS-risks of certain materials or apparatus in use are considered to be too high to be controllable or tolerable, these items should be phased out or their use avoided from the outset. After 20–30 years of debate, experts and the public have concluded that the risks of nuclear fission are too high and that nuclear reactors should accordingly be shut down. At its inception, nuclear technology was largely praised and in no way seen as problematic. With most generic innovations in modern history it has usually been the other way round. They are treated with great suspicion in the beginning, but then become assimilated, selectively integrated, and finally the public becomes accustomed to the erstwhile innovation. Examples of this are railways and car traffic, synthetic fibres, television or the many suspicions which genetic engineering is contemporarily subject to.

A problem which has so far been given little attention are emissions from impure low-quality materials in utility goods (Braungart/McDonough 2002 45–67, 92–102). For example, when a hairdryer or electric mixer is switched on, gas-chromatographic analyses reveal that such gadgets can give off a wild mixture of chemicals, in more than just harmless traces, originating either from the electric motor inside a gadget or from its plastic casing. Without oversimplification, it can be said that at the root of this problem are cheap imported materials of low quality and high impurity, if it is not indeed the entire product which has been imported from some country with low standards. Globalisation here has the effect of replacing high-quality, and more costly domestic materials with low-quality cheap imports. In this case, however, this is not caused by WTO trading rules but by a lack of domestic standards regarding low-quantity emissions of hazardous substances in materials and electric motors which are used in utility goods.

All utility goods are going to be disused sooner rather than later. Impure materials undermine recyclability. Downcycling them cannot even be recommended because accumulation of pollutants in the reprocessing chain is no more acceptable than is accumulation of contaminants in the food chain. Not all materials are food, but materials are feedstocks for some subsequent process, which is not that different after all.

An example of design for disassembly is a recyclable semiconductor plate. It is designed of two or more different plates which can easily be separated when the appliance is disused. The first plate is a conductor plate made of a very thin foil of thermoplastic material. Conventional plates made of one piece can only partially be recycled because imprinted elements contain bromine as a flameproofing agent. Even though there are two or more plates, the entire module incorporates only 20 per cent of normally required materials. And these, after disassembly, are easy to recover and recycle.

Another design idea of late could be addressed as design for virtuality. A standing example of this is a virtual net-based answering machine that can be operated from any telephone by using a PIN. In mobile phones, virtual answering machines are state of the art. They fully replace conventional stand-alone answering machines, or gadgets which are integrated in single desk-top telephone sets. This at last is one example of 'dematerialisation', an almost real one, since there is of course still a large central server carrying out the job for all concerned. This is why the virtual, while indeed being advantageous, must not be overestimated.

Product design for environment and metabolically consistent choice of materials certainly need some framing regulation, especially on environmental standards. But that which promotes them best is entrepreneurial chain management. Those who are in the best position to implement chain management are the producers of complex end products, notably (1) manufacturers of utility goods, (2) manufacturers of vehicles and (3) house-builders, i.e. architects and their clients. It is they who put together all of these hundreds and thousands of different parts from different intermediate manufacturers and materials processors. End producers are the key manufacturers in that they determine by selective demand what kind of product or material is preferred. Just as they engage in negotiations on price and time limits, they can complement the materials side with EHS requirements applying to the quality, i.e. ecological consistency of the materials and intermediate products under negotiation. This also holds true for large retailers. Seen from the side of their counterparts, they are the gatekeepers who have to be passed in order to obtain a place in the range of goods which is on public offer. It is encouraging that quite a number of end producers and merchants have consciously chosen to act as key chain managers – though, of course, demand cannot act in lieu of supply. One can only buy which is in supply. But key manufacturers, in contrast to private consumers, can exert immediate pressure on suppliers to come up with new supply items with improved ecological properties.

4.9 MATERIALS REPROCESSING AND WASTE MANAGEMENT

4.9.1 Regime Rules

Before going into details of TEIs in the realm of materials reprocessing and waste management, the basic rules of a resource-efficient and materials-consistent waste regime will be outlined. These are:

1. Design for disassembly, partial reuse, and recycling
2. Utilisation of pure high-quality materials
3. Refurbishment and reuse of long-lived high-quality parts
4. Industrial symbiosis on the basis of combined production processes
5. Recycling and downcycling of products and materials

6. Composting/fermentation of biotic final waste
7. Thermal treatment of final waste, including hazardous waste
8. Depositing of mineralised, chemically non-reactive waste only. For the remaining types of waste: the closing down of dumping sites.

Regime rules (1) and (2) seem to be of the utmost importance. They lie upstream in the product chain and a product's life cycle, before reprocessing actually starts. Products and their materials content ought to be reusable or recyclable in the first instance. If handled as waste they ought to be treatable and renaturable without too much effort, and disposable without causing harm. Downcycling and thermal treatment ought to be avoided as much as possible. This can only be achieved by a corresponding design of products, including utilisation of high-quality materials.

It is the designers and producers, particularly of end products, who are in a position to assure attainment of these goals rather than reprocessors and waste managers who can do nothing more than responsibly deal with what is handed over to them. If products and materials were properly designed, many of today's recycling and waste problems would be significantly reduced.

4.9.2 Refurbishment and Reuse of Products and Parts

The most widespread and certainly best known examples of reuse, actually circulatory use of products, are deposit bottles, plastic crates and gas bottles, i.e. containers made of plastics, glass and metal. These are long-lived materials which can last through dozens and hundreds of reuses. Glass bottles, though, may only endure 15–25 times as a result of mechanised cleaning. Another example of reuse are the casings of disused telephone sets which are refurbished and then integrated in new sets.

Refurbishment of long-lived parts and reuse of these in products or production processes has a certain tradition among second-hand dealers. A new example of this was given in the chapter on building, where long-lived construction parts made of reinforced concrete can be refurbished and reused. In traditional building, reuse of wood beams, ashlar blocks and bricks was current practice. There has also always been a market for second-hand spare parts of machines and vehicles.

Machines and entire production plants can have a useful second life by being taken over by domestic manufacturers or manufacturers in developing countries who can well do without the latest-generation model. A particularly striking example of reuse are control devices such as speed meters, water meters or electricity meters. Eighty per cent of these can be used a second time at the same level of accuracy and quality. In the context of current take-back and recycling policies, automobile manufacturers are thinking anew about how to reuse parts of automobiles and motors instead of sending them back for materials reprocessing chain-upwards. Refurbishing and reuse of product parts can save raw materials and primary materials in a range of about 40–80 per cent. This includes corresponding amounts of natural resources, process energy and auxiliary chemicals. There are, of course, limits to reuse, similar to the limits to modular design. Life cycles of products quite often tend to be relatively short. A newly calibrated old water meter can easily be reused, not, however, the speed meter from an older car in a new one. So the time perspective of reuse is normally rather limited.

Where applicable, however, reuse should be given preference to materials recycling and downcycling, because finished parts represent a higher level of structuration than crude input materials. It is important to systematically identify those objects where refurbishing and reuse makes sense, and to integrate these useful second-time objects into the life cycle of a product. Furthermore, gathering and refurbishing of used parts has to be organised, which is by no means a trivial matter. Last but not least, there has to be active marketing of refurbished parts.

4.9.3 Industrial Symbiosis on the Basis of Combined Processes

Follow-up use is different from reuse in that byproducts and waste products from production processes are going to be used rather than disused products or parts thereof. Also follow-up use is indeed a long-standing practice, and has recently been relaunched on ecological grounds under the heading of industrial symbiosis. The principle of industrial symbiosis can be described as cascadiç re-entry of waste products or byproducts in subsequent production processes on-site and inter-site.

Among well known examples of industrial symbiosis is the secondary use of steam from power plants for heating of buildings and greenhouses in surrounding areas, whereby the greenhouses can also absorb some of the CO_2. In Kalundborg, Denmark, four big companies and a number of small businesses utilise each other's residual products in a network on the basis of bilateral contracts with freely negotiated prices. They are the Asnaes power station, Gyproc, a plasterboard producer, Statoil refinery, Novo Nordisk, a pharmaceutical and biotechnological group, and greenhouses and fish farms. Among the residual products exchanged are waste water and cooling water, steam, heat, gas, sulphur and gypsum.

In the context of the recent environmental discourse, industrial symbiosis has become closely related to the idea of zero-waste production flow, which is achievable if all outputs are inputs for something else. Hence projects such as the Zero Emissions Research Initiative of the United Nations University aimed at 100 per cent recovery of the carbon dioxide emitted during the brewing of beer. The idea is that of an inter-site industrial symbiosis where waste streams from brewing, aquaculture, fish processing, greenhouses and algae production will feed on each other.

The principle was formerly also known as combined production processes, a fundamental principle in the chemical industry ever since it came into existence. The combined processes principle was also at the core of production planning in centrally planned socialist economies. There are certainly economic and ecological benefits to be experienced from combined production or industrial symbiosis, but there is also evidence for undesirable inflexibility or lock-ins (for example difficulties in eliminating chlorine chemistry). Once such a structure of combined processes has been installed, it is difficult to change one element without repercussions upon the others. So, in practice, industrial symbiosis also has its limitations.

An example of cascadic use was given by Braungart (2002). Animal residues, as far as not used for human food, are an almost ideal feedstock into five follow-up production processes. The first and most valuable is to produce meal of animal protein as a feeding stuff for animals. A second best use is the production of amino acids for the chemical industry. Third, animal residues can be used to produce fertilisers and synthesis gas, and fourth, they can be put into fermenters to produce biogas. Fifth, which is the last, and least desirable option, it can finally be burned, thereby obtaining heat and electricity. The various steps represent different levels of structuration, from low to high entropy, or to put it in other words, from higher to lower organic molecules and then to smallest inorganic molecules. Organic materials especially should of course be kept at the highest possible level of structuration. To incinerate animals for fear of infectious content, without any intermediate selection and use, is an admission of total defeat in the face of agroindustrial processes which are seemingly out of control.

4.9.4 Collection and Separation of Waste Materials

Waste materials which are not subject to immediate combined uses have to be collected, either separately or in mixed containers. If in mixed containers, waste has to be separated in order to be properly reprocessable.

Collection is distinguished according to client-delivery systems and collector-pick-up systems. In the end, all systems are a combination of decentral bringing and more centralised picking up of waste. Materials reprocessing then starts with separation. Neither collection nor separation of waste presently represent mature systems. Important changes can still be expected.

Consider collection and separation of residential waste. How many bins are there in a household? How many would be required? Where are the bins located? Future systems of household waste management may continue to separate paper, and also batteries and other hazardous household waste. In contrast it may not be necessary to separate glass, plastics, metals, earthenware, wooden materials and bio-waste from urban households. In principle, all waste of this kind can be put in one container.

A technology which can deal with the one-bin-for-all approach is the dry-stabilate method. It was developed and launched by Herhof in 2001 and is being adopted by a number of municipalities in Germany and Italy. The bins are picked up as usual, and the content is dried in a central plant through an automated biological process. Within six days microorganisms eat up the leftovers in the mixture, thereby creating heat which dries the waste. The remaining dry waste is then shredded, whereby all of the useful content materials such as plastics, glass and metals are separated, even in rather small quantities, and supplied to reprocessors. The residual substrate, the final dry-stabilate, just consists of small pieces of plastics, paper or textiles. Its energy content is equal to that of brown coal. It can be used in any suitable combustion pro-cess or as an additional fuel in refuse incinerating plants combined with heat and power generation. Only the remaining inert slag or ashes have to be disposed off, if they are not used as an aggregate material in building. The dry-stabilate method can save waste management companies, public and private alike, considerable investment sums, while citizens can benefit from much lower waste charges. One would not have to bother any longer with those various bins. The use of such bins is an example of unpaid work to the benefit of companies which themselves work on the basis of imbursement for work done. Its value has always been more of an educational rather than of a substantial nature.

In waste separation, there is a general trend towards semi or fully automated sorting plants. Both municipal waste and vehicles and appliances can be taken back by industrial companies or their joint recycling subsidiaries. Different types of waste and different purposes certainly require different sorting plants. In the past 10–15 years a series of components of automated waste separation have been developed. An important part of this is automated materials recognition by optical, sonic and other high-tech sensor systems (I/4.5.4.3). Other parts include sophisticated mechanics, integrated and controlled by equally sophisticated software-and-actuator systems. There are various demonstration plants in most industrial countries. The development of disassembly plants for vehicles, appliances, e-scrap, old furniture, other bulky waste, disused shoes or textiles is heading in the same direction. The bottleneck for the time being is adoption by main users, notably municipal or regional waste management companies. In any case, the men with the ill-fitting gloves will not have to stay at the conveyor belts forever.

4.9.5 Recycling and Downcycling

In the early days of the ecology movement recycling was one of the core responses to the problem of overcoming wastefulness and environmental deterioration. Today, any assessment of recycling will be more differentiated and will also include criticism, particularly with regard to downcycling. Materials have different degrees of recyclability. Glass and earthenware can in principle perpetually be recycled at almost 100 per cent, as is the case with metals presupposing that they are relatively pure and do not contain too many other metals of a different kind. By contrast, fibrous materials such as paper and textiles, or fibre-like materials such as plastics (polymer chains) are of limited recyclability. When reprocessed, the fibres or molecular chains break into shorter pieces. This results in a material of lower quality. So these materials can be recycled only 3–6 times, and primary fibres have to be added to the cycleware in order to maintain quality.

In practice, any materials recycling is in fact downcycling, though at a different speed and within different time frames depending on the particular material. Aside from the decline in quality, downcycling quite often also involves an accumulation of hazardous substances, similar to the accumulation of contaminants in the food chain. A telling example is chemical solvents. In most cases, where specific purity requirements do not apply, solvents are nothing other than mixtures of recycled waste chemicals. Downcycled materials tend to accumulate and enrich the pollutants handed down from primary materials. The problem would of course largely be reduced if primary materials were completely free of hazardous ingredients. Envisaging 100 per cent recycling, another way of saying zero waste, is nonetheless a vain hope, besides being unnecessary. The important thing is metabolic consistency, i.e. to re-integrate the industrial metabolism into nature's metabolism, not to cut it off and live in a self-contained world.

Materials recycling begins with in-process recycling, or microrecycling, in a closed production loop. Typical examples of this, which do not even entail particular reprocessing, include re-entry of residual dough in baking, or re-entry of excess metal in smelting and casting, or the reuse of filling and cover materials in the production of quilted blankets and jackets. This type of in-process reuse results in a 90–95 per cent reduction of production waste which would otherwise occur.

In-process recovery of valuable waste content, by contrast, requires some more effort. A case in point was the silver which flowed off with the effluent from photo developing baths. Only in the 1980s were methods introduced to recover this valuable ingredient. Examples of recovery of hazardous waste content, which is nonetheless valuable, include in-process recycling of solvents, sulphuric acid, chlorine and other auxiliaries in the chemical and pulping industries.

Inter-site recycling, or macrorecycling, sends materials from production processes and disused products back to the foot of the production chain where it becomes secondary raw material. In every case, this must not be of lesser quality than primary materials. For example, top quality amino acids and peptides can be obtained from waste feathers from poultry slaughterhouses and waste down. The job is performed by *Fervidobacterium pennivorans* (the literal translation of which is feather guzzler) and *Thermoanaerobacter keratinophilus*, or by enzymes thereof. This resolves an expensive waste problem and replaces chemical hydrolysis.

Recovery of secondary raw materials by macrostructural recycling is organised on local, national and international levels. Certain green fundamentalists demand that waste management be kept within local or regional boundaries. There is no sensible reason for this. Cycleware represents a quite normal category of products, as recycling represents quite normal production processes. These work best, i.e. find themselves an optimum position within the manufacturing chain, when they are mediated by a free market like any normal production process. The only acceptable exception to this rule is hazardous waste. But this is regulated, even at the international level by the Basel Convention on the Control of Transboundary Movements of Hazardous Wastes and their Disposal from 1989. It is illegal to export hazardous waste to destinations which are not equipped to deal with it.

The long-term life cycle of the recycling industry is taking off. Among OECD countries the Netherlands, Belgium, Austria, Switzerland, Denmark, Finland, Germany and the USA now reach quotas of recycling and composting of municipal waste in the range of 25–30 per cent. The other countries do worse, especially the Mediterranean countries, Ireland and Japan (OECD 2001b 7.3). Recycling quotas of useful materials such as paper and glass are nonetheless on a steady rise in most countries, having now reached 50–70 per cent in Japan, Denmark, Austria, Germany, the Netherlands, Sweden, Switzerland, yet only about 40 and less per cent in the US, France, Belgium, Italy, Spain and the UK (OECD 2001b 7.4A).

Recycling quotas of metals, which are valuable, are in general much higher. The example of aluminium is particularly impressive. In the EU, it is recycled to 80–95 per cent. The share of secondary aluminium is now at about a quarter of all aluminium, up from very little 20 years ago. The highest recycling quotas are obtained in vehicles, construction, machines and the electrical goods industry. Even aluminium from composite packaging materials is recycled to 40–72 per cent.

In all cases, secondary raw materials and cycleware save significant amounts of natural resources, energy, auxiliary chemicals and fresh water. In most cases, recycling also saves money, except with plastics, where recycling is often more expensive than producing primary material from cheap petroleum.

High recycling quotas can only be achieved with the help of advanced methods of recovery. The same or similar technologies which are applied to sorting and separating municipal waste also help to separate various factions of waste materials which were collected through different channels, and to recover or extract the valuable factions thereof. For example, e-scrap has been a problem until recent developments in recognition or detection technology. An average nation's e-scrap is easily worth several billions of euros a year because of its valuable metal content. An advanced sorting and separation plant can recover rare metals such as gold, silver, copper, platinum and aluminium from e-scrap by colour detection of mechanically crushed pieces. Such a plant first shreds, crushes and separates the various factions of e-scrap mechanically. Crushing, by the way, is in ever more cases done by water as an environmentally benign auxiliary. A high-pressure waterbeam can crush carpets, floor materials, fibre compound materials, tyres, etc. Also metals can be cut quite neatly by such a beam. The small pieces then pass an optical detector which recognises metals according to their specific light reflection, i.e. colour. These pieces are then removed by targeted jets of compressed air.

Recycling of building materials and construction waste is now also being more systematically organised. The construction industry and the building materials industry are setting up cooperation networks, creating technical infrastructure and employing new machinery aimed at reprocessing building rubble of any kind, including steel, aluminium, wooden materials and parts made of plastics. This not only saves vast amounts of primary building materials, but also the high cost of large volumes of waste disposal. Officials say it might be possible to cut present levels of construction waste, which have come down already, by half within 10 years.

Some of this recycling, however, would have to be examined critically. Is the quality of materials acceptable, or are bricks and similar materials tainted with residues of plastics and chemicals? What is chipboard made of? If it is made from waste furniture, this certainly saves a number of production steps, for example 65 per cent heat and 45 per cent electricity in comparison to the production of conventional chipboard, but it may also incorporate a number of hazardous chemicals with which the furniture was treated and coated.

Similar things can be reported about the recycling of plastics, textile fibres or leather. Disused tyres, for example, can be seen in questionable reuses such as construction elements in sonic barriers along roads, possibly covered with soil, and whatever other tragicomic inventions there may be. It would be preferable to recycle tyres by recovering their soot content. This can be done by depolymerisation of old tyres in liquid tin below 500°C. In this way the creation of pollutants is largely avoided. The process results in fine soot which can again be used in the production of rubber for new tyres.

Plastics are recycled, actually downcycled today on a regular basis. Downcycled plastics are certainly acceptable in many cases, even though not all of these satisfy good taste. After two or three downcycling loops, the waste plastics can be burned or liquefied anyway as a fuel which is equivalent to the petroleum it was originally made from. It must be ensured, however, that such downcycling does not involve dissemination of too many hazardous ingredients. Among the more blatant cases of this was leather shred which was recycled as floor covering in a riding school. Soon thereafter, the riding teachers fell ill. The reason was that the continuous stamping of hooves on the leather floor covering created a lot of dust which contained significant amounts of hazardous chemicals such as chromium, used in the tanning and dyeing of the leather. Purity of materials is indeed an outstanding requirement of design for environment.

Recycling does not represent a virtue in itself. It is useful insofar as it increases resource efficiency, particularly of those resources which are scarce or difficult to come by. Most of the more valuable materials tend to have a considerable ecological footprint or materials rucksack. Increased resource efficiency through recycling thus helps to minimise human invasiveness into natural ecosystems. Non-invasiveness itself, however, is by no means an uncontroversial criterion. The crucial aspect is that materials be metabolically consistent, pure and of the highest possible quality. Recycling of impure low-quality materials, by contrast, represents an EHS problem in itself.

4.9.6 Treatment of Solid Final Waste

When products made of primary or secondary materials are disused and will no longer be recycled, then, according to current practice, they are either deposited in a landfill or burned in an incineration plant before the resulting slag and ash is deposited. In historical perspective, waste dumping in landfills or at sea is a rather primitive production practice comparable to gathering, hunting and fishing in the wild. It was only in the 19th century that civilised town dwellers stopped dumping their bodily and other waste in their backyard or even in the street in front of their door. Since then, waste dumping has been concentrated in sites outside the town gates. This no doubt made an immediate difference to town dwellers, but in historical perspective it might not be all that significant.

It has been argued that landfills, in terms of square kilometres, only cover a very small part of the entire land. This is a negligent point of view, like saying 'don't worry about small doses of poison', the more so since landfills are not concentrated in one single place but widely scattered over a territory. Even if all deposited materials were easily bio-degradable within an appropriate time horizon and did not set free gases and toxic substances upon decay, it is the sheer amount of waste, and valuable materials content, which makes dumping a rather dumb option. In reality, much of the waste is environmentally problematic, and geological conditions in certain locations actually seem to preserve rather than to decompose waste.

So waste regime rules (6) to (8) from I/4.9.1 must be strictly maintained. These state that waste for which there is no better use and which can in no useful way be recycled, must be treated before being deposited. Treatment can be biological for the biotic faction of waste, i.e. composting or fermentation, preferably inside a closed plant or in a biogas plant, in both cases with recovery of the ensuing biogas. While biogas from biomass cannot in general be considered to be an optimal eco-consistent approach, it can make sense in certain special cases such as agricultural biogas plants. Another such special case is treatment of biotic final waste, including co-fermentation of final bio-waste from various sources (such as sewage sludge, manure, residues from food processing, and the biotic faction of residential waste). The solid product is a valuable substrate for soil melioration and fertilisation.

All of the remaining final waste, hazardous and non-hazardous, must be treated thermally. The conventional way of thermal treatment is incineration with combined heat and power generation. This reduces the volume of waste and renders its residues chemically inert. Where applicable, waste and hazardous waste can alternatively be incinerated in industrial blast furnaces, for example in the production of pig iron. This, however, requires waste of high energy content such as sewage sludge, sludges from coating, dyeing, varnishing, contaminants-charged activated coke or plastics granules. In such cases waste is a valuable substitute for regular fuels. There are no quality problems with the iron and no additional problems with emissions control. The waste owner can even sell the waste (for less though than the price of regular fuel) rather than having to pay waste fees to the incineration plant.

A special problem with the burning of sludges of any kind is that they are wet and have to be dried before incineration. When done by heating, this consumes large quantities of energy and is thus cost-intensive. Two alternatives of late are solar and mechanical drying. Solar drying takes place on the concrete or asphalt floor of a transparent, glasshouse-like hall. A push-turn device running on tracks continually revolves and turns the substrate over. Solar drying does not pose hygienic or odour problems. Energy requirements are very low at 25 kW per ton of dried up water. Solar drying thus is cost-efficient, though not as cheap as mechanical drying of sludges. This is done by press-wringing. Wet sludge from sewage plants or other industrial processes (such as phosphate sludge from the automotive industry, or pulp and paper sludges, hydroxide sludge or sludges from grinding and varnishing) enters a cylinder-like mechanical press that wrings the water out of the sludge. The trick with this method consists of special filters and membranes.

Alternatives to waste incineration are various approaches of flameless thermal treatment with recovery of valuable content materials. The latest innovation of this kind is the dry-stabilate process as described in I/4.9.4. Earlier technologies of the 1980s–90s are based on pyrolysis, i.e. the thermal cracking of chemical compounds. One such pyrolytical approach is the smouldering-burning process as developed by Siemens. It carbonises waste in several successive steps at relatively low temperature, whereby metals and glass can be recovered. Something similar is achieved by the Italo-Swiss Thermoselect plants.

After such treatments, there is always some kind of inertised slag and ash, which is chemically non-reactive under ambient conditions. Mineralisation in this form is a kind of re-naturation which seems to be acceptable according to the present level of knowledge. It is important that the depositing of non-treated, non-selected, non-mineralised waste is ruled out. As a consequence, all conventional dumping sites, even if equipped with sealing layers, covering, drainage and aeration pipes, would successively be closed down. There would have to remain some landfills for well sorted, not-yet-recyclable mineral construction waste and for the slag and ash of waste incineration if this cannot be processed and recycled as an aggregate material to mineral products. Such a waste regime is a major step towards closing the industrial metabolism and towards its reintegration into the metabolism of nature. It is furthermore one of the issues on which environmentalism and conservationism concur.

Compared to prevailing practice, complete treatment of final waste still is far from being a widespread reality. In contrast to hazardous waste, which has to be incinerated in special plants in all industrial countries, municipal waste still tends to end up in some landfill. Only in Denmark, Switzerland and the Netherlands has the percentage of deposited waste been reduced to below 25 per cent. In the UK, Ireland, the USA, Germany and the Mediterranean countries it is still above 50 per cent. Also incineration, not to mention flameless thermal treatment, is far from being common practice. Only France, Denmark and Switzerland have incineration rates of about 50 per cent, Japan even 70 per cent. In most other countries, including the USA, the UK and Germany, the percentage of incinerated waste still is below 20 per cent (OECD 2001b 7.5) – which leaves ample space for innovative initiatives.

4.10 EMISSIONS CONTROL (AIR, WATER, SOIL/SITES)

Many of the environmental technologies aimed at end-of-pipe treatment of exhaust gases, downstream treatment of effluent water and ex-post remediation of contaminated sites were developed as early as the 1960s–80s, so not many frontline innovations in this realm can be expected in the future. Nonetheless there are some. And there are many new generation variants, new system components and incremental innovations of established technological paradigms.

4.10.1 Exhaust Air Purification

The latest, and probably most important, innovation in exhaust gas purification is CO_2 sequestration. This has already been described in I/4.2.3 in connection with the clean coal strategy. Exhaust air purification in vehicles has been discussed in I/4.7.3 in connection with survival strategies for internal-combustion engines.

In power plants and industrial furnaces the key components of air purification are removal of airborne particles, desulphuration, denoxation and removal of volatile organic compounds (VOCs). Airborne particles such as dust and soot are removed by electric filtration. This has reached the stage of a mature technology with some incremental refinements such aerosol microfiltration. This can be carried out by a corona-aerosol separator which especially filters out oil aerosols and ultra-thin particles in the range of 10–2.5 micrometres, some even below 1 micrometre. The particles are electrically charged up to 10,000 volts when passing a nozzle plate, and are then captured electrostatically. In this way 99 per cent of ultra-thin particles are eliminated. This is of particular relevance in metal processing where chemical lubricants and coolants are involved. In addition to improved exhaust air quality (which is actually cleaner than ambient air) the new process increases energy efficiency by a factor of 10 in comparison to older generation electric filtration.

Desulphuration of flue gas is conventionally done by a wet process which includes limestone powder, ammonia or sea water. The typical byproduct of desulphuration is gypsum. Dry desulphuration as an alternative is more efficient and less expensive. The dry process, though, needs a special desulphuration additive which activates the limestone in the flue ash. The additive is recovered from industrial residues of thermal power plants and furnaces in steel production.

In a number of production processes, advanced membrane-type filtration as developed in the chemical industry (I/4.5.4.2) has successfully been applied to off-gas purification with recovery of auxiliary agents. In this way hazardous or valuable substances such as vinylchloride, hexane and dichlor-methane, can be removed, or recovered respectively. Similarly, benzene can be separated from gasoline gases, resulting in less toxic gasoline at the point of use. This is done by pervaporation which is less energy-intensive and cheaper than comparable conventional methods.

Nitrogen oxides in exhaust air are nowadays routinely broken down with the help of catalysts, such as the three-way catalytic converter in internal-combustion motor cars. Catalysts also help to dissolve VOCs and CVOCs from solvents. But as these are more complex there are various approaches. One of late is the use of porous burners as described in connection with clean-burn technology in I/4.5.4.1. In a flameless thermal oxidiser, hazardous VOCs and CVOCs in an exhaust gas are converted into CO_2, water and hydrogen chloride. This occurs in the central reaction front inside the loosely packed porous ceramic body. The method can be applied to off-gases from process and waste streams, and also to site remediation on- or off-site.

Additional approaches are UV light and biocatalysis. The UV approach has for example been implemented by the Phoenix™ apparatus. This is a 3-module apparatus developed for application in the printing industry, but easily applicable elsewhere. Module 1 pre-filters dust and steam, module 2 is a catalyst, and module 3, the centrepiece, is a UV oxidation unit which decomposes VOCs into water and CO_2. Biocatalytic removal of organic compounds from off-gases with the help of microbes is in principle as effective as flameless oxidation and UV oxidation. However it involves longer reaction times. This may be compensated for by lower cost.

4.10.2 Sewage Water Purification

In effluent purification, 3-phase plants with integrated ammonia stripping have been the best available technology for quite some time, although still not implemented everywhere. Phase 1 is for sedimentation. In phase 2 the sewage water is mechanically recirculated and aerated. Phase 3 is the biological phase where microbes transform the organic waste content.

New components in this overall framework, similar to previous realms of materials processing, include advanced membrane-type filtration processes, application of combined UV and ozone treatment, ionisation and significant improvements by newly bred microbes in the tertiary biological process.

New membrane-type filtration technologies of various kinds are especially applied for removal of organic compounds. Ceramic membranes, for instance, have proven to be effective in treating sewage water from the production of isopropanol and ethanol. Cellulose-stainless steel-membranes seem to be best at eliminating dyes from sewage water.

Ultrafiltration utilises low pressure at one side of a semipermeable diaphragm, causing substances at the other side to pass through. The diaphragm will differ according to the specific purpose. Ultrafiltration has turned out to be very suitable for purifying sewage water from dairy production and other processes with a high degree of organic pollution of sewage water. This results in a 25 per cent reduction of sewage water, 75 per cent less CSB, 60 per cent less nitrogen and 75 per cent less energy requirements. When applied to the waste water from the cleaning of chemical tanks, whether stationary or mobile, ultrafiltration results in a 95 per cent reduction in salts and chemicals residues in the water compared to earlier methods. Even the parameters of heavy metals, AOX and CSB are all close to detection limits. Special gels are also used for filtration purposes. They absorb high-molecular substances. Apart from large plants, advanced filtration technology is also usable in small decentralised units

Water has normally been disinfected with large amounts of chlorine, for example in swimming pools. Chlorine gas, however, causes irritation of skin and eyes, and one of its dangerous byproducts, trihalogenmethane, is carcinogenic and mutagenic. A healthy clean alternative to this is electrolysis of saline solution (brine). An equally advantageous way of doing away with chlorine is water purification and disinfection by a combined application of ozone and UV light. A further approach to water disinfection is ionisation of silver or other agents.

Biocatalytic purification with the help of bacteria and other microorganisms appears to be the field where innovation activities in sewage water treatment are most highly concentrated. New breeding generations of special GM bacteria come with much higher decomposition rates of organic pollutants. As with diaphragms and membranes, special purposes require special microbes. For example, purification of fatty sewage water from food processing (fish and meat industry, large kitchens, producers of ready-to-serve meals) is a task to be left to thermophilic microorganisms above 50–60°C. They deliver a hygienic output and do it highly efficient.

Effectiveness and efficiency of the bacteria can be further increased by their combination with some already existing technical element. For example, brown coal coke (activated carbon) can be used as drop feed for bacteria. The coke, which is porous, provides a much enlarged reaction surface. This enables quick throughput even of heavily contaminated industrial and dumpsite sewage water in larger volumes, if necessary repeatedly, with subsequent normal purification. The combination of coke and bacteria results in higher performance, shorter treatment and significant cost savings. Similarly, bacteria can be combined with advanced membrane-type filtration. In this combination, the bacteria form the main element of some multi-layer filtering arrangement.

With the development of advanced filtration technology and biological treatment, the linear 3-phase arrangement of water purification was overlaid by a number of feedback-loops between the three processes. This has in turn led to the development of sequential reactor plants. This means that biological treatment (tertiary process) and sedimentation (primary process) occur in one reactor. This saves an entire reactor, and is nonetheless reported to deal more flexibly with changing demand.

An approach with a certain green appeal is plant-growth purification stations. These let the sewage water simply pass at low speed through some moist biotope which is composed of an appropriate combination of plants which absorb or transform the contaminants in the water. Such stations can indeed be effective. But they require extensive knowledge and know-how. They can deal with small amounts of residential sewage water, less so with highly concentrated industrial and urban effluents. They can, furthermore, only treat limited amounts within a given period of time. So plant-growth purification stations might be useful on a small scale and in rural areas.

In addition to central purification plants there are now also small purification devices for decentral applications. For example, there is the prototype of a domestic sewage water purification tank, for use in private households and small businesses. It consists of a set of catalysts and aluminium electrodes fitted inside a small tank. When powered up, the fatty particles shed their negative charge and begin to clump together. The hydrogen which bubbles off one of the two electrodes carries the coagulated fatty matter to the surface. There it can be skimmed off. The resulting water is clean enough for any non-drinking purpose. This is of interest to restaurants, canteens and food processing firms which must operate under special hygienic conditions. The tanks are expected to reach the market by 2005.

Another approach to implementing decentralised small purification units is microfiltration through biomembranes. These partly hold back and partly transform pollutants, and eliminate germs. Such a unit is intended for use in isolated houses without access to a sewerage system, or to avoid the expense of connecting scattered houses to this system. The unit itself can be placed in the cellar, whereas the gas exhaust would preferably be located on the rooftop. The purified water can be used in the washing machine, or in the garden, or be discharged into the draining ditch. There is no sewage sludge, and energy demand and maintenance requirements are low.

4.10.3 Site Remediation

Site remediation refers to the clean-up of contaminated industrial brownfields such as abandoned production sites, waste deposits, landfills and deposited mining overburden. Contaminated brownfields are a major nuisance in industrial development, as a dirty heritage of metabolically inconsistent production, as well as an effective deterrence to redevelopment and follow-up investment on brownfields. Public and private investors shy away from brownfields because of the environmental and health risks involved, and especially because of the high amount of capital which has to be devoted to brownfield remediation.

Among the conventional approaches was to entomb affected soil in a concrete coffin built around the zone from all sides and below. This would appear to be a monstrous idea originating in the construction business rather than a solution to the problem.

The conventional effective approach is thermomechanical decontamination. Affected soil is placed in rotary kilns where the pollutants are cracked thermally. This has to be done off site if the kilns are stationary. It can also be done on site in mobile kilns. The soil has ultimately to be removed in any way. Thermal treatment off site can be more profitable due to the larger kilns, but involves more and longer transport. Thermal treatment on site can only use smaller kilns but does not involve long-distance transport of soils. The big disadvantage of thermomechanical decontamination, besides being very expensive, is that the soil obtained is a biologically dead mineral substrate. It takes many years, if ever, before such soils, which are profoundly burned out, recover to a normal state of fertile soil.

Chemical decontamination applies in special cases such as accidents on construction sites, or transport accidents on railroads and motor roads, also on sea (oil spills). Depending on the substances involved, various chemical agents are applied which neutralise escaped hazardous substances.

A special case is chemical neutralisation of acid mining overburden with the help of Bauxsol™ or 'red mud', an Australian invention of late. Bauxsol™ is made of caustic leftover from bauxite refinement, diluted in sea water. It is applied in the form of fine-grained powder which is added to metals-laden wastewater or soil. It is a robust and simple method, relatively cheap and effective.

Biological soil remediation then is more particular again. An approach with long-standing traditions is phytoremediation, i.e. growing on a site special plants which absorb, transform and neutralise pollutants in the soil. The difficult thing is to identify the right species for a specific task. For example, *Arabidopsis thaliana (GM)* eliminates the toxin methyl mercury in contaminated soils. The plant transforms ionised mercury into its elementary state. This is then released into the air (which is not problematic because occurrence is in small traces only which become widely dispersed and reabsorbed). Phytomining is limited to surface contaminations since the catchment depth of the plants depends on their roots. A general disadvantage of phytoremediation, at least from a landowner's or investor's point of view, is the relatively long time it takes to do the job. But if the right species are chosen and combined, phytoremediation is effective and cheap.

Microbial bioremediation is, of course, also under intensive development. In the case of oil-spills on land, also on sea after tanker accidents, there is now a series of microbial agents commercially available, as an alternative to chemical agents. On land as on sea, however, the microbes need to be given the time which is necessary to absorb and digest the pollutants. Neither on highly frequented traffic ways nor under conditions of stormy weather on sea is this normally the case. Under normal conditions, microbes can be effective crackers of organic contaminants and absorbers of inorganic pollutants. At the Idaho Chemical Processing Plant of the American Department of Energy a bacterium, *Deinococcus radiodurans (GM)*, has been tested which even transforms radioactive chemical waste on site.

A viable alternative to thermomechanical treatment is high-pressure aeration of contaminated soil which has been mixed with biological and mineral additives. The contaminated soil has to be removed as in a thermomechanical process. Suitable bio-additives are compost, mulch, straw, chaff, while suitable mineral additives are crushed rubble or perlite, which is porous and thus creates inner surface expansion for bacteria. Together with nutrients and water these additives are mixed into the contaminated soil and the resulting substrate is piled up in a heap of 4–5 m height, the size which is necessary to be economic. The substrate, according to content, can be kept stable or must be constantly turned. In addition, air is injected into the heap under some pressure, because the microorganisms need sufficient quantities of oxygen in order to efficiently decompose the different hydrocarbons that represent the contamination. The decontaminated soils are immediately reusable. The bio-additives result in biologically viable soil, self-recovering within a relatively short period of time.

This is supported by new highly efficacious methods of planting greenery which were developed in the construction of traffic ways and strip mining. One such method is to splash water from a hose filled with seeds, fertilisers and wood fibres from residual timber. This method is of particular advantage on steep or otherwise difficult ground.

In strip mining, there is a wealth of experience with possibilities of recultivation and reforestation of shut-down mining areas and slag heaps. This knowledge is also applicable to otherwise excavated land (such as newly developed sites, new traffic ways), or desertificated land, or farming fields which were devastated by over-intensified agroindustrial use, or areas which were clear-felled and then abandoned. Depending on local geological and climatic conditions, some soil transfer, special combinations of plants and a quite long time horizon are all necessary to successfully set up a self-supporting process of plant covering in specific successions of vegetation.

There are now also ecosystem services whose business is to restore damaged habitats by implementing a near-nature succession of plant and animal societies (biocoenoses). For example, erstwhile farmland in the American Midwest will by itself not revert to prairie again because previous tillage has created homogenous soil which is good for certain crop monocultures such as maize, not however for plants living in the wild.

Since soil quality, soil moisture and regional water conditions are closely intertwined, such services are simultaneously linked to regional water management. If a water catchment area is covered to more than 20 per cent with asphalt roads, car parks, residential and production sites, the waters draining away from these areas transport a high content of certain bacteria which are harmful to the life in natural surface waters, including health hazards in bathing areas. In rural areas it is frequently recommended, because cheaper and ecologically more consistent, to restore previously drained wetlands or to construct artificial percolation polders rather than to catch the drain water in pipe systems and purify it in plants.

4.11 MEASURING AND MONITORING

Metrology refers to methods and instruments of detection and measuring of substances, in an environmental context especially toxins, contaminants and pollutants. Its importance tends to be underestimated. Measuring and monitoring is certainly not an end in itself, and does not by itself bring about efficient and ecologically consistent TEIs. But metrology-based analyses and environmental monitoring involves a large number of methods and instruments which serve as aids to identifying physical, chemical and biological content, and to spotting irregular occurrences. These analytical aids are indispensable control devices, both within industrial processes and on different territorial levels. Without them, much of the technical progress would not be possible from the outset, particularly not in those areas where the main frontline of today's progress is to be found, i.e. on micro- and nano-scale levels.

4.11.1 Metrology

State-of-the-art metrology most often includes spectroscopic and chromatographic analyses. While these methods are to a degree almost traditional nowadays their potential does not seem to be exhausted yet. Spectroscopy is based on the fact that each atom and its isotopes have a characteristic spectrum of electromagnetic radiation which can be broken down into their constituent colours or wavelengths.

Recent examples of new devices based on spectroscopy include the ion mobility spectrogram synthesiser. This measures within seconds organic contaminants in waste wood, such as PCP, lindane, DDT, as well as some inorganic ones such as chromium-copper salts. This is crucial to the recycling of wooden materials and reconstruction of old houses. It replaces time-consuming and expensive laboratory analyses.

Mass spectrometry is an approach which identifies molecules on the basis of atomic weight. Since everything has a specific weight, as it has a specific radiant spectrum, mass spectrometry, if applicable, can be used for identifying literally anything. Beyond chemistry, applications are also to be found in medicine (for example antibodies, thus detecting diseases), the military (detecting chemical and biological warfare agents) and in environmental monitoring (detecting pollutants of any kind). There are now also very practical mobile devices the size of a briefcase. Unfortunately, at more than hundred thousand euros these highly sophisticated devices are very expensive.

Chromatography is a process by which single components of a compound or substrate are adsorbed, and thus separated from the compound or substrate, by a passing stream of gas or liquid, the liquid in most cases being a highly pure organic solvent. Various substances are adsorbed at different rates (speed). These differentials permit the identification of the presence of certain substances in a substrate and measure their volume. One application of it is chemical fingerprinting by isotope markers. This is helpful in polluter control and may also contribute to identifying the origin of food and feedstuffs.

Another category of metrology is that of optical, particularly laser-based, analytical devices. These measure dispersion, reflection and absorption of light. These also characteristically vary according to substance. Laser-based devices can measure the flow and composition of materials. This allows non-contact probing and replaces hazardous chemicals in conventional analyses. Metrological laser devices can identify and measure metals as much as organic substances and microorganisms.

Among new commercial devices is one which measures the flow of sewage water, also in mobile applications. This is not only of in-process relevance but also to control compliance with regulation and avoid fines. Conventional measurement methods are cumbersome. They include expensive construction measures and unfavourable sedimentation. Inaccuracy can be as high as 50 per cent. The new laser-based method is highly precise and comparatively cheap. An approach called LIPAN, i.e. laser induced plasma analysis, can quickly and on site identify particular hazardous substances in waste wood, for example inorganic wood protection such as boron, fluorine-containing biozides or arsenic. The presence of such substances determines whether or not the waste wood can be recycled or must expensively be incinerated as hazardous waste. LIPAN is highly precise and replaces time-consuming and expensive off-site laboratory methods.

Beyond lasers, there is a variety of other optical, photo-acoustic and sonar methods. A new 32-channel array detects UV-B radiation by counting photons with a high degree of precision. Photo-acoustic sensors detect gaseous airborne contaminants such as carbon oxides, organic hydrocarbons, alcohols and xenon. A method which combines sonar and radar at the same time (SAD-RASS principle) can be used for automated mobile remote measuring of airborne pollutants. Its vertical reach is 250 m, covering a space of up to 30 m above ground. All this advanced metrology is quick and replaces time-consuming and expensive laboratory analyses.

Membrane-type dialytical analysis uses an in-line concave fibre membrane combined with a gas-dialytical penetrometer. This quickly detects a great variety of substances depending on the specific acceptor solution which is filled into the device as reaction agent.

Electrolytic analyses are applicable to liquid-state substances and substances dissolved in a liquid. Nitrate monitoring is one such case. Electrical analyses, by contrast, measure specific electrical resistance of substances. Mercury can thus be detected by a gold-plated sensor chip which measures changes in electrical resistance when mercury combines with the gold to form an amalgam. Both electrolytic and electrical analyses are simple, quick and cheap. They substitute for time-consuming and expensive photometrical processes.

Different metrological approaches certainly have specific strengths and weaknesses, but some are also interchangeable to a certain extent. Printing, for example, requires moist agents containing the alcohol isopropanol. Its evaporation during the printing process results in significant VOC emissions. To date isopropanol input has been optimised, if at all, by the method of density measurement. Much better results are obtained either by spectroscopy or by bioelectric sensors or by measuring sonar speed. Either one enables reduction of isopropanol input by 30–50 per cent.

An outstanding field of advanced metrology is biosensors, which have already been discussed in chapter I/4.4.3 in connection to chemistry, biotechnology and food processing. These examples included DNA-chip technology, optical biosensors, as well as enzymatic and cell biosensors. All of these can of course be applied to environmental measuring and monitoring.

Biology has developed a special new subdiscipline on natural bioindicators. Most often these are lichens, indicator plants and small organisms which display specific reactions upon exposition to specific pollutants. Miners in the past used canaries to indicate dangerous concentrations of firedamp. So bio-indicators are not exactly an innovation, but there are systematic new findings on specific reactions of particular species to pollutants.

4.11.2 Monitoring Infrastructure

During the past few decades, national environmental authorities have installed a new infrastructure of territory-wide environmental monitoring, sometimes overlapping or in combination with ground stations which serve weather forecasting. These ground stations cover a variety of environmental pressure and state indicators which have become standardised such as concentration of specific pollutants in the air (dust, sulphur, ozone) or in waters (COD, BOD). Results are published on a regular basis in environmental data reports and have in this way also become established as part of everyday news content.

As metrology as well as information and telecommunications technologies advance, new features and systems are added to the existing infrastructure. One example is monitoring of ultra-fine soot particles which pose a significant health hazard. Hitherto it was technically not possible to detect and measure ultra-fine soot outside the laboratory. Only recent developments have made it feasible to produce small devices which automatically detect soot particles in the range down to 2.5 micrometres and transmit the data to a central computer.

New developments in sensor technology and wireless data transmission can considerably extend the territorial reach of environmental monitoring systems. The elements of a wireless sensor network are composed of motes (i.e. remote sensors and radio transceivers). These are small interconnected monitoring devices. They are powered by small batteries and equipped with a small processor and memory chip as well as with tiny sensors and a small radio transceiver to broadcast data to nearby neighbour motes, or to receive and pass on data from these respectively. Such wireless sensor networks can cover vast territories without needing to be wire-connected to the electrical or communications grid. Of course, motes can be applied to monitor not just environmental data on light, humidity, pressure, heat, or to monitor movements and living conditions of wildlife. Motes can equally monitor the movements of humans, just as in George Orwell's world of Big Brother.

The ground-based systems are complemented by satellite systems stationed in orbit. Satellite-based earth observation, i.e. global environmental monitoring, particularly of forests, soil, urban agglomerations, water, sea and air is becoming ever more established. This has links to weather forecasting and early detection of volcanic eruptions. Satellite-based monitoring of environmental ground conditions, in combination with GPS, can also significantly contribute to precision farming in the field (I/4.3.5).

4.12 MAIN FINDINGS

4.12.1 Summarising TEI Trends

If one tries to recapitulate the TEIs discussed in the preceding sections, a number of broader trends and transitions can be identified.

In the realm of energy, to begin with, there is a marked trend towards de-carbonisation through a double transition, one from carbon fuels to hydrogen, the other from carbon fuels to fuelless energy. Related to this is a transition from furnaces, vessels, combustion engines and similar combustion technologies to electric motors powered by fuel cells of any size and purpose. Remaining combustion processes include ever more clean-burn technology, involving higher proportions of natural gas which, if not exactly benign, is at least less harmful. Whether the age of carbon fuels, the coal age, will definitely come to an end within the coming decades or whether there will be some coal age II side by side with the rising solar age remains to be seen. There is an undeniable niche for clean coal, notably in zero-emission central power plants with CO_2 sequestration. The purpose of these power plants is to produce hydrogen by steam reformation as much as to generate electricity. New sources of electricity generation (decentral micropower) are leading towards distributed power generation, though this probably will have to include a certain number of large central stations. Over-centralised power generation and one-way distribution will thus be replaced by an integrated two-way-flow grid management.

In the realm of natural resources, the Rio process has helped to develop and introduce new sets of resource regime rules in order to put an end to traditional over-exploitation of resources and deterioration or even outright destruction of ecosystems. There are at least four major transitions:

- transition to comprehensive water management (surface and groundwater)
- transition to low-impact mining
- transition from clear-cutting to sustainable forestry
- transition from large-scale open sea fisheries to aquaculture, i.e. fish farming, also with transgenic varieties, and offshore ranching.

Aside from the fine statements of intent there is, however, in many places considerable resistance, or inertia and a lack of political will in these matters. So it is not clear how far and at what pace the new sets of resource regime rules will further be implemented in what countries.

In agriculture there are three trends all of which contribute in their way to replacing conventional over-chemicalised and otherwise environmentally non-adapted agriculture. One is organic farming. Its original intention was to offer a general alternative to agroindustry, while in fact it is in the process of becoming another agroindustry itself. The other two trends relate to ecological modernisation of agroindustry, on the one hand by precision farming in the field and within closed-loop greenhouses and agrofactories, and on the other by agrobiotechnology in the form of GM crops.

Moreover, there are two additional trends regarding agricultural products. First, in addition to growing food and feeds, agriculture will produce more biofeedstocks and fibres for industrial manufacturing. Second, current innovations also indicate a trend towards fuel crops and biofuels. This, however, seems to be artificially pushed and would not have a genuine place in an ecologically consistent scenario. The exception to this is biogas from those biowastes that cannot be used as feedstuff or industrial biofeedstock.

In the chemical industry, biofeedstocks are on the rise, replacing petrol to a certain extent. So there is a partial transition from carbo- and petrochemistry to phytochemistry. A second transition relates to biotechnology. The chemical industry is to a considerable extent in the process of transforming itself into a biotechnical industry and, in the field of inorganic materials, into a nano-industry. Biotechnical production processes under ambient conditions substitute for environmentally problematic physico-chemical production processes at high temperatures and high pressures, as well as for processes which involve the use of hazardous chemicals. This connects with a third major trend in the chemical industry, that of benign substitution, i.e. a systematically pursued reduction and elimination of hazardous chemicals in production processes and products of the chemical industry.

In the realm of materials and materials processing, which is itself in fact simply another field of chemistry, the same or very similar trends and transitions as in the chemical industry can be identified. Dry processes, biotechnical processes, plasma treatment and nanotechnical surface working are among the TEIs which will reduce or fully eliminate the present need for hazardous chemicals and high-temperature high-pressure processes.

In addition there are new composites and compound materials with unheard of properties. These regularly contribute to increases in eco-efficiency. Among the trends regarding the choice of materials are the following:

- Increasing share of secondary raw materials (from materials reprocessing).
- Increasing share of cycleware in end products.
- Synthetic fibres replace natural fibres, as natural fabric replaces synthetics, and both synthetic and natural fabric may be combined.
- Biotic compound materials replace plastics and wood in the interiors and furnishings of buildings, vehicles, casings, furniture, or similar.
- Near-nature solid materials with properties of clay, wood, straw, or similar, replace steel and concrete building materials in certain applications.

- Parts made of plastics replace parts made of metal in machines.
- Aluminium is the construction metal of choice, allowing more energy-efficient designs and being ideally recyclable.
- Composite materials substitute for metals and reinforced concrete in vehicles and buildings.

In construction and building these trends have become fully established. Growing shares of construction rubble are recycled, with a tendency towards complete recycling. Beyond materials, almost all TEIs in building and architecture revolve around the energy design of buildings, to which insulating materials are crucial. Architects include ever more elements of solar thermal heat and photovoltaics in the energy design of houses. The propagation of local firewood or other biofuels will probably turn out to be a green fad. Architects remain reluctant regarding the possibilities of geothermal heat.

As for vehicles, a belated though final breakthrough of clean car propulsion substituting for internal-combustion engines is underway. For the moment, the race is led by PEMFC-powered e-cars which are currently being launched in test markets and will probably be mass-produced from around 2010. A transition to hydrogen-fuelled internal-combustion engines, as a competing approach, cannot be precluded, though this seems to have become less probable by now. At the same time, as a defensive counter-tendency, there are incremental innovations aimed at making internal-combustion engines ever more fuel-efficient.

Moreover, all cars are becoming subject to a complete system of product stewardship with take-back and recycling of disused vehicles. An important missing element is clean aircraft propulsion. Aside from some desktop drawings there is no serious effort to develop a hydrogen-fuelled aircraft. Latest-generation airships as freight carriers are equally being ignored. The transition to high-speed maglev trains has been blocked in Europe by fast trains such as the French TGV and German ICE trains.

In the realm of utility goods the emphasis of TEIs continues to be on power-efficiency of household and office appliances. Another focus is on benign substitution, i.e. reduction and elimination of hazardous substances in product materials. Tendencies within the domestic markets of industrialised countries towards non-hazardous pure high-quality materials are however being counteracted by cheap low-quality imports. This spoils somewhat the strong trend towards increased recycling and use of cycleware in utility goods. Principles of design for environment are being increasingly adopted and are thus becoming a self-evident part of professional product development (which is also true for vehicles and building).

In materials reprocessing and waste management there is a new trend to refurbishment and reuse of high-quality parts, although this represents only a small part of the picture. A trend which is not to be overestimated either can be seen in moves towards industrial symbiosis on the basis of combined production processes, i.e. any byproduct or emission becomes input for some other process with the ultimate goal of accomplishing zero emission. Sorting and separation of municipal waste is gradually becoming semi and fully automated. New sorting and separation technologies will enable a return to just one bin for everyday non-paper household waste. The long-term trend to recycling of materials is strong and progressing further. There are, however, increasing problems with materials purity and downcycling This will have to be dealt with by stricter design principles and standards regarding materials requirements, and also by learning more about optimum rates of recycling.

TEIs regarding final and hazardous waste almost exclusively represent methods of biological and thermal treatment with integrated recovery of valuable content-materials. Although political trends are slow and unclear in many countries, technical innovation in this field clearly heralds a long-term transition from dumping and directly burning wastes to cascadic recycling and final biological or thermal treatment of all wastes. Only mineralised, chemically non-reactive waste, i.e. slack and ash, will be deposited in land-fills, if at all, so that most dumping sites can successively be closed down.

End-of-pipe purification of exhaust gases and downstream purification of sewage water – which despite all integrated problem solving remain neces-sary to a certain extent – are increasingly carried out by advanced physico-chemical methods. Biotechnology has a growing share particularly in water purification and also in site remediation (where site remediation is actually carried out and not delayed).

Finally, environmental measuring and monitoring is adopting advanced metrology such as genetic, molecular and cell biosensors, as well as advan-ced physico-chemical technologies such as spectroscopic, mass spectromet-ric, chromatographic, membrane-type dialytical, electrolytic, laser-based, other optical and sonar technologies. These are highly precise and can regis-ter traces in the micro and nano range. The new measuring technologies are applied within a network of monitoring ground stations which regularly record data on environmental indicators, enhanced by satellite-based monito-ring systems. The general tendency is towards integration and an ever closer meshing of these monitoring systems.

Some of the broad trends and transitions which have been identified are cross-sectoral, thus representing true key technologies in that they apply in several sectors and many segments at the same time. This is particularly the case with regard to

- the transition from carbon fuels and combustion technology to hydrogen, fuelless energy and fuel cell technology across all sectors, segments and operative functions

- the spread of biotechnology in mining, agriculture, chemistry, materials processing, reprocessing and recycling, emissions purification and site remediation
- applications of micro- and nanotechnology in energy, chemistry, materials processing, vehicles and utility goods
- optical, laser and sonar technologies in process control across all sectors
- a growing importance of membrane technology in chemistry, energy, materials processing, reprocessing and recycling
- benign substitution in chemistry, materials processing and reprocessing, materials design, buildings, vehicles and utility goods
- design for environment regarding end products (building, vehicles, utility goods), i.e. design for energy-efficiency, disassembly, recyclability and use of metabolically consistent, pure high-quality materials.

4.12.2 TEIs Tend to be Upstream Rather than Downstream

Two central findings emerge from the present investigation. First, TEIs cannot be denied having a huge potential to significantly contribute to solving all of today's anthropogenic environmental problems. For each environmental problem one or several TEIs can be identified which represent a solution. Whether successful problem-solving on the basis of these TEIs will in fact be achieved is a question of political will and coordinated endeavour rather than technical feasibility. Finance is in no case an insurmountable problem either. Ecological modernisation thus gives a valid orientation, also including, in addition to TEIs, complementary adequate regulation and financing as well as building up a co-directional knowledge base, attitudes and political will. To make this convincing to the public, policy- and decision-makers alike has been one of the main reasons for carrying out this research into TEIs.

The second central finding is that most TEIs, and also the most important ones in terms of structural impact, tend to be upstream in the manufacturing chain and upstream in the life cycle of products or technologies. This is all the more true if energy technology is considered to be upstream to any other manufacturing step in that it is a basic input component in every single production function (I/3.3.4). While a number of academic and political experts, especially those whose focus is on consumer-oriented action, may be surprised by the upstream tendency of environmental innovation, others have certainly been aware of this fact. They may, however, be surprised about the full extent of this tendency as it has emerged from this investigation.

The database underlying this book may not be particularly suited to sophisticated statistical analyses. The main reason is that some of the datasets, or entries, are at the level of technology domains or families, whereas others are at the level of kinds and cases. Some contain just one example, others half a dozen. Some are of particular structural importance, some are not. By and large, however, this can be expected to even out so that coarse percentage proportions give a reliable orientation. Truly significant findings are methodologically robust, i.e. they show up in the same or a similar way regardless of the specific statistical method applied.

In terms of integrated solutions versus add-on measures, 85 per cent of TEIs represent integrated solutions, 15 per cent environmental add-on technology. This is hardly surprising, serving to confirm what was to be expected. (Recycling was accounted as 'integrated', though this might be controversial. Technologies of measurement and monitoring were not included.) Of the 85 per cent integrated solutions, 49 per cent can be said to be driven mainly by ecological motives so that these are TEIs by prime intention, whereas in 36 per cent of the examples ecological motives, though these may also be considered, cannot be said to be the main reason behind that innovation.

As can be seen in Figure 4.1, more than a quarter of the datasets (27 per cent) are relating to energy, including vehicle propulsion and energy design of buildings (not however innovations related to energy demand of appliances). The picture would be even more pronounced if the statistics counted all single cases within datasets. The number of TEIs in the realm of energy is matched only by 24 per cent of technologies relating to the extraction, processing and reprocessing of materials, which accounts for 34 per cent if agriculture is included, and 44 per cent if TEIs in the realm of the chemical industry, 52 per cent if emissions control is added to this.

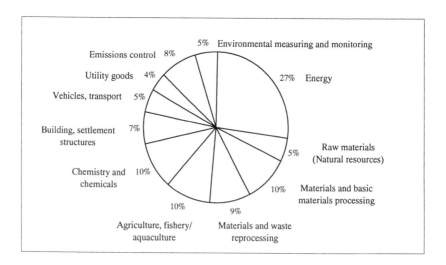

Figure 4.1   TEIs according to realm of innovation (n = 298)

We can try, as shown in Table 4.4, to ascribe energy technologies and emissions control to the sectors where they actually occur within the vertical manufacturing chain. We can thus make a distinction between end products on the one hand, and intermediate products and primary or base products, both representing pre-products, on the other. We can furthermore distinguish user or consumer behaviour from producer practices. We again exclude TEIs regarding off-site measuring and monitoring. This results in the categories of (a) primary or base production and materials, (b) materials processing and intermediate productions, (c) end products/productions and (d) user behaviour or consumer practices as shown.

Table 4.4 TEIs according to chain position

Primary productions or base products e.g. raw materials, primary fuels, power stations, agriculture, forestry, base materials (steel, cement, pulping, tanning, etc.), secondary raw materials from recycling, including related add-on purification technology	Pre-products and pre-producer practices 71%	44%
Materials processing and intermediate products e.g. metal and surface working, paper making, wood processing, furniture, textiles, production of cycleware, dyeing/coating, food processing, including related energy technology (e.g. industrial furnaces, stoves, burners) and also including related add-on-purification technology		27%
Final productions, end products Buildings, vehicles, utility and consumer goods, including related energy technology (e.g. car propulsion, house heating, electricity demand of appliances), also including related add-on-purification technology as well as producer practices		25%
User behaviour, consumer practices		4%
n = 288		100%

In this way the upstream concentration of TEIs emerges even more markedly. Primary productions and base products represent the biggest slice with 44 per cent, materials processing and intermediate products 27 per cent, making up 71 per cent of the TEIs in primary and intermediate productions. Final productions and end products represent 25 per cent. Most of this relates to building and vehicles rather than office and household appliances and consumer goods. By contrast, innovative practices regarding consumption and private user behaviour (such as soft driving, car-sharing, leasing instead of buying, avoiding overheating of rooms) at 4 per cent do not count for much. Even if this figure were doubled or tripled, the finding and its message would basically remain the same: TEIs are upstream rather than downstream. The important broad trends and transitions as summarised above are occurring upstream anyway.

As regards the life cycle stage of the TEIs as shown in Table 4.5, 3 per cent are just an idea on paper, 10 per cent are in an early stage of research and laboratory demonstration, 26 per cent in a more advanced stage of development, i.e. 36 per cent in the development stage. 35 per cent then are at market launch or shortly thereafter, or otherwise being introduced to regular practice. 16 per cent are experiencing growing adoption, though this does not always represent an impressive take-off. The remaining 10 per cent represent mature technologies in a rather late stage of structuration and diffusion.

Table 4.5   TEIs according to stage of life cycle

Idea, concept on paper	3%
Early stage of research and laboratory demonstration	10%
Advanced stage of development	26%
About market launch or introduction, or shortly thereafter	35%
Experiencing growing adoption	16%
Mature stage	10%
n = 289	100%

The distribution in Table 4.5 is not surprising either. Ideas and early experiments are normally not communicated to a broader public, just as technologies in a late stage of their life cycle are normally not subject to public attention since nothing particular is occurring any more. It is nonetheless important to have this documented empirically, because it confirms one of the basic recommendations that can be drawn from life cycle analysis: true progress which includes structural change, particularly regarding a change in the consistency of the industrial metabolism, requires a change in path; i.e. it requires the development and implementation of new technologies rather than the modification of long-established mature systems already in place.

As is shown in Table 4.6, most TEIs do indeed bring about change in metabolic consistency. It was not possible to determine exactly how big each contribution is, just as it is not possible to be specific about the exact extent to which TEIs contribute to environmental ease or improvement (something beyond 'considerably' or 'insignificant'). To make specific statements on various environmental impacts would require an entire technology assessment project on each case, and even then results usually tend to be debatable. But we can in most cases clearly say whether qualitative properties of products and processes are changed or whether an innovation is merely about reducing unit-specific quantities involved.

Table 4.6   TEIs according to ecological consistency and efficiency

Consistency improvement without efficiency change, or efficiency unclear, or even slightly decreased	16%
Consistency change in the sense of lesser degree of inconsistency without efficiency change, or efficiency unclear, or even slightly decreased	8%
Consistency change in the sense of lesser degree of inconsistency combined with increase in efficiency	12%
Both consistency improvement and efficiency increase	41%
Mere efficiency increase without actual change in metabolic consistency	23%
n = 281	100%

In about a quarter of the TEIs there are significant efficiency increases without a change in metabolic properties. Typical examples of this include reuse of parts and recycling of materials such as solvents or sulphuric acid, or increased fuel-efficiency in internal-combustion engines.

Three-quarters, however, involve some change in consistency. The biggest share, at 41 per cent, is that of examples where there are both consistency improvements and efficiency increases, for example latest-generation solar cells, fuel cells of any kind and many biotech applications of where there is simultaneously less or no environmental pressure and higher yield. The percentage of such double-surplus TEIs is nevertheless lower than was hypothetically supposed, whereas the number of cases in which there is less inconsistency rather than really benign improvements are more than was expected. Typical examples of a lesser degree of inconsistency without efficiency change include exhaust catalysts, transmutation of nuclear waste, or HCFC-22 as a halocarbon replacement for conventional CFCs.

Replacement with SF_6 then is an example of a lesser degree of inconsistency combined with some efficiency increase, because on eco-balance SF_6 helps to save CO_2 emissions. Further examples in this category include GM crops tolerant of agrochemicals. In general this is about incremental process innovations which help to reduce hazardous auxiliaries or materials-content, in certain cases simultaneously increasing output.

An example of benign consistency improvements without efficiency changes are hydrogen-fuelled internal-combustion engines. Previous generations of solar cells, solar heat and wind power, though clearly clean and metabolically consistent, came with less efficiency than conventional like technologies.

5. Conclusions: Upstreaming Environmental Policy

5.1 PARADIGM SHIFT FROM UPSTREAM TO DOWNSTREAM

Besides providing evidence of the huge potential of technological environmental innovation, the second central message from the above findings concerns a paradigm shift from downstream to upstream in the product chain and technology life cycles. This implies a parallel shift in the emphasis of policy: from regulation to innovation, and from final use to base and intermediate production. Environmental policy would again have to focus on industrial production as was the case in the early days of environmental policy. At the time, however, the policy pattern was quite bureaucratic as the political actor constellation was highly confrontational. First add-on technologies merely deferred rather than resolved the problems. This was one of the reasons then why a shift set in from production to the consumption and demand side. This shift finally included everybody and attributed the same level of importance to everything, devoting an almost absurd degree of attention to plastic bags, deposit bottles and numerous dustbins while not being anywhere near serious enough about hydrogen, coal and agriculture.

What makes the difference today is that we are now dealing with metabolically consistent and efficient TEIs, including advanced add-on technologies. A return to basics would not mean abandoning the wealth of instruments and intervention methods which has been developed, but would mean reshifting priorities. Action has to be refocused onto those industrial operations where large environmental impact actually occurs – in energy, raw materials, in agriculture, in chemistry and base industries, partly also in building and vehicles; not, however, in consumer goods and user behaviour.

5.1.1 From Consumer Orientation to TEIs in Core Productions

The fact that most environmental impacts are determined upstream in a life cycle and caused upstream in the manufacturing chain puts final consumption in a somewhat paradoxical role. This ecological paradox of consumption is as follows: On the one hand, expectations of high and still rising levels of affluence are indeed among the main driving forces behind the ongoing growth of industrial production and the large volumes of materials turnover; on the other hand, however, the immediate contribution of consumer behaviour to the industrial metabolism is rather low, about 5–15 per cent to gauge it generously. This is because most of the environmental pressure, the big ecological footprints or materials rucksacks of consumer society, occur during the different steps upstream in the product chain and are determined by the basic design of a technology in the early stages of its life cycle. In contrast, consumption in service businesses and private households entails final steps downstream in the product chain (Huber 2001 407–411).

Therefore, if we are really concerned with environmental impact, changing consumer behaviour cannot lead very far. Again consider the case of the automobile. Not using a car at all is a non-option, open only to some young social experimenters and a few old mavericks. What remains for the user is soft driving which can save 15–20 per cent of gasoline consumption. Similarly, with heating, putting on a pullover instead of turning on the thermostat can save 6 per cent fuel consumption for each 1°C in the range around 19–23°C room temperature. Such savings are certainly significant, but the two cases represent rare exceptions to the general rule that at the level of consumption normally very little can be saved, and structurally next to nothing can be changed.

Consumers may save on water, assuming that they live in a region where water is scarce and where it consequently makes ecological sense to save on water. Consumers, under urban conditions, however, cannot change the physical quality of the water they tap and how that water is extracted, processed and reprocessed after use. TV viewers can switch on a TV set, zap around, and switch it off. They cannot decide which construction materials and communication technology go into the TV's manufacture. Air passengers can decide to book or not to book a flight, theoretically, but once they are aboard they do not have any influence on the ecological effects of flying, neither on how aircraft are flown nor how the aircraft and its engines are designed and produced. But it is precisely these aspects which determine the environmental performance of air traffic. If a product as such is inherently problematic, as is the case with carbon-based fuels, being an efficient eco-manager and a good consumer does not help much. What really helps is changing the technostructure and its metabolic properties, upstream in a technology's life cycle and upstream in the manufacturing chain.

In recent years, many activities have been launched to appeal to the green-minded consumer, to create eco-markets from the demand side and to lure consumers into spending more on organic produce or on clean electricity, or buying green and ethical investment funds (Vergragt 2000, OECD 2001d,

Bouma et al. 2001). These initiatives are no doubt useful to some extent. The more effective targets, though, continue to be in production: the broad-based large-scale development of environmentally benign new technologies and newly designed products which would render obsolete the distinction between a normal product and an ecological product.

Today's unfocused environmental policies which experiment with eco-taxes, emissions trading, green labels, and which proclaim that everybody would have to contribute their share, cannot change the basic and simple truth that important environmental problems and solutions are placed chain-upwards, and what has to change is technology, i.e. the products and processes which industry is based on and which in fact determine the industrial metabolism.

5.1.2 From Final Demand to Technological Innovation and Key Manufacturers' Chain Management

Orientation towards the upstream direction of innovation life cycles and product chains represents a shift from final demand to technological innovation and key manufacturers' chain management. One can take it as a strategic guideline that environmental priority targets are first and foremost to be set in research and development, and secondly they are to be found in production; by contrast in trade, services and private consumption, while there may also be some interesting environmental innovations, there are few really important ones.

In the long run, we cannot afford to hold on to environmentally damaging technologies or products; and final demand by itself cannot bring about innovation. Innovation comes from innovators, from creators and suppliers of technology, from producers of products. They may need a degree of political support as well as legal and financial framing, and also, at a later stage of an innovation and diffusion cycle, some demand pull by the green-minded consumer (II/9.9).

To put it simply: do not worry too much about consumers, services and trade; look to designers and producers, to engineers, chemists and biologists. Do not appeal to airlines; talk to the designers and producers of aircraft. Do not charge users of electricity ever higher prices and taxes; check the power business and the technology it uses. Do not bother car drivers and house owners; work out an agreement on innovative standards with car makers and producers of heating systems and impose environmental performance standards which are demanding, but achievable. Do not promote extra-profitable premiums on organic produce; make sure farmers, forest industry and fishers adopt practices that make any produce ecologically sound and healthy.

Markets and demand side impulses are effective controlling agencies, good at selecting and organising optimum cooperation. But they are not creative and never produce anything. That is why environmental awareness and will-building, regulation, green marketing and business management remain ultimately pointless in the absence of a prior strategy of technological environmental *upstream* innovation to give a common focus to environmental policies.

The starting point thus would be primary supply rather than final demand. What final users and consumers can do is to 'vote' at the point of sales. This is certainly a selective signal to retailers and producers, but quite often rather irrelevant in face of the facts of technology and production. It is similar to legitimising politics at the ballot box. One has the choice to vote for this or that party or politician, but one cannot 'vote into being' parties and politicians that do not exist. One has to accommodate oneself with the ones that are 'on offer'. It is only in the dynamics of politics itself that politics can structurally be changed whereby certain institutions and persons are better placed than others to take the lead.

With technological environmental innovation it is quite similar. Final demand certainly has some selective influence, but this is less decisive than commonly shared views about market economy and consumer society (which represent sediment of theoretical teachings and political doctrines) would have us believe. If there is a position at all from where important demand power can be exerted, then it is the big manufacturers of end products such as buildings, vehicles, appliances, furniture and textiles, and also demand by large retailers, who are in a position to effectively implement supply chain management. This is none of a user's nor of a government's business.

Manufacturers of complex end products and large retailers are roughly in the midway between base productions and final consumption. In this chain position impulses from the supply side and the demand side can really be fed back to each other. For once they are gatekeepers who select which items are rejected and which ones pass through. This, on a higher level of influence, is basically the same type of selection which is exercised lower down by final buyers. Second, and more importantly, key manufacturers and large retailers, in contrast to external end-users and consumers, in fact communicate and cooperate with their suppliers within a production network. This can simultaneously be activated as an innovation network (II/9.2–3). Key manufacturers and their suppliers, as well as manufacturers and large retailers, talk to each other. They bargain for prices, quantities, dates, quality and properties of products to be delivered. And this quite simply is the way in which supply chain management works, notwithstanding the fact that this can be enhanced by some computer-based chain management information system.

5.2 GOVERNMENT POLICIES

5.2.1 Policy Regimes: Bureaucratic Command-and-Control versus Coordination and Context-conditioning

A shift from downstream to upstream also entails a shift in prevailing policy regimes. Two patterns of government intervention, regulation and sets of policy instruments are at hand when dealing with environmental problems: on the one hand is a regime of bureaucratic command-and-control, and on the other, a regime of coordination and contextual control.

Each type of regime includes the three components of (1) actor constellation, (2) policy style and (3) policy determination, i.e. goal-setting, institutional arrangements and choice of instruments. The underlying spectrum can be characterised by (1´) an actor constellation that is confrontational versus cooperative, (2´) a policy style of statist command-and-control versus co-ordinative networking, and (3´) a choice of policy instruments that are directivist versus contextual.

The command-and-control pattern usually emerges from a background of confrontational actor constellation. It makes use of compulsory directives and control procedures, and economic incentives such as fees and fines, taxes and subsidies which are familiar from centrally planned economies. The co-ordinative pattern of contextual control presupposes a more cooperative climate among actors involved and makes use as much as possible of civil law and private contracting. This results in a preference for policy instruments such as liability regulation, voluntary agreements, mediation (negotiative conflict settlement), informative tools of disclosure, comparative eco-balances, or benchmarking on environmental performance over time. In addition, this policy type may also apply incentives such as subsidies (particularly during research, development and market launch), and disincentives such as taxes, if the latter clearly discriminate in favour of desired alternatives.

There has been some shift from administrative dirigisme to coordinative context-conditioning since the 1990s. But an overall shift from the one to the other one is not exactly what happens nor what a strategy of TEIs necessitates. Both policy patterns do have their advantages and shortcomings. The bureaucratic approach is effective in the short term, notably if there is urgent need for action. It is equally effective in overcoming insecurity about legal conditions and dissolving collective prisoners' dilemmas (how to share burdens and rewards, costs and benefits). In the longer term, however, it reproduces the adverse effects of any bureaucracy, including cost-inefficiency and obstruction of change and innovation. Change and innovation, in turn, represent the biggest advantages of a regime of coordination and contextual control which, however, only begins to deliver on its promises in the longer term. This becomes disadvantageous in cases where the cooperative climate among the actors involved breaks down.

So the two policy patterns cannot substitute for one another at will but have most often to be combined. For innovation purposes, however, a regime of coordination and context-conditioning is clearly superior to command-and-control. Ideas, shared visions, ingenuity and innovation do not occur by ordinance. Creativity and innovativeness cannot be prescribed. Science and technology are primarily cultural phenomena, the self-expression of a utilitarian society with a predominantly instrumental mindset. Modern culture actualises itself primarily through scientific discovery and technology, to much greater extents than it formerly did when the frontline of cultural evolution was in the arts, literature and classical music. In a superficial analysis, technical innovation seems to serve the interests of financial capital accumulation. In the end, however, capital is the catalyst which serves the evolution of technology and technical productivity which is the basis of material wealth.

If technological innovation is to be politically fostered, the best way to achieve this is by coordination and suitable framing, communication and organisational support. This is feasible if there is a cooperative actor constellation among government, industry, academe and other research institutions, and, if relevant, social movements. The ensuing policy style would then be coordinative networking based on open and inclusive communication among the aforementioned participants. This network approach would apply continually to all of the stages of a policy cycle, i.e. agenda-setting, programme formulation including goals and instruments, implementation and evaluation.

The role of politics and administration in such a setting includes three main aspects: the enhancement of the national innovation system, the setting of tough environmental performance standards while refraining from imposing cumbersome administrative procedures, and the provision of well-balanced financial support during research and development until market launch. Table 5.1 gives an overview over an integrated multi-level approach to a regime of coordination and context-conditioning for TEIs.

5.2.2 Enhancing the National Innovation System

An environmental policy aimed at fostering TEIs will more than hitherto have to take the form of S&T policy, or innovation policy, or structural or industrial policy, or however else it may be called in different countries and political contexts. In any case, more elaborate and better equipped policies of technological research and development should come to the fore, policies that promote, communicate, mediate, and regulatorily and financially pave the way for TEIs.

Table 5.1   Integrated multi-level approach to a regime of coordination and context-conditioning for TEIs

	Invention, research	Organised development	Structural unfolding and diffusion in	
			Production	Use
Pivotal actors. Main addressees for policy	Individual inventors, research institutes, small start-ups, large companies.	Start-ups, smaller firms, large companies, cooperation between large and small.	Growing role of larger companies, or growth of smaller ones.	Corporate and private users.
Structuration, selection	Creating competing prototypes, demonstrating practicability.	Working out competing designs for becoming dominant until/during full market launch.	Standardised products/services and production processes.	Routinised use practices.
Policy tasks	Variety generation. Encouraging and enabling innovative initiatives.	Selective strategic development support.	Creating level playing field. Enabling chances of new entrants.	
Action strategies	Issue-related research programmes. Specified inventors programmes.	Public-private innovation networks. Strategic niche management.	Appropriate regulation.	
Patenting	To be streamlined. If necessary subsidised.	No more patenting subsidies.		
Lawmaking, standards-setting	Strict proactive environmental performance standards.	Setting necessary technical standards in time.	If possible, doing without administrative procedures (approval, permission, licensing). If unavoidable, ought to be simple, swift, flexible and efficient.	
Subsidisation	Well-funded broad-range basic research. Plus special-purpose seed money (public, foundations). Demonstration programmes.	Still some seed money. Co-funding of development programmes; partial investment subsidies.	Market launch support. Partial investment subsidies/ favourable investment credit facilities.	

Table 5.1 continued

Investment and capital market		Increasingly private venture capital.	Regular invest-ment via the stock exchange and banks.	
Taxation	No special tax concessions. Instead, favourable conditions of offsetting depreciation against tax as well as building up investment reserves.			
Market for the innova-tive items		Potent test-users. Some government guaranteed demand. Pilot markets.	Approachable lead-markets (lead users).	
Entrepr.ship and labour market	Flexible arrangements for setting up new firms, including flexible employment. Cross-sectoral career opportunities for high-potentials.			
Culture	Taking risks, trying out new ways, being at the forefront of progress, as a value of high esteem, rewar-ded in prestige, career promo-tion and extra income.	Non-adverse attitude towards business and making money, notably on the part of academe, political parties, associations and social movements.		

Public-private innovation networks, i.e. task forces, agencies or similar, each with its own focus on a special technology, can act as promoters, change agents and initiators of R&D cooperation. They communicate, stimulate, promote, co-finance, help to organise and coordinate; in brief, they enhance specific research and development activities within the national system of in-novation, Europe-wide at the EU level (II/10.2). The interorganisational cross-boundary networks to be initiated include researchers and S&T institu-tions, professional associations, industry and government officials from the local to the national and supranational level.

Certain innovations originating from within large corporations may not need external cooperation. Normally, however, large corporations also have additional benefits from national and international networking, particularly from building close ties between their product development divisions and ex-ternal research laboratories. With complex innovations, furthermore, more broad-based cooperative research arrangements are required, thus public-private innovation networks. These can be set up by the government, for in-stance the ministry of S&T and/or the environment and/or economic affairs.

They need in any case official government approval and support, including government funding or co-funding.

During the last 20 years there has been a clear shift away from individual R&D projects (from 85 per cent down to 33 per cent) to collaborative projects (from 15 per cent up to 67 per cent) including corporate, governmental, academic and private partners (FMEL 2002 43). Setting up interorganisational cross-boundary research and development networks has actually become a common practice in the US, Japan, the EU and its member countries. Sometimes they are purely governmental and rather bureaucratic such as Japan's MITI, sometimes the networks are purely industry-run, and in the US it is frequently the Pentagon which acts as central national planner and the driving force behind the development of new technologies.

What needs to be done is to transfer this practice to the development and introduction of TEIs, and to avoid all too firmly established planning bureaucracies. The history of nuclear power is a telling example of a government-driven, and highly problem-laden innovation process which was rigidly pushed forward without any timely self-correction. Similarly, French 5–10 years 'planification', or 'concerted actions' in Germany, or strategic planning departments in industry, are examples of how institutions get stuck on their own beaten track. Their fate is either that they are soon ignored because things are moving on anyway, or that they find themselves on the wrong track, usually the track of yesterday's tomorrows, failing to keep pace with the choice of futures of the day.

Innovation and research are first about variety generation, competing paradigms and alternative approaches. Development and introduction into regular practice is about successive testing, proving and selection. Paradigmatic pluralism in society must not to be planned off, and innovation contest must not be abolished. So simultaneous support for different paradigms which compete to become the leading design should actually be encouraged during early and also later stages of development, until the time has come for government, markets or science, to take further decisions.

Even well-informed bureaucracies do not know enough to make such decisions alone. The markets, i.e. investors, producers and users, do not know for sure either. Nor do researchers and designers. Such a process has to be communicated and organised according to what it is in reality, i.e. a collective process of searching and learning. Even all these elements working together can err. But a firmly installed government planning bureaucracy is more likely to go wrong than a multitude of public-private innovation networks with a specific task, and a time-horizon which is certainly long-term but nonetheless limited (II/9.7).

Kemp et al. (1998 183) have proposed an innovation policy which they call strategic niche management. They start from the observation that suc-

cessful innovations in history were brought about by a combination of entre-
preneurs or system builders and the availability of niches for application. A
niche in this meaning is different from the ecological niche or evolutive
niche of something which encompasses the entire evolutive potential
throughout a life cycle. In Kemp et al. a niche means a first, rather small
space where something new may have its chance to take hold, a field of first
convincing application that will trigger further adoption on a broader basis:

> Niches are instrumental in the take-off of a new technological regime and the fur-
> ther development of a new technology. Apart from demonstrating the viability of
> a new technology and providing financial means for further development, niches
> help to build a constituency behind a new technology, and to set in motion interac-
> tive learning processes and institutional adaptations – in management, organisa-
> tion and the institutional context – that are all-important for the wider diffusion
> and development of the new technology (Kemp et al. 1998 184, Kemp 1997 288).
>
> The creation of a market niche or protected space may be an important step-
> ping stone for the further evolution of a radically new technology, e.g. an energy
> technology: it helps suppliers better to understand user needs, to identify and solve
> specific problems, to achieve cost reductions in mass production, and perhaps
> most important, to create a constituency behind the new product, finding support
> from other actors (Kemp 1997 307).

A niche can be identified by a single entrepreneur, or by any other type of
individual or collective actor. The authors themselves envisage the government, or some agency set up by the government respectively. The reason is
that the authors conceive of strategic niche management as

> the creation, development and controlled phase-out of protected spaces for the de-
> velopment and use of promising technologies by means of experimentation, with
> the aim of (1) learning about the desirability of the new technology and (2) enhan-
> cing the further development and the rate of application of the new technology
> (Kemp et al. 1998 186).

In practice, strategic niche management comes down to designing smaller or
larger field experiments and test uses with a new technology, for example
hydrogen-fuelled public busses in the mid-1990s, and then trying to communicate the findings related to the experiment, and finally, if evaluations were
positive, expanding the experiment into a market niche (Hoogma et al. 2001).

5.2.3 Regulative Instruments: Setting Strict Performance Standards

In the environmental literature, the term regulation is most often used in the
threefold sense of including (a) general lawmaking, (b) administrative ordinances and procedures such as registering, approval, permission-granting,
licensing, etc., and (c) setting environmental performance standards such as
critical limit values. Overgeneralised statements about regulation can be con-

fusing, because the three aspects can include each other in quite different ways.

A regime of command-and-control tends to rely on administrative procedures of a dirigiste and interventionist character, hampering and impairing initiative in research, development and production alike. The latest object lesson regarding this is biotechnological research in certain EU countries. Standards tend to be prescriptive in technical detail, referring to some judicially defined 'proven practice' or 'best available technology' or 'state of knowledge' (which tend to be yesterday's best knowledge and technology).

A regime of contextual control, by contrast, will in general try to minimise administrative procedures. If unavoidable, it will prefer, say, registering to formal approval and permission granting. It will try to keep procedures as simple, swift, flexible and efficient as possible, rather than as encompassing and detailed as possible. As for standards-setting, a regime of contextual control will try to avoid prescribing technologies and details thereof. Instead it will concentrate on certain technical standards which are necessary to enable reliable individual planning and development activities, and environmental performance standards.

Many TEIs discussed in this book confirm an insight of recent social-science environmental research. The insight is that it is not just one policy instrument which leads to successful environmental innovation. There needs to be an appropriate mix of instruments depending on the special case. Similar to cooking, the mix must be well chosen and balanced, and must not be overdone in that it includes too many ingredients. There is one element in such a mix which always tends to be present and which has proven to be of particular importance. This proven core element of effective environmental regulation is the setting of tough performance standards. This may also involve ruling out unwanted substances, materials, products or processes. An appropriate time-frame must be set for achieving these standards. The standards shall preferably not prescribe application of particular technologies, just performance goals, normally limit values, with regard to certain technical and ecological parameters. Strict environmental performance standards are an indispensable and very effective guide to developing TEIs (Ashford et al. 1985, Ashford 1993, Kemp 1997 317, Hemmelskamp et al. 2000).

This is not surprising since performance standards are technical by nature and represent an integral element of the rules of a technological regime (II/6.3), in contrast to, say, liability claims or taxes on resource consumption and emissions. Strict environmental performance standards also give the first movers who are willing to proactively meet those standards a competitive advantage which is hard to catch up with (Porter/van der Linde 1995). Scientific evidence about the limitations of using resources and sinks, critical loads and exposures, remains relatively questionable, so that standards may have to be changed as new findings indicate different limitations. Nonetheless, determinate performance goals are requested in order to orient actors and keep planning and development processes going.

5.2.4 Financial Support: Seed Money, Venture Capital, Level Playing Field at the Market

Financial support has different functions at different stages of a life cycle. In early development stages there is no real alternative to public research funds, unless an innovation comes entirely from within industrial corporations. With generic systems innovations, however, this is normally not the case. Furthermore, industry tends to hold its own research capacities rather small while outsourcing ever more research and development activities which are not yet very close to market launch. This should be seen as a welcome tendency because it contributes to the sharing of knowledge and know-how within national and international networks.

Beyond regular funding of broad-range basic research, special-purpose seed money has to be provided to fund promising TEI developments in academe and industry. This seed money can come directly from the government, or from special foundations, or through those public-private development networks, no matter by whom they were originally set up. The purpose of donating seed money is to stimulate the generation of variety.

In later stages of development, public agencies or public-private networks should still have funds available to grant venture capital (risk capital) until market launch. In contrast to seed money, which is donated, venture capital ought formally to be treated as an investment which may yield a return in later years if successful. Within the frame of a co-investor model the capital-granting agency becomes co-owner of shares or otherwise documented property of a new technical development. A related different practice is the co-funding model, i.e. to provide favourable investment credit facilities, or securisation of investment borrowing or export activities. Even though there will be no return in many cases, and development loans have to be written off, there will be profit in other cases.

Should the investment prove to be a success, the practice ought to be to sell off shares to co-owners or the public in following years. In principle, private venture capital and regular investment via the stock exchange should of course contribute as much as possible. The purpose of public venture capital, after all, is not to reintroduce nationalised industries but to promote innovation for the common good. That private co-owners would profit more than others is no contradiction. Whoever is prepared to take initiative and risks, should have some advantage, the more so since it is society in general which will enjoy the utility of TEIs.

Degressive development subsidies represent an alternative to temporary public venture capital (co-investor model) and development credit (co-funding model), and actually a regular practice today; i.e. public agencies continue to donate money, but ever less the more a development advances and reaches the threshold to market launch. The principle is to decrease subsidies step by step until full market introduction is achieved. In effect this may not be too different from granting external credit or internal capital as a co-owner. The actual difference is return foregone to the public purse.

Public venture capital may in certain cases be unnecessary because industry and the stock exchange are so convinced by an innovation that they provide all the venture capital which is necessary by themselves. In certain other cases doubts and risks may be perceived as being too high to go it alone. Then public money has to take some of the risk, preferably in an arrangement of mixed public-private investment. There may also be cases in which industry is not interested at all because it is on the defenders' side and does not want to see challenging innovations come up. Then public agencies have to go it alone for a certain period of time and make advance concessions to new entrants.

When it comes to market launch, government or public-private agencies have still to provide some support. A market does not come into existence by itself. Building up a new market is a precarious process. The best a government can do to build up markets for TEIs is appropriate lawmaking and suitable standards-setting. Government can furthermore create demand, for example by itself guaranteeing to purchase a significant amount of a newly introduced technology or product. The American military and the Japanese government have never hesitated to do this. An alternative approach is to guarantee selling prices to make new technologies competitive against well-established like technologies. This is another form of subsidisation, currently practised in many European countries in order to introduce photovoltaics, wind power or biogas. Unless shielded in this way, they could not become competitive against power from conventional sources.

Shielded sectors among otherwise competition-exposed sectors is normally seen as undesirable. But with technological succession the situation is different in that it tends to be a competition between established defenders in a strong position and challengers who are still weak in their infancy. Promoters of a new technological regime and defenders of an old one do not play on the same level. Defenders of old regimes can rely on high cost-efficiencies, i.e. low per-unit costs, whereas innovators unavoidably have to operate at high per-unit costs. Techniques, practices and brains in old regimes are readily available and practice-proof, not least because they are not innovative any more, whereas new technologies still have to demonstrate their supposed utility, connectivity and reliability (II/8.3). Established regimes have a wide constituency whereas new ones still have to gain one. So the justification for selectively and temporarily promoting and protecting promising innovations is that it levels the playing field so as to encourage new entrants to take the field against established oligopolists.

Creating a level playing field at the market launch certainly needs careful investigation case by case. With regard to industrial nation-building in the 19th and 20th centuries, shielding young national industries was a strategy widely chosen by newly industrialising nations. But there were cases of success such as Germany and Japan, and cases with mixed results such as Russia, or even failure in a number of developing countries. All of them have finally decided to connect themselves to international market competition and give up artificial protection of their domestic markets because there was seemingly too great a lack of endogenous potential. With new technological regimes it is in principle much the same as with newly industrialising nations. If there is not enough endogenous potential, an artificially levelled playing field can even be counter-productive in that it pushes or conserves inefficient and unconnective structures. Furthermore, measures taken should not over-impair interaction with the selective environment. If a new technology is subsidised in order to be price-competitive, selection through market demand should be kept intact.

Creating a level playing field entails the use of financial instruments. Today's environmental policy has devised quite a number of these instruments such as fines and fees, bonuses, deposits, administered prices, tradeable emission permits, ecologically motivated expenditure taxes, tax reliefs and direct subsidies. Some of these, such as green taxes, while they have been praised by environmental economists have more often been abused in practice. For example, a tax on electricity from whatever source, and on petrol and gas, while exempting coal and biomass, is not an 'eco-tax' but a staggering example of a mischievous, mislabelled hotchpotch which was created under the influence of conservative industrial powers, including mining unions.

Another bad example is to grant regenerative power an unlimited demand at a high selling price, both of which are forced upon oligopolistic electricity suppliers. This represents an administered price in combination with enforced contracting. While this certainly satisfies anti-monopolist resentment, it simply adds fuel to the flames. The oligopolists will of course raise prices to compensate for the extra costs enforced upon them, resulting in reduced overall productivity of the economy. Moreover, there is no incentive at all to reduce unapproved-of unclean energy in the energy mix provided.

Financial instruments are important control factors indeed, similar to laws and administrative procedures, but they cannot 'control into existence' that which does not yet exist, and they cannot make up for a lack of specific technical and environmental performance standards. If there is a financial instru-

ment at all which can effectively contribute to bringing into existence something, i.e. contribute to innovation, then this is simply to hand out money to
the innovators: research funds, seed money, venture capital, preferential
credit, securities. But this has to be targeted very carefully, appropriately in-
strumented and evaluated at regular intervals. One should keep in mind that
all of the aforementioned financial incentives, including subsidies, belong to
the arsenal of centrally planned economy. Normally they do more harm than
good and should therefore be treated like hazardous substances – be avoided
if possible, and if unavoidable be used in well measured doses only.

In general, a policy shift from bureaucratic command-and-control to
coordinative context-conditioning is naturally wedded to the market model of
the economy – though a simplistic idea of 'the markets know' so 'let the
markets decide' is clearly a myth removed from reality and bound to fail as is
now known from any economics of social and ecological externalities. What
would be needed in addition to the original price-related economic market
model is a general model of push and pull factors, i.e. 'supply' and 'demand'
factors in societal subsystems and their interactions with one another. Understanding functional push-pull dynamics would greatly help to develop appropriate, possibly non-directive, policies.

A policy instrument is not automatically compatible with market dynamics simply through its being a financial instrument. On the contrary. Most
instruments which treasuries have at hand date back to redistributive feudal
practices structured by status and privilege, thus alien to free and open market economies. Those which seem to be relatively more acceptable than others are precisely those that could be envisaged when trying to foster TEIs.
They include, in ascending order of efficacy and economic desirability the
following:

- Emissions trading. This in fact represents an appendix to centrally planned
emissions contingents. It comes at a high transaction cost, entails lengthy
administration and its innovative effects are overestimated.
- Discriminatory expenditure taxes, i.e. taxes on resource consumption, as
well as taxes on emissions. These are discriminatory in that they draw a clear
line between approved-of and disapproved-of items, thus making for a clear
selective impact, for example between desired fuelless power and hydrogen
(which would be tax-free), natural gas and perhaps biomass as transitionally
semi-desired fuels (charged with a lower tax rate), and coal, petrol and uranium as undesired fuels (charged with a higher tax rate). By contrast, taxes
on electricity consumption or energy use in general are non-discriminatory
and thus counterproductive.
- Direct subsidies degressively accompanying the upstream stages of an innovation life cycle. These are the measures proposed above, such as regular

funding of research, donating seed money, providing venture capital, preferential credit and securities.

- Phasing out environmentally counterproductive subsidies and tax reliefs. These include tax reliefs and subsidies in over-intensified agroindustry, open sea fisheries, as well as subsidisation of coalmining and of kerosene in air traffic.

Much of the money which is channelled into promoting TEIs will not have an immediate pay-back. But the few paying investments, success stories of contemporary industrial history, will more than make up for those many little failures. Seed money, venture capital and introductory subsidies for approved-of TEIs have to be seen as strategic investments. Strategic investment has of course to be measured and its outcome has to be evaluated, but penny-pinching controllers must not be allowed to spoil initiative. A high die-off rate is completely natural for inventions, development initiatives and market launches of new single products, even though this hurts the brave. It is better to pay for highly qualified, forward-looking personnel who maintain the knowledge and skill base of the time than to pay for counterproductive subsidisation of backwardness.

PART II

Innovation Life Cycle Analysis

6. Innovation: Definitions and Distinctions

6.1 WHAT IS AN INNOVATION? AND WHAT IS INNOVATION?

Innovation can be seen as a process by which an actor (an individual or a group of individuals) brings into existence something new which evolves further and becomes socially more diffused. The word innovation also refers to the thing that is being innovated. According to the US Patent and Trademark Office, a patentable innovation is 'a creation that is novel, non-obvious and useful'. Anything can become subject to an innovation process if it is structurally somehow new and does not yet exist in its emerging genotype or phenotype, or at least is not yet known to the members of a social system at the time when it occurs to them. Anything can be created and further developed and be spread among ever broader groups of actors in the evolutive course of a life cycle: ideas, words and usages, manners and other behavioural patterns, expressive styles, values, rules, institutions, regulations, monetary and financial practices, work practices, and scientific and technological paradigms.

The term innovation applies neither to natural changes in the geological environment and to biological evolution and reproduction, nor to plants or animals, nor to the birth and life of human beings. It applies, however, to biological changes in organisms which have come about through human cultivation, breeding and genetic modification.

Innovation is among the central determinants of structural change. Theories of social change, and theories of innovation and modernisation imply each other. They belong to the same family of evolutive paradigms. Evolution equals 'creative destruction', as Schumpeter put it. It imposes, in the words of Darwin, a permanent struggle for life, a continual competition to occupy available niches, i.e. opportunities of coming into existence and evolving. Hence, any structural change has its pros and cons, its promoters and opponents, its winners and losers. Innovation is a source of cooperation and competition, of alliances and conflicts.

243

6.2 TECHNICAL AND 'SOCIAL' INNOVATIONS

In most cases, the term innovation is applied to technical innovation or prod-uct innovation. The emergence and diffusion of other new things in society can of course equally be seen as cases of innovation. Diffusion theory, after all, was invented by cultural anthropology and sociology. All non-technical innovations are usually referred to as 'social' innovations. Furthermore, there is growing awareness that technical innovations are not an isolated phenome-non but occur in a conditioning societal context. This view was put forth dur-ing the 1960s and early 1970s in the socio-technical systems approach by Fred E. Emery and others from the London Tavistock Institute of Human Re-lations. The socio-technical view continues to be valid, though by now, more than a generation later, the meaning of what is 'social' in a socio-technical approach may need some further specification. To start with, innovations can be distinguished according to the functional subsystem of society in which they occur as shown in Table 6.1.

Table 6.1   Innovations according to society's functional subsystems

Societal Subsystem		Kind of Innovation
A	Operative system of manifest action, esp. technical and bodily activities (work) of production/consumption.	→	Technological innovations, including practices and behavioural routines, as well as organisation of cooperation and physical transfers under con-ditions of divisional specialisation.
B1	Economic system	→	Economic innovations, i.e. price-related mone-tary and financial innovations, including mecha-nisms of allocation and distribution, as well as monetary and financial institutions.
B2	Ordinative system	→	Legal, institutional and administrative innova-tions based on law and formal authority, e.g. re-gulation, from standards-setting to market order.
C	Formative systems such as politics, media, education, arts, science, religion, value base.	→	Formative innovations, i.e. any 'social' and cultural innovation other than above.

It is implicit in the systematics given in Table 6.1 that innovations concern-ing technologies or products represent a special type of societal innovation. So technical innovation is not different from 'social' innovation but is in fact a special form thereof. Technical innovation *is* societal innovation. Further-more, since social systems are interdependent, technical innovations are usually not just technical but imply certain corresponding co-directional changes in other societal subsystems. This is all the more true, the more the innovations under consideration constitute generic systems innovations of major importance entailing a potential for structural change. To understand the embedded reality of technology and TEIs, it is import to see how different system levels relate to each other. To this end, Table 6.2 shows the systematics of Table 6.1 in a slightly different way.

Table 6.2   Technological innovations, economic and ordinative controls innovations, and formative innovations

Effectuative Innovations	Operative Innovations	Technological innovations = Performative methods and practices of human and technical work, particularly in industrial production/consumption.
	Controls Innovations	Economic innovations = Monetary and financial mechanisms, including any price-related transactions.
		Ordinative innovations = Legal and regulatory as well as innovations in public administration and business management.
Formative Innovations	Formative Innovations	Political and cultural innovations of any kind other than the above.

Production activities are conditioned and controlled by economic and ordinative factors, as well as, taking a more comprehensive view, by political and cultural factors. For example, least-cost planning is an economic method for selecting among technical choices. A similar function is fulfilled by setting regulatory environmental standards. It apparently makes sense, on the effectuative system level, to recognise that regulatory (ordinative) and economic innovations are representing innovations in their own right which are as important as technical innovations. But it should be kept in mind that controls innovations are not effective by themselves but become so through processes of selecting and controlling specific operations, especially technologies and practices of producing and using goods and services.

Similarly, environmental awareness, as a cultural innovation, is an indispensable precondition for the success of any strategy of environmental innovation. But awareness, of itself, does not alter the industrial metabolism. Awareness, however, directly contributes to altering economic and ordinative controls conditions as it has an immediate influence on technology.

6.3 TECHNOLOGY AND TECHNOLOGICAL REGIME

Technological innovation, it can now be said, is innovation in the operative subsystem of society. A technology is a method of achieving a specified operative purpose by special operative means such as tools (instruments, materials, machinery, equipment, infrastructure) and practices. The set of principles and characteristics that make up a technology is referred to as a technological paradigm (Dosi 1982 158, Dosi et al. 1990 75–113).

Technology is a body of formative and effectuative knowledge, especially know-how, but necessarily also some theoretical know-why as well as know-what-for to make sense (also Freeman 1987 235, Kemp 1997 7). When talking about a technology, for example computing, one tends to think of physical things, typically some apparatus such as a computer, which is a product. Products – as much as production processes and services – are specific instrumental manifestations, apparatus-like implementations of a technology. Products, processes and services tend to be marketable, commercialised applications of a technology.

Any technology is based on, or represents, a respective set of human skill, know-how and further knowledge from wider domains. Even in cases of advanced automation, a technology originates with and is operationally dependent on personnel who are qualified to different degrees. The knowledge base and specific technology-related qualifications and skills of persons are an integral part of a technology.

A particular variety or form of a technology, which has successfully been established as the pre-eminent model of that technology by staying the test of innovation competition, is called the dominant design of that technology, as introduced in the writings of Utterbeck and Abernathy, Clark, or Teece. A dominant design in this understanding includes the dominant shape of tools (instruments, materials, machinery, plant equipment, infrastructures), not, however, dominant practices. A dominant design thus represents the key tools and design features of a technology.

The specific design of tools as well as of dominant practices are in turn guided by operative or technical regime rules, i.e. formalised or informal guidelines on how to do what and when. Examples of such regime rules are technical standards regarding size, shape, weight, or speed limits for motorised traffic, or limit values of emission standards, or dosages of agrochemicals and crop rotation in agriculture.

Since any technological activity is part of a structure of division of labour, i.e. a division of operative and other effectuative functions, technology is subject to various forms of cooperation, increasingly being investigated under the heading of network analysis. Aspects of organisation that directly relate to organising operations, i.e. cooperation, are thus also an integral part of technological innovation and a technological regime. The notion of a technological regime which builds up in the course of the development and the subsequent establishment of a technological innovation was introduced by Nelson and Winter (1982). 'Regime' refers to the fact that any technical solution to an operations problem is not just a set of tools but necessarily also includes some further 'practices and organisational settings' (Hoogma et al. 2001).

In the present text a technological regime refers to the ensemble of dominant-design tools (instruments, materials, machinery, plant equipment, infrastructures) and practices as governed by specific regime rules, plus all of the knowledge base and skills involved, plus all related and regularised cooperation (networks of any kind) among researchers, developers, producers, users, regulators, investors and may be some stakeholder groups.

Similarly, in Kemp (1997 269) a technological regime includes all the knowledge involved, skills, routines, regulations, patterns of working and living, social norms and commonly shared preferences. Kemp, Schot and others think of a technological regime as

> a relatively coherent complex of scientific knowledge, engineering practices, production technologies, materials and product properties, skills and procedures, and institutions and infrastructures that make up the totality of a technology. A technological regime is thus the technology-specific context of a technology which pre-structures the kind of problem-solving activities that engineers are likely to do, a structure that both enables and constrains certain changes. ... A technological regime combines rules and beliefs embedded in engineering practices and search heuristics with the rules of the selection environment (Kemp et al. 1998 182, Rip/Kemp 1998, Kemp/Rotmans 2001 7).

Starting from the distinctions made in II/6.2 one can distinguish between an operative notion and a more encompassing, broader notion of a technological regime. The operative notion would include tools, practices, operative regime rules, technical and environmental standards, structures of cooperation and the entire technological knowledge base. It would not include institutional arrangements, market controls, finance, administrative and managerial procedures. All of the latter aspects would be part of a broader all-effectuative concept of a technological regime.

6.4 THE QUESTION OF ECONOMIC ENDOGENEITY OF TECHNOLOGICAL INNOVATION

The term technology makes explicit reference to the cognitive, to the knowledge, logic and sense constituting a thing. According to Romer (1990) technology can be considered as 'targeted knowledge', as 'a set of instructions' to achieve something physically. Any material 'hardware' without its infor-

mational 'software' of knowledge would be a nonsensical and useless phenomenon. Hence, it is inappropriate to consider knowledge and machinery as
economic factors which can be substituted for one another, as is postulated in
most models of neoclassical new growth theory. Technical and organisational knowledge and skill on the one hand, and machinery and infrastructure
on the other, are complementary non-substitutional aspects of technology.
Technology thus represents one single factor, much in the same way as the
factor work incorporates the physical work *and* the mental activity, qualifica-
tion and experience that are required to carry it out.

Human physical work and machine work can partly substitute for each
other. This is possible indeed because both technology and work are operative factors, not economic ones. Operative factors and economic factors,
however, cannot substitute for each other. Economic factors can only be substituted for other economic factors, especially prices involved, such as the
price of labour and the price of physical equipment. The confusion between
operative factors and economic controls factors, the consequent confusion
between production functions and price-based financial repartition functions,
and the attendant confusion between operative productivity and financial
profitability, is a long-standing economic fallacy. With technology there
must be a non-economic utility function which fulfils a human idea, need, or
similar motive and practical purpose. There is in fact neither a utility function
nor any production function in the economy. The only functions which are
present in the economy are such as profitability, rentability, monetary interest, return on investment, or cash-flow functions of various kind.

Neoclassical economic growth theory has been criticised for treating technology as a factor exogenous to the economy. So the latest generation of new
growth theories have now adopted technology as well as knowledge and education as endogenous economic factors (Diebolt/Monteils 2000, Lucas 1988,
Romer 1990). A similar attitude can be observed in alternative approaches
such as evolutionary economics and institutional economics, albeit that
technology there figures partly in its own right, within the broader framework of a 'techno-economic system' (Kemp/Soete 1992).

Seen from a functional, systemic-evolutive point of view, the supposed
economic endogeneity of technical progress appears to be rather inappropriate. Technology is technology, primarily connected to and originating in
technology. It is an operative factor, not an economic one. Similarly, a particular 'techno-economic' system is no more real than a 'socio-technical', or
a 'politico-scientific' or a 'cultural-economic' system. These are just partial
views taken for special-purpose analyses. They may be useful within a particular context, but should not be reified. In empirical principle, there is an
operations system (of manifest physical actions by humans and their technologies) separate from the economic system, separate from the ordinative

system, separate from formative functions. A company, for instance, is neither a 'techno-economic' system nor simply an effectuative system. It is an institutional microcosm encompassing *all* societal functions in the sense of operations (technical), controls (economic and ordinative), and formative factors (political and various cultural functions).

Science and technology, as much as real work (not wages), are in fact exogenous to the economic system. Neoclassical economics could be criticised for paradigmatic colonialism, i.e. for imposing utilitarian economic modelling upon everything, rather than keeping within the limits which are drawn by the fact that the economy is just one of a number of societal subsystems, and actually not the dominant one as was supposed in the classical and Marxist thinking of the 19th and early 20th centuries. No societal subsystem could be said to be the most important one. To come back to the organism analogy, one would not consider circulation more important than motion or digestion. Money is not the only social medium that makes the world go around. Any subsystem is a system sui generis in that it fulfils specialised functions that are not and cannot be fulfilled by the other subsystems. By complementing each other, the subsystems maintain their niche within the entire system, and thus also maintain the niche of the entire system within its environment.

There are certainly interactions and interdependencies between the operative (technology-based) and the economic (money-based) subsystem, as there are interactions and interdependencies between other subsystems. Basically, however, all of these systems follow their own rules and 'laws'. Technology follows technological logic, just as the economy follows economic logic.

It does not make much sense to attempt to describe and analyse the industrial metabolism in economic terms. Technology is not a function of the economy, nor is the economy a function of technology. In genealogy, technology originates in formative functions, as do economic and legal controls mechanisms. Once it has been created and is evolving in its own right, technology develops from within itself, i.e. from the operative experience, skill, vision and knowledge of humans. New technology, then, by way of systems interaction, offers new opportunities (and new risk) for investors, as much as it may require new regulation.

Regulatory and financial controls factors are part of the selection environment of technological innovations (Nelson/Winter 1982, Kemp/Soete 1992). But legal standards or prices are in no way a primary cause of technological evolution. Prices, however, are partly a result of technical performance, i.e. productivity gains in the course of a technology's learning curve. Prices are thus indeed a controls factor, a selective secondary feedback factor, reinforcing or weakening the further development and diffusion of a technology.

7. The Innovation Life Cycle

7.1 APPROACHES CONTRIBUTING TO INNOVATION LIFE CYCLE ANALYSIS

Innovation life cycle analysis builds on a number of traditions neighbouring and partly overlapping each other. Among these underlying traditions are

- the anthropological study of cultural diffusion
- the sociological theory of diffusion of innovations as initially developed by Tarde and Simmel in the late 19th century, then more fully developed by Lazarsfeld, Merton, Coleman or Rogers in the 1950s–70s, and continued by others up to the present day (Rogers/Shoemaker 1971, Rogers 1995, Lopes/Durfee 1998), including
- theories of opinion leaders, change agents and innovation champions, and their particular application to a business context in the theory of innovation promoters by Witte and Hauschildt
- the theory of economic development as founded by Schumpeter and numerous other authors who have contributed to the literature on long-wave innovation cycles of structural importance. Related to this are
- the approach of evolutionary economics, especially by Nelson and Winter
- life-course analysis of 20th century industrial products and processes as in Abernathy, Utterbeck, Clark, Williamson
- and last but not least the empirical research on S-curve-shaped growth patterns of technology diffusion by Marchetti, Grübler and Nakićenović at the International Institute of Applied Systems Analysis.

In order to avoid misunderstanding it should be pointed out here that the life cycle of a technology or product is different from product eco-balances which try to assess the total environmental impact of a product from cradle to grave. Such eco-balances are also often called product life cycle analyses. In the context of this book this could be confusing since a life cycle in its proper meaning refers to the historical learning curve and diffusion curve of an item, figuratively most often represented in the form of an S-shaped or bell-shaped curve. Environmental product life cycle analysis should therefore indeed be called product eco-balance, or be renamed as product chain analysis.

7.2 THE LIFE CYCLE MODEL: STRUCTURATION, IMPACT, ACTOR FUNCTIONS AND DIFFUSION

Schumpeter's model of an innovation cycle continues to serve as a frame of reference for most scholars. The model distinguishes three stages: (1) invention, i.e. conceptualisation and creation of a new thing, referred to as R&D in an industrial context, (2) innovation, i.e. introduction into the market, alternatively named commercialisation, and (3) diffusion, i.e. growing adoption by users and the public in general up to the point of market saturation.

The model especially fits a perspective of mass production and marketing of a standardised physical product used or consumed in mass quantities. The product itself is seen as an application of prior technological progress which is then applied in a company-specific R&D process towards the specific production and marketing goals of that company. It seems that after almost a century of ongoing development the model is in need of some further differentiation, since it tends to amalgamate quite different aspects.

To begin with, the term innovation (as the process rather than the thing which is being innovated) shall encompass the whole process of structuration and diffusion of a novel thing from the beginning through to the end of a life cycle, not just a particular stage thereof, nor a special sub-function within the innovation function. It is not appropriate to restrict the term innovation to the special stage of main unfolding and diffusional take-off of an innovation, i.e. market introduction and mass customisation.

Diffusion generally signifies the process by which an innovation is adopted. According to diffusion theory, and different from a Schumpeterian view, diffusion of an innovation does not start sequentially after it has been introduced into the market. Diffusion starts right at the beginning of any innovation life cycle, i.e. with the idea or perception of an innovation and the communication processes accompanying it. Diffusion then continues throughout the different stages of a life cycle. An innovation's potential for diffusion, i.e. its niche, may be more or less actualised, and may itself evolve.

With regard to these questions we follow the approach of diffusion theory, albeit diffusion theory itself is also in need of further differentiation too: The stage of structural evolvement of an innovation and the degree of its diffusion rather than being identical, represent two different aspects of an innovation's life cycle.

Schumpeter intended his model to apply to technological systems change, i.e. to the introduction of new technological paradigms or entire new product families rather than to incremental innovations of things already introduced and well-known. Underlying this view is Schumpeter's analysis of business cycles, especially long-term Kondratievs (long waves) constituted by the introduction of new key technologies such as the steam engine, chemical dyes and fertilisers, or electricity. Schumpeter's focus on long-wave structural change probably counts among his greatest achievements in innovation and modernisation theory. However, the question of an innovation's structural importance needs to be treated separately from the question of evolutive stage, diffusional stage and life cycle functions.

So we can proceed by resolving the Schumpeterian model and the model of diffusion theory into several distinct dimensions. An innovation and its evolutive life course, as will be discussed subsequently, can be laid out in four dimensions. These are interrelated but should be kept apart in order to properly analyse their interplay:

1. The structural impact of an innovation, i.e. its impact on structural change.
2. Stages of structuration of an innovation. The structure of an innovation refers to what it consists of, what its form is, and how it fits into and connects with different elements in its environment. An innovation's structure in this sense determines the structural impact an innovation has.
3. Different actor functions in a life cycle (creation, production, use, finance, regulation, communication/coordination).
4. The degree of diffusion of an innovation, i.e. the reach and frequency of its utilisation by populations of adopters.

7.3 STRUCTURAL IMPACT

7.3.1 System Innovations and their Evolvement by Incremental Innovations and Modifications: High versus Low Impact

An innovation's impact on structural change refers to the distinction whether an innovation is considered to be of great importance, i.e. high structural impact, or whether it is of minor importance, i.e. low structural impact. An innovation of high impact induces far-reaching changes of the status quo. It has ample evolutionary potential so that it can be called a fundamental innovation (Schumpeter's Basisinnovation), or systems change, key innovation, or similar. An innovation is of minor importance or lower impact when it induces smaller and perhaps more short-lived changes of the status quo. A low-impact innovation is usually referred to as an incremental innovation, or a modification, the latter actually conserving rather than changing an existing structure. Innovations are thus categorised according to whether they represent a high-impact scenario or a low-impact scenario of structural change.

Ever since Schumpeter tried to circumscribe the introduction of a new technological system, to be treated as different from ongoing technical progress based on already established technology, a variety of further terms have been suggested. Probably the most widespread is the distinction between radical and incremental innovation. Calling an innovation radical signifies that it is breaking new ground, introducing, in the words of Dosi, a new technological paradigm the evolvement of which opens up new evolutive paths. A radical innovation introducing a new technological paradigm is characterised by relatively high uncertainties, in contrast to the normal incremental progress of a technology which follows a well established trajectory with less uncertainty.

Another suggestion is generic innovation (Clark 1987 34). The meaning of it again comes very close to Schumpeter's Basisinnovation. A generic innovation brings about a new technological system or a new product family such as rail traffic, the automobile, assembly lines, computers, etc., thus new fields of venture, long-term investment, new markets and occupations.

Impact and frequency of innovations are inversely proportional to each other. Major generic system innovations are rare, and their chances of occuring are limited to a rather narrow window of opportunity that may open up only once every several decades. Middle-range and minor incremental innovations, by contrast, are what occur normally and relatively often in greater numbers, although no truly significant innovation is an everyday phenomenon. That the frequency and impact of innovations are inversely proportional to each other is a self-evident consequence of the nature of a system's learning curve on its evolutive life course (II/7.7).

By definition, incremental innovations of major or minor significance do not constitute life cycles of their own. Their raison d'être is fully identical with the life cycle of the (sub)system or component they belong to. What constitutes a life cycle are the major, high-impact innovations on the (sub)system or component level. Depending on the case under consideration, however, such a life cycle will often be a subsystemic integral part of the life course of the whole system, or the supersystem it belongs to, whose general trajectory it follows and whose main characteristics it shares.

7.3.2 Systems, Components and Parts

It can be useful for certain purposes to return to the distinction between a system as a whole and its different elements on different levels. With regard to technical systems we can speak of functional components (= technical subsystems) made of assembled and single parts (= sub-subsystems or basic elements). In the same way, Henderson/Clark (1990) have introduced a distinction between architectural innovations relating to the entire set of configuration principles of a complex product or technological system, and modular innovations relating to different components within that system.

Furthermore, many technical systems need an operational infrastructure which also consists of functional components and lower-level parts. Radical innovation then describes a generic system innovation as well as the innovation of new structures within a generic system. Consider computers, both hardware and software. Computing machines and programming languages represent a complex system innovation or generic innovation, leading to a family of product varieties such as mainframes, personal computers, laptops and handhelds. Typical functional components of computing systems are different kinds of drives and memories (internal/external, ROM/RAM/DRAM/ MRAM, SIP, electric/magnetic), screens (tube-based or LCD, low/high pixels), entry devices (keyboard, mouse, touchpad), telecom interfaces, operation software and application software.

Or consider the automobile which has an infrastructure of roads, fuel provision and maintenance services, and is made up of a number of functional components such as an internal-combustion engine or an electric motor, gear system, brake system, coachwork, chassis, suspension, steering, electrical system, etc., all of which are by themselves quite complex subsystems and can again be broken down into assembled or single parts (sub-subsystems). While the motor car as we know it at present represents a vehicle system already at a late stage of its life cycle, its components and elements may still undergo radical changes on the basis of major generic innovations.

7.3.3 Successive Generations of an Innovation

The evolvement of computers and their functional components were closely linked to the innovative steps that led from radio valves or tubes to wired transistors and then to semiconductors which enabled the construction of microprocessors on silicon chips. Steps like these can be described as radical component innovation, whereby the miniaturised circuits on silicon wafers represent a true key innovation which today has been incorporated into almost everything. Since its introduction the 'chip' has followed the course of a typical efficiency-increasing learning curve from the 8, 16, 32 and 64 bit generations of microprocessors to 286 and 386 processors, to a series of Pentium processors I, II, III, IV. This learning course prototypically exemplifies successive steps in a process of major incremental improvement – major, because the successive additions to the quantity of computing capacity enabled qualitatively new dimensions of computer applications.

Successive variants of a technology or product which occur within the frame of its learning curve of structuration, and which introduce technically improved and, more importantly, structurally changed properties on the level of the system or its components, are called generations. Examples of contemporary technology include, again, the generations of mainframes and personal computers in general, or hard drives in particular. First-generation (1G) hard drives were introduced in the mid-1950s. They were made of 50 platters and weighted a ton. Later on they were improved by thin-film induction heads using a single element to read data from and write data to the disk. The breakthrough to second-generation (2G) hard drives was achieved around 1990 with AMR read-heads (anisotropic magneto-resistive), being more sensitive to weaker signals from smaller bit-carriers. Storage capacity increased year after year and reached one gigabit per square inch. 3G hard drives with multi-layered GMR read-heads (giant magneto-resistive) which use processes at the quantum level were already possible in 1997. Storage capacity is now at about 30 gigabytes per square inch and is bound to reach 120 gigabytes in the near future.

With mobile phones, there have been two generations so far, and a third one in the pipeline. 1G mobile phones from about 1970–90 were based on analogue voice transmission, whereas 2G mobile phones available since the early 1990s are based on digital encoding enabling limited data services such as text messaging and wireless access protocol (WAP). These are now being further enhanced for higher transmission rates and are thus referred to as 2.5G. True 3G mobile phones of the near future will offer full Internet access and videotelephony based on high-speed, always-on data connections.

Similarly, three generations of photovoltaic cells can so far be distinguished. 1G solar cells are those state-of-the-art black and blue-coated panels made of crystalline, polymorph or amorphous silicon with an efficiency grade of 12–16 per cent, already quite near to the technical maximum of about 25 per cent. 2G, close to market launch by now, are made of different materials (metal compounds). Their advantage is not a better energy yield, but considerably reduced materials requirements (50–100th of volume, less purity) and fewer production steps. 3G solar cells of the future, now under research and early development, will almost certainly be made of organic polymers in combination with inorganic nanomaterials (I/4.2.7.1).

In oil refinery, to quote an example of process innovation, there was a succession of three technology generations of cracking crude oil. 1G cracking was thermal. Transition to 2G cracking towards the end of the 1940s and during the 1950s brought catalytic cracking. 3G practised since around 1970 is hydrocracking. Another example of generation succession is the substitution of steel for cast iron and wrought iron between 1870 and 1930.

The examples demonstrate how new generations of a technology maintain the specific purpose, function and use, and build on the cumulative knowledge base of the technology. They can nonetheless radically change the technical system as such or certain components therein.

7.3.4 Synopsis of Attempts to Define High-impact Innovations

Further attempts to coin terms which describe technological innovations of major structural impact include breakthrough technology (Angel 1994), defining technology (Bolter 1984) and key technology, the latter representing a general usage. These terms too have a meaning similar to Basisinnovation in the Schumpeterian sense. Whereas breakthrough refers to the aspect of breaking new ground or heading for new frontiers, the terms defining or key postulate that certain new technologies become the starting point for a greater variety of new products and processes in a broad range of applications. Key technologies in this sense are pervasive in that they are taken up in a large number of sectors and applications, and introduce structurally new technical systems, components or product groups.

Table 7.1 summarises the categories used for describing the structural significance of technological innovations.

Table 7.1   Synopsis of categories referring to the structural impact of an innovation

Authors	High impact Major significance	versus versus	Low impact Minor significance
Schumpeter 1911, 1939	Basisinnovation	Ongoing technical progress	
General usage	Radical innovation,	Incremental innovation	
Kemp/Soete 1992	System innovation	Drop-in-innovation	
Mensch 1979 Albach 1994	Basisinnovation	Improvement innovation	Pseudo-innovation
Huber 1995, 2001	System innovation	Unfolding innovation	Modification
Dosi 1982, Dosi et al. 1990	New technological paradigm		
Clark 1987	Generic innovation	Incremental innovation	
Angel 1994	Breakthrough innovation		
General usage	Key innovation		

Applying the categories discussed so far to real empirical cases can reveal that they do not always represent clear-cut definitions. Real cases of innovation often have more features to them and demand additional specific characteristics. But the orientation given by the general categories discussed here so far is by and large correct and relatively close to empirical reality.

7.3.5 The Tree of Technology: The Role of Disruptive and Sustaining Innovations in Ongoing Modernisation

Christensen (1997) and Ashford (2001) describe the structural impact of major innovations according to whether they are disruptive or sustaining. The notion of disruption represents another way of reasoning about what Schumpeter called an innovation's structural impact of creative destruction. Railway and motorcar were disruptive to horse and carriage, which is an example of a succession contest between new and old technologies. Today, there are new wireless telecom technologies emerging such as smart antennas, mesh networks, ad hoc architectures and ultra-wideband transmission. They promise to be more effective and many times more efficient than conventional technologies. If they were to become successful, they would clearly be disruptive to conventional telecom technologies, even to their most advanced 3G varieties. This is an example of a niche contest between several new technologies for becoming the dominant design (II/9.4).

By contrast, new electronic road and car controls systems represent a generic innovation that introduces a new component of traffic infrastructure and which represents a sustaining addition to the existing system whose overall evolution is cumulative. This can be seen from the fact that the existing system has not been replaced and has time and again incorporated new structures.

Computers, another example, were first thought to be disruptive, for example in a transition from paper to paperless, and from traditional literacy of reading, writing and calculating to some sort of allegedly brainless artificial intelligence. Soon, however, computers proved to be a highly sustaining key technology by keeping and contributing to technical progress in almost every field and by creating new options and an enlargement of access to information for almost everybody. A new 'electronic divide' was identified between those with and without computers and access to the Internet. Schumpeter's categories, it would appear, need an additional specification: A generic system innovation which represents a new technological paradigm, 'creative' as it is, does not necessarily come with a 'destructive' impact but can be sustaining, which indeed is quite often the case.

Innovations of the type Schumpeter had in mind represented generic industrial system innovations, laying the foundations for new industries, markets, vocational occupations and enlarged lifestyle choices, occurring in a rhythm of long waves of about 4–6 decades in length. Such waves of technological innovation can be disruptive to predecessor technologies which traditionally fulfilled a comparable function, but new technology as such does not necessarily imply the replacement of older ones (Freeman/Soete 1999). Disruptive innovations seem to have occurred more often in the 19th and the first half of the 20th century, whereas later innovations have increasingly tended to become sustaining or cumulative. The reason is that those long waves of industrial revolution step by step replaced traditional production practices by modern technology, and the more the historical great transition, as Karl Polanyi put it, from tradition to modernity was advanced, the more modern science and technology established itself as a coherent operative complex.

Today's technological innovations are part of a family tree of modern science and technology to the ongoing growth of which they contribute. That is why new key technologies today mostly add to the existing structures of technology and industry rather than replacing them. The family tree of modern science and technology, or, as one can also say, the tree of industrial sectors and markets, point first to the overall structure as a whole: a tree of sustained modernisation with ever more new branches blossoming and adding to the whole. While there are new branches growing, the tree keeps its roots. The number of new branches which are added to the common stock is greater than the number of old branches which die off whether through natural causes or through being killed off by the competition.

When relating innovations more specifically to their historical role in societal modernisation, two cases ought to be distinguished:

a) the historical transition, called industrial revolution, from traditional pre-industrial society to modern structures of management, finance, markets and industrial production, with modern technologies and products replacing traditional ones,
b) today's transition from now long-established industries, that have become old industries, to high-tech society, as part of an ongoing process of self-modernisation of modernity, with the substitution of the newest high technology for older-generation industrial technologies.

Consider a number of cases, for instance long-distance roads, a case in which traditional structures were totally modernised without having been replaced. Roads represent an infrastructure common to both traditional and modern societies. The road as a generic infrastructure may be complemented by other traffic infrastructures (rail, air) but it will not be replaced. It will nevertheless continue to undergo certain major or minor component innovations – new construction principles, new materials, multi-lane motorways, and there is now a new infrastructure of ICT-controls systems being added to the motorways. Similarly, pulping and paper making has been mechanised, chemicalised and partly automated, but the materials and methods of pulping and paper-making continue to remain basically the same as in ancient times. Similar statements could be made on other natural fibres such as wool and its use in the manufacture of textiles, and similarly also on most agriculture and breeding during industrialisation. Those things have been mechanised, chemicalised and partly automated, but the fundamental methods and the basic metabolism involved have in principle remained unchanged and have become integral to modern structures too. This is at least still the case, it may of course change in the future.

A typical example illustrating now the transsecular transition from traditional pre-industrial to modern technostructures has already been mentioned, i.e. the motorcar replacing horse and carriage, or the steam boat replacing the sailing ship. These are rather spectacular cases of disruptive innovation, similar to the replacement of horse power and human drudgery by steam-powered and electricity-driven factory machines. Less spectacular, though equally important and disruptive, was the transition from wood as a fuel and material to coal as the main fuel, and iron as a materials substitute for wood. In addition to coal, further fossil fuels were introduced in the course of time, especially mineral oil and natural gas. Though leaving coal with a comparatively smaller market share within the modal split of fuels, these new fuels came as additional ones. Their effect on coal was complementary or cumulative rather than disruptive. A rather late example of the transition from traditional to modern production is the assembly line, representing an advanced stage of mechanisation by putting together taylorised elements within the organisational frame of the assembly mega-machine. That was definitely disruptive to the remnants of traditional crafts practices in factories, but – and this is important to note – it was very sustaining to industrial manufacturing.

An example of more recent processes of ongoing self-modernisation of modernity, i.e. the transition from old-industrial technologies to even more advanced newest-generation high technology, is the transition from workers to robots and operators, i.e. to semi and fully automated processes of materials processing, assembly and shipping. This has proven to be disruptive to old-industrial workers, the ones with the big spanner in their strong fist, many of whom have recently been removed to the newly created museums of industry and labour, whilst technicians and engineers together with a few clean-collar workers have taken over on the factory floor. Similar examples include the substitution of electronics for mechanics in a number of fields, such as the pocket calculator having replaced the slide rule and clumsy mechanical devices, the electronic watch having replaced the precision-mechanical watch, the PC having replaced the typewriter. The old mechanical industries exposed to that change, and failing to adapt and assimilate in time, underwent difficult crises of structural change, or went bust and have disappeared. The overall effect on the technostructure in place, however, rather than being disruptive, has been sustaining, enlarging and enriching, thereby adding to the family tree of modern technology. Similar earlier transitions had replaced steam power by electricity or gasoline (Freeman/Soete

1999 59–79). The final replacement of steam as it is still used in thermal power plants (steam-driven turbines) is yet to happen.

Another replacement by self-modernisation which is entering the arena today is the decarbonisation of the energy base through the phasing out of fossil fuels. High-tech industrialisation will almost certainly do away with carbon-burning in favour of ecologically sustainable clean energy (I/4.2). Smokestack industrialisation was modern in its time but has by now become an old-industrial epoch, a closing chapter of modernity. High-tech clean energy sources are an example of generic system innovations that are disruptive to one branch of old industrialism, in this case fossil energy, but obviously add to and sustain the entire operative system of advanced high technology.

The impact of generic system innovations can be generalised in this way: disruptive as they may be to traditional pre-industrial and old-industrial approaches, they sustain the evolution of ever more advanced technology of later stages of modernisation. In this sense, all of the Kondratiev or Schumpeter long waves, even though some of their achievements have partially been replaced in the meanwhile, have brought about some lasting generic innovations which have added to the tree of technological modernisation rather than fundamentally challenging it:

- Mechanisation around the beginning of the 19th century
- Railway, steam ship, telegraph, dyestuff, fertilisers around the middle of the 19th century
- Electrification, telephone, motorisation, taylorisation (assembly line), synthetics around the beginning of the 20th century
- Mass-traffic in motorcars and aeroplanes, automation, mass-media (radio, TV), pesticides, plastics since around the middle of the 20th century
- Computerisation, TC-networks and the Internet, photonics (lasers), specialty chemicals, add-on envirotech, transgenic crops towards and around the end of the 20th century
- Already underway, but still to come on a large scale as epoch-defining technologies: more transgenics, clean energy, more photonics, microtech which combines electronics, optics and mechanics in a range of hundredths to thousandths of a mm ($\approx 10^{-6}$ m), and nanotech for creating miniature parts and materials surfaces in the order of ten to hundred thousandths of a mm or even less ($\approx 10^{-9}$ m).

To be sure, any life cycle in time and space is bound to reach its end. So will processes of ongoing modernisation of society in general and technology in particular. Since, however, we are dealing with a transsecular life cycle which has been pursuing its path for centuries now, do not expect to see that end coming anytime soon, not at least for endogenous causes.

7.3.6 High-tech versus Low-tech Innovation

As awareness of the importance of technological innovation for national wealth and power has grown, politicians have got keen on it, and economists and statisticians have come up with measures that indicate the state of technological modernisation and innovativeness.

One approach is to count new products, machinery, equipment and buildings as a percentage of all products, or the ratio of new ones to older ones. The results are replacement rates. They deliver information on whether an industry's or an entire economy's vested capital is new or old. But whether it is really up to date may be a different question. The latest collection of the season is certainly new, though not necessarily innovative. Replacement rates do not properly disclose whether the newly bought items represent latest-generation technology or yesterday's innovations in some modification.

A different approach is that of counting patents, pending and granted, from locals and foreigners, at home and abroad. Patenting practices, though, differ considerably across the world. That is why statistics on patenting can be difficult to interpret. So another measure has become the key indicator for measuring levels of technological innovativeness: expenditure on research and development.

All of the then new industries of the 19th century were based on scientific research from the beginning, the chemical, pharmaceutical and electrotechnical industry as much as machine and rail construction or motor vehicle manufacturing. Engineering as a genuine field of science with a number of faculties of its own grew up with those industries. Most of the research was still done outside the factories, in traditional and technical universities or in comparable academic research institutes.

As the transsecular process of science-based technisation gained further momentum in the 20th century, it became a decisive advantage in business competition to be the first to have marketable R&D results. Ever more of the bigger companies started to have their own scientific R&D departments, whose importance to their industry grew, and so also grew the share of personnel and means they absorbed. It has become an established fact that technologically advanced industries or products incorporate a high degree of R&D. They are R&D-intensive. Conversely, the more R&D an industry or product line incorporates, in-house or by way of outsourcing, self-made or acquired, the more technologically advanced it is (OECD 1996).

There has been talk about high tech for thirty years. For about twenty years now it has been defined in terms of R&D intensity. This means R&D expenditure above the average of about 4 per cent of turnover of companies or industrial sectors. According to Grupp (1998 208), true high tech, the one he calls leading-edge technology, is characterised by an R&D expenditure of over 8.5 per cent relative to turnover, whereas the rest of high tech, called high-level technology, ranges from 3.5 per cent to 8.5 per cent. Examples of today's leading-edge technologies include aircraft and spacecraft, advanced optics, precision instruments, pharmaceuticals, computers and transgenics. Typical high-level industries, in contrast, are automobiles, special machine tools, cameras and many chemical products. Products below the 3.5 per cent threshold of R&D are usually referred to as low-level or lower-level products such as steel and bulk chemicals, agro-products, food, textiles and furniture. High-level tech, then, is actually medium tech, representing yesterday's leading-edge achievements, though still constantly being upgraded by incorporating elements of leading-edge technologies.

Grupp calls the leading-edge and high-level technologies 'Schumpeter industries', whereas the lower-level technologies are deemed 'Heckscher-Ohlin industries', referring to neoclassical market models by Heckscher and Ohlin in which prices and cost competition alone are decisive, while supply differentiation by product quality or product innovation do not play a role. Distinguishing Schumpeter markets from Heckscher-Ohlin markets, handy as it seems to be, may not be fully appropriate. Even though it is true that conditions roughly approaching the Heckscher-Ohlin type of economic cost reasoning tend to prevail in mature industries in their retentive stage, lower-level industries cannot necessarily be said to have become cold spots for innovation. There is some innovation in lower-level technologies too, for example innovation of new components or new materials, and if not for endogenous causes than certainly for exogenous impulses. There are always innovative trickle-through processes, top down or inside out, i.e. some restructuring also of old and very old industries by incorporating and assimilating new elements and components of later and latest technologies of the day. Such processes can easily be seen in agriculture where every wave of technological innovation caused some innovative restructuring, for example by successive steps of mechanisation and successive generations of agrochemicals, up to ICT-based precision farming and transgenic seeds which are today's leading-edge features on their way to the farm.

7.3.7 The Modal Split within Functional Clusters

Products serve purposes. The recent shift in environmental policy from product-orientation to purpose-orientation may not be the great discovery its promoters think it to be (I/3.4.3), but it certainly remains useful in its downstream ramifications. It may help to restructure product stewardship, trade structures and producer-consumer relations in general.

Whenever a general purpose, i.e. a functional cluster, can be identified it entails a modal split. The term modal split is familiar from transport. The modal split there is the choice between ship, rail, tram, bus, bike, car, truck, air, etc. Analogously there is a modal split of fuels, building materials, fibres or food. Examples are given in Table 7.2.

Table 7.2 Modal split within different functional clusters

Functional cluster	Modal split
Fuels	Wood, peat, charcoal, hard coal, brown coal, mineral oil, natural gas, uranium, hydrogen, ...
Motors and engines	Steam, internal-combustion Otto, Diesel, Wankel, hot gas Stirling motor, electric motor, powered externally or by battery or fuel cell or wind-up device, propellers and jet propulsion, ...
Transport and traffic	Ship, rail, tram, bus, bike, car, truck, air, ...
Materials	Wood, metals, clay, stone, concrete, glass, plastics (poly-ethylene PE, polypropylene PP, polyvinylchloride PVC, epoxy resins, polyester, polystyrene, ...)
Fibres	Wool, silk, cotton, bast (linen, flax, hemp, jute), cellulose (viscose, triacetate), synthetics (polyamide = nylon, perlon, polyester, polyacryl, PVC, ...), carbon fibres, glass fibres, ...
Food	Cereals, vegetables, fruit, nuts, roots, dairy products, meat, eggs, fish, ...
Housing	Stand-alone family home, terraced house 1–2 floors, multiple dwelling, large house with many flats on 3–6 floors, residential block, high-rise building
Further clusters	... ...

A modal split refers to the existence of different available alternatives which render the same or a similar service. Any such modal split represents a situation of niche competition – sometimes temporary, sometimes lasting; sometimes fierce, sometimes, and for the most part, relatively mild. This depends on the degree of actual substitutability and complementarity of different alternatives fulfilling a same function.

7.4 STAGES OF STRUCTURATION

A life cycle can also be called a life course. A course follows a path, a direction, or a trajectory, and may change its path and direction. The general evolutive direction of any life cycle can be said to progress from ill-defined openness and fluidity of an emerging system structure to well-defined specifity and maturity, as illustrated for 20th century industrial product and process innovations in Figure 7.1 by Clark (1987 106).

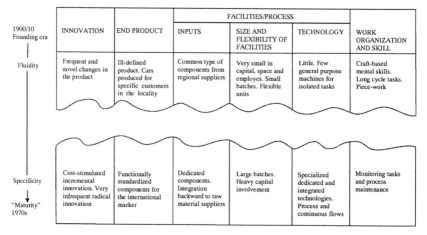

	INNOVATION	END PRODUCT	FACILITIES/PROCESS			WORK ORGANIZATION AND SKILL
			INPUTS	SIZE AND FLEXIBILITY OF FACILITIES	TECHNOLOGY	
1900/10 Founding era ↓ Fluidity	Frequent and novel changes in the product	Ill-defined product. Cars produced for specific customers in the locality	Common type of components from regional suppliers	Very small in capital, space and employes. Small batches. Flexible units	Little. Few general purpose machines for isolated tasks	Craft-based mental skills. Long cycle tasks. Piece-work
Specificity ↓ "Maturity" 1970s	Cost-stimulated incremental innovation. Very infrequent radical innovation	Functionally standardized components for the international marker	Dedicated components. Integration backward to raw material suppliers	Large batches. Heavy capital involvement	Specialized dedicated and integrated technologies. Process and continuous flows	Monitoring tasks and process maintenance

Source: Clark (1987 106).

Figure 7.1   The life course of technologies from creative fluidity to mature specificity

There are various ways of categorising subsequent stages of a technology life cycle, for example two stages as in Schumpeter (invention and innovation), or three stages as in Mensch (1979) and the OECD's Oslo Manual from 1992 (starting with the development of a new product, followed by incremental improvement in performance, and coming to an end with just minor technical or aesthetic modifications). In Grossmann (2001) there are as many as seven stages. All of these models basically converge. This is apparent from the fact that they are all represented by an S-shaped learning and growth curve, or a bell-shaped curve of growth *rates*, increasing until inflection point, decreasing thereafter (Figure 7.2).

Following some common usage, the structuration cycle of a technological innovation will be described here in five particular stages as shown in Figure 7.3:

1. Early formation by initial invention or discovery, through research, experimental exploration or making things on a handicraft basis
2. Advanced formation by organised development in successive steps
3. Accelerated or prime unfolding of an innovation
4. Maturation, and
5. Retention or decline, phase-out, replacement.

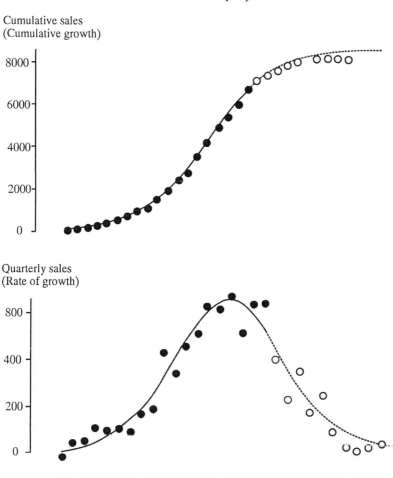

Note: At the top, growth of cumulative sales in Europe for Digital's minicomputer VAX 11/750. The S-curve shown is a fit carried out in 1985. The dotted line was a forecast at that time; the little circles show the subsequent sales. ...

Source: Modis (1992 58).

Figure 7.2 Representation of a life cycle as an S- or bell-shaped learning and growth curve

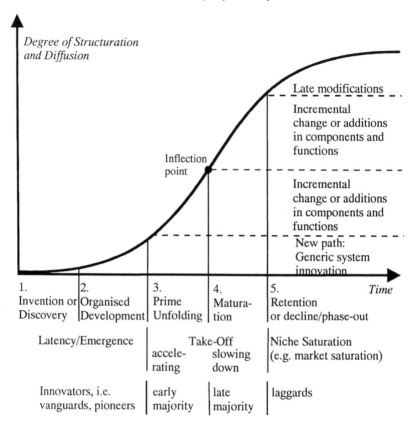

Figure 7.3 The innovation life cycle

The S-shaped life cycle as shown in Figure 7.3 represents a simple and robust model. As an analytical tool it should be seen as a heuristic metaphor,
not as a blueprint for cloned realities. The model, though, is not taken out of
thin air but based on the evidence of many hundreds of empirically documented cases of real innovation processes all displaying the characteristic
S- or bell-shaped form in manifold variants (flat or steep, short or long, sym-
metric or asymmetric, complete or fragmentary/unfinished). In all those cases
where an innovation is allowed to follow its path, i.e. where it is not diverted
or disrupted by dramatically changed internal or external factors, its trajectory comes close to the general model. (For a systematic overview of examples see Modis 1992, and all of the writings by Marchetti, Grübler and Nakićenović.)

The learning curve of a structuration cycle represents a transition from before the process to after the process. On the basis of systemic interrelations as

explained in II/6.2 such a transition is not just technical but includes co-directional structural change in connected societal functions; i.e. it is related to co-directional readjustments, if not co-transitions, of economic and ordinative controls systems as well as of political and cultural impulses.

A life cycle can be understood as a process of actualisation of evolutive potential, from first ideas and concepts via step-by-step development on to prime unfolding and maturation. Evolvement has the meaning of unfolding of the structural and functional potential of an innovation. It can also be described as a step-by-step dialectic of realisation of possibilities of a potential, resulting in a reinforced or rearranged structure of remaining or ensuing possibilities, some of which are in turn realised again, and so on.

In the course of its evolvement, an innovation ceases to be innovative and becomes an assimilated part of the stock of existing and established structures. Qua definition one could say, it is the inflection point of the curve that marks the transition from innovative to reproductive; but it may also be appropriate to say that it is the beginning of the retentive stage which marks the end of innovativeness.

Distinguishing stages of a life cycle is a necessity in that the ensemble of qualitative and quantitative characteristics of a system changes over time. This is what evolvement is all about. Distinguishing, however, four or five instead of, say, three or six stages represents verbal convention which relies on experience rather than distinct categories with clear-cut boundaries. Basically, the learning curve of an innovation is an ongoing process of structural unfolding, not necessarily in a continual smooth flow, not necessarily unchanged in momentum and direction, but in effect nonetheless progressing.

As far as successive stages of a life cycle are defined, any of these can be subdivided or put together according to the special subject and purpose of an investigation. For example, (a) invention and (b) organised development have been categorised here as two periods in the formative stage of a life cycle. There is no need to start a controversy over whether this represents one or two stages. Invention and development are called formative stages because they are about shaping emerging realities according to ideas and interests on the basis of knowledge and know-how, whereby there are in turn important additions to, or restructurations of, the knowledge base of an innovation in the course of basic learning processes during the formative stage.

7.4.1 Early Formation by Initial Invention or Discovery

Schumpeter referred to the first step of an innovation life cycle as invention. Quite often, though, something new happens to be (re)discovered rather than invented, or something going on may just be perceived for the first time and considered to be new. Whether invention or discovery, the stage of early for-

formation starts at that moment when someone gets an idea of something new and begins working it out and communicating it.

In some of the life cycle models the process of getting inspired with some novel thing is treated as a stage in its own right. This is because many ideas do not make it so far as to become a structured venture of systematic exploration, experimenting and first prototype designs. New ideas, it is said, are a thousand a penny. As a rule of thumb, only one out of ten ideas will seriously be submitted for further investigation. Out of these, only one or two end up in small projects. And from these, just one or two are selected for becoming new developments and, perhaps, new market launches. Not every launch, finally, is bound to become a success. As a result, the survival rate of new ideas may even be less than one in a thousand.

An idea may be brilliant. Or just erroneous. At the beginning one does not know yet. That is why we cannot consider any new idea to be the beginning of an innovation. And that, at the same time, is why we can define a viable invention as an idea, the feasibility of which can be demonstrated or made plausible. So, early formation by invention or discovery is an evolutive stage that includes having an idea and exploring its feasibility, working it out to stabilised experimental demonstration, first model, or laboratory or garage prototype. Invention certainly begins with an idea or a significant discovery, but it continues its existence by exploring and working out the real structure of the new thing in detail. Thus it also lays the foundations for the technological regime to come (and its evolutive potential at that time), and it ends with technological demonstrations and prototypes, which in turn are the beginning of the next stage of organised development.

In most of the important cases, the stage of invention ends in applying for a patent by disclosing detailed information about the innovation to the patent office. If there is a pending patent on a novel item, this signals the end of successful invention and the beginning of advanced formation by organised development. A purely legal definition of invention, however, would not do because there are many new practices that are not patentable or not worth the effort of patenting, and because patenting also occurs in later stages of structuration. There may also be cases in which inventors do not want or do not have the means to have their invention patented. Usually, though, they want to, because patenting comes with advantages both for the patent holder and for the common good, in spite of the fact that only two out of hundred patents are applied commercially.

Patents grant the holder a limited period of 20 years during which the holders have an exclusive right to make use of an innovation, i.e. it is protected against unauthorised replication or use by others so that possible benefits can be appropriated by the patent holder individually. This creates a strong incentive to inventiveness as so far as inventiveness is rooted in utilitarian rather than idealistic motives. When patent protection expires, the innovation and all official information about it becomes part of the public domain so that everybody else may reproduce it without authorisation. Today, as industrial R&D has become complex and expensive so that new technology needs much pre-financing before it earns money, individual and corporate inventors tend to licence their invention to someone else as soon as possible, because this creates a cash flow right from the beginning, or they try to bring their patent into joint ventures with solvent partners. According to a recent estimate by Baumol (2002), about 20 per cent of the total economic benefit of a patented innovation is appropriated by inventing firms, whereas 80 per cent spill over in those various ways to the (inter)national economy at large.

Invention and later development often come in rival approaches, representing competing varieties, if not competing paradigms. Variety generation is a typical characteristic during the early stage of formation, particularly when an idea's time has come and 'is in the air' to those who are receptive enough to make out which might be the next thing worth exploring. Variety generation ought in general to be encouraged during formative stages of any kind. Selective realities will then make felt themselves anyway.

Inventors are the vanguard or pathfinders of an innovation, in contrast to later development pioneers and further adopters and followers. Inventors tend to be creative visionary people, though this is not a precondition. An invention can be made by an individual or a group, within academic research or industrial corporate laboratories, or by amateurs and private enthusiasts, who may form associations for the advancement of their novel paradigm. While inventors in the past were quite often non-experts, nowadays they tend to be well-trained specialists. They continue, however, almost certainly to be outsiders to those communities who represent established like paradigms.

7.4.2 Advanced Formation by Organised Development

The second stage in an innovation life cycle is organised development. It is still a formative stage, but one in transition to regular production and use. Development is about putting an idea or new principle into regular practice. The innovative venture is transformed into a coordinating endeavour of organised practical research and development, which is aimed at making an invention operational, by working out how to use and produce it in a regular way. This usually takes place within an industrial setting. Further aims are to connect it to existing realities beyond technology and production methods. These include patterns of use and social behaviour, regular finance, market supply and demand (sorting out cooperation and competition), development of new regulatory standards and regime rules. This organised development also includes cooperation between firms and authorities, or compliance with existing regulation, or, if relevant, dealing with cultural (non)acceptance.

Organised development is the time of the pioneers. Their pioneering mission, to retain that metaphor, is building roads and bridges into the new territory, taking possession of it, and making it ready for regular use by those who will follow. The participants are now exclusively professional, forming an (inter)national innovation network of academic, industrial and government agencies (II/9.3).

The pioneers run pioneer companies, in the order of 10–100 collaborators. During the recent new-economy hype they won some fame as start-ups. Another term calls them development-stage companies. Some of them are listed at the stock exchange. This is possible to the degree that they attract venture capital on the basis of 'market fantasy', and cooperation with some market heavyweights, i.e. well-established corporations acting as lead users, or otherwise contributing to funding long-term development works.

Organised development entails ever more production activities in statu nascendi. Technology or product development can thus also be seen as an intermediary or link between creation and production. Development continues to contribute know-what and know-why to the knowledge base of a novel technology, but its main contribution now is increasingly new know-how, i.e. knowledge about how to use and produce the new thing.

In the more advanced stages of R&D, a product design becomes a production design too, or merges with a production design. Organised development thence is also about systematically mingling science with technology, and technology with manufacturing. According to a certain usage, an innovative process is seen to go from laboratory stage (invention) via technology stage (beginning organised development) to engineering stage (late organised development). An innovation is thus step by step being structurally evolved in shape, scope and size. In this process, more new patents relating to the innovation are registered. Not least, development costs greatly increase with every step taken, somehow in mirror image to the die-off rate of innovative ideas. A marketable product easily costs several thousand times the amount of a first laboratory desk-demonstration prototype.

Organised development represents a four-to-eight-step process of pioneer production, including pilot production plants which grow step by step in size and volumes, cooperation with pilot users or test users, and, towards the end of the development stage, also introduction on pilot markets or test markets, or test use under regular everyday conditions. The development process leads from small-scale one-off laboratory prototypes (end of invention) to regular production of a dominant-design 1G product on a scale that is large and cost-efficient enough to enter regular production and marketing. At the end of organised development there ought to be a preliminary regular product introduced on test markets and adopted, or at least tried out, by test users. Test users are easily approachable customers. They may be believers in progress, or have a personal fable for certain kinds of products, or be contracted users, or may otherwise regularly cooperate with a producer, for example medical doctors cooperating with the pharma industry, or airlines cooperating with producers of aircraft, or electricity companies cooperating with the suppliers of power plants, or automobile producers cooperating with development-stage companies in hydrogen production and fuel cell technology.

Details vary of course from case to case, depending on industrial sector and the type of technology involved. Development of new drugs, for instance, according to a release by the Boston Consulting Group, represents a six-step process from target identification and target validation to screening, optimisation, pre-clinical testing and regular clinical trials. The process may take 12–15 years and cost almost a billion euros, with the trials being the most expensive followed by target validation and identification.

A comparison with life cycle models by other authors may be useful for further clarification. Dodgson (2000 57) distinguishes between the two classical formative stages of science-driven research and technology-based development, which he again subdivides in two substages each:

(1) In the research domain there is, following a commonly shared view, (1a) basic research and (1b) applied research. Basic research is about generating new knowledge, understanding things in theoretical principle and model. Applied research then is about generating new knowledge with a practical aim, and developing tools, instrumentation, etc. Both basic and applied research have to rely on external sources of knowledge in addition to own findings, hence the case for cooperation or networking. The output of both substages consists of research reports, bench-top demonstrators and patents.
(2) In the development domain there is (2a) the advanced engineering domain of experimental development. It is about demonstrating technical viability, eliminating technical uncertainty, and choosing actual technologies and materials. The output of this is further demonstrators coming with new specialised know-how. Activities are then followed by (2b) the development domain per se. This is about translating demonstrated principles into new products, resulting in defining designs and marketable model products.

It can be argued that basic research (1a) is not yet part of the invention stage of an innovation life cycle. Basic knowledge, up to now at least, is public knowledge to be shared commonly for any potential purpose. It is thus part of diffuse latency conditions of any innovation. Innovation then entails some specific application as this determines the direction of applied research (1b). So Dodgson's applied research (1b) and advanced engineering domain (2a) are both part of what is called invention in this manuscript. The term 'advanced' can, however, be as misleading as 'engineering' since both are justified in comparison to previous research, but strictly speaking jump ahead to the development domain (2b). The latter is in accordance with the notion of organised development.

Weyer et al. (1997 35) see any major technological paradigm passing through the three stages of (1) origin in small informal groups of vanguards and pioneers, where the 'socio-technical core identity' of an innovation is determined, followed by (2) stabilisation through developing a dominant design within increasingly exclusive expert networks of academic, industrial and government agencies, and finally (3) carrying through the novel development to regular production and customisation with wide use. The three stages in Weyer et al. by and large correspond to the stages of (1) early formation by invention, (2) advanced formation during organised development, and (3) beginning prime unfolding as discussed below.

Organised development comes with a die-off rate probably as high as that of inventions, i.e. they die at a supposed rate of about 90 per cent, without having had the chance of reaching the more advanced state of being launched at market or otherwise being introduced for regular use.

Some other new technology may have to endure a long latency period before its time has come to emerge on a broader basis. The fuel cell was in principle known for 150 years before it finally had its chance under today's conditions. The hologram was invented in the 1940s, but at the time there was no source of coherent light sufficient to create such images. Only with the development of lasers in the 1960s–70s did such sources of coherent light become available (*MIT TechRev* Jan 2003 88). In motorcars, a smooth-running automatic gear system called continuously variable transmission had been known since 1900 when it was first built into a French Fouillaron, and around 1950 also in Dutch DAF cars. It never made a lasting breakthrough, however, because materials and parts available at the time were not yet advanced enough to stand the test of practice. Only today are such materials available (*Economist* 1 Sep 2001 65).

7.4.3 Accelerated or Prime Unfolding of an Innovation

Prime unfolding of a new technology tends to be closely interrelated with the diffusional take-off of a respective innovation. Accelerated unfolding in the course of a take-off necessitates a big enough market, in the beginning some kind of niche market, which it is hoped will become a lead market with a certain potential for future growth (Kemp 1997 288, 307). Users who are adopting a new technology or product – lead users opening up lead markets – are important trendsetters at this stage. Quite often it is the test users from the organised development phase who stay with the new technology and act as lead users during take-off. Feedback between users and producing suppliers continues to be important, whereby that feedback is now no longer just direct exchange of experience, but increasingly takes place by way of market interaction, i.e. dynamics of supply and demand.

The focus of structuration now shifts from product development to process improvement, servicing efficiency and specific cost savings. Gains from economies of scale are being made. Some incremental redesign of the product may also occur, introducing different materials, new single components or even, after a while, major innovation in the form of 2G products (II/7.3.3).

If the case in point is a complex generic system innovation, prime unfolding and diffusional take-off is accompanied by a rapid growth in related equipment, necessary infrastructures, introduction of complementary products and various services.

Marketing and finance come to the fore. Investors are now receiving regular return on their investment. Market growth is accelerating and per-item profit margins are growing along with it. The number of competitors is consolidating. Some of the pioneers have not managed to make it. Some others are establishing themselves as market leaders. A territorial expression of this is the formation of new regional clusters, or the reinforcement of already existing regional clusters connective to the new productions (II/10.1).

Close cooperation between producers, users, designers/developers and regulators as well as investors is as crucial during this stage of accelerating prime unfolding as it has already been during the stage of organised development before regular market entry. That is why important networks during prime unfolding include cooperation between producer-supplier, producer-(lead)user, as well as producer-regulator and producer-investor. In preparing market launch and starting take-off it is clearly producers who are the pivotal actors in the innovation arena.

7.4.4 Maturation

Maturation is the stage of later unfolding after the inflection point of a life cycle learning curve. Not only is the pace of market growth rates slackening (which is a diffusional aspect), but innovativeness is also fading. Efficiency gains tend to become smaller now. They are compensated for by systematic efforts to rationalise production, sales and cost management, including layoffs. There is no further major redesign, no more launch of new product generations. At most there may be a last one, bringing some further incremental change and modifications in various details rather than a major redesign of the system and its components.

Product, production process and use practices have become assimilated by many participants in many places and functions. So the accommodation of a mature innovation to outer systems is reaching some relative optimum during late unfolding. The diffusional niche of a technology also tends to reach its limits. Market saturation comes perceptibly closer.

Patenting can be relevant throughout all of the stages of a life cycle. Naturally, more patenting, and more patenting of key innovations and major improvements, occur during formation in research and development as well as during prime unfolding. During maturation patenting becomes less, or becomes simply patenting of minor incremental alterations rather than of key innovations and major incremental changes. During later stages of maturity and retention, obtaining patents is no longer an important issue. Patents may in fact have expired some time ago.

7.4.5 Retention, or Decline, Phase-out, Replacement

When structural unfolding has approximated the limits of its potential, i.e. marginal utility, the learning curve of an innovation, then no longer being novel, enters into a final stage of retention, or decline, phase-out and replacement. A retentive stage is characterised by mere modificational alterations, usually accompanied by a situation of final niche saturation.

A comparison may again contribute to further clarification. The model of the industrial product life cycle by Williamson (1975 215) distinguishes first an early exploratory or formative stage (including first steps of low-volume batch production and early market entry), second an intermediate evolvement, and third a mature stage. The formative stage in Williamson corresponds to early and advanced formation by invention and organised development in this text. Intermediate evolvement parallels prime unfolding during take-off and beginning maturation. The mature stage then includes advanced maturation and retention. Product life cycle models are obviously fully compatible with diffusion models in general. They continue, however, to confuse stages of structural evolvement with stages of diffusion and with the life cycle functions of creation, production and use, which they describe as being sequential, not simultaneous and feedback-looped as they in actual fact are.

7.5 DEGREE OF DIFFUSION

The degree of diffusion describes the degree of realisation of a niche potential, or the degree of niche penetration of an innovation among various groups of adopters, especially users. A diffusional take-off signifies the introduction of a newly developed thing into everyday operative practice by a critical mass of adopters. What is usually considered here is market entry and market penetration of a commercialised product or service. Under conditions of a market economy, commercialisation constitutes the economic side of niche actualisation (Rosenberg et al. 1992).

Processes of market penetration, i.e. adoption of a commercialised product by users or consumers, have been most widely investigated. In principle, any innovation is subject to adoption, or 'imitation' in the words of Gabriel Tarde who was the first in the 19th century to have a closer look into the dynamics of social diffusion of innovations. The idea of 'imitation' led to epidemic models of diffusion, i.e. adoption as a process of contagion or proliferation.

In a life cycle with fairly full niche saturation the process of diffusion represents a statistical standard distribution (bell-shaped) as shown in Figure 7.4, from innovators or vanguards/pioneers (2.5 per cent) to early adopters (13.5 per cent), early majority (34 per cent), late majority (34 per cent) and laggards (16 per cent).

Innovators and early adopters characterise the diffusional stage of emergence of an innovation (comp. Figure 7.3 on p.266), whereas early and late majority are the two adopter groups during the stages of first accelerating, then slowing down take-off, and the laggards finally coming in during niche saturation.

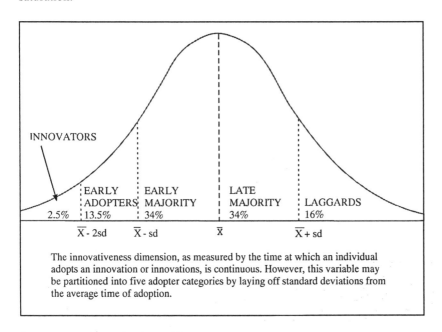

The innovativeness dimension, as measured by the time at which an individual adopts an innovation or innovations, is continuous. However, this variable may be partitioned into five adopter categories by laying off standard deviations from the average time of adoption.

Source: Rogers/Shoemaker (1971 182).

Figure 7.4   Categories of adopters in the diffusion of an innovation

In contrast to symmetrical logistic functions as shown in Figure 7.4 there are models where the inflection point comes earlier or later along the curve, the course before and after inflection not being symmetrical, as in the Gomperts model and Bass model. (For a discussion of those models see Kemp 1997.) There is no need to go into further details here. As always with modelling, its validity and reliability depend on the categories and variables in the function and on the assumptions made. Whereas the categories can be accepted in this case, the assumptions are far from being realistic: constant rate of adoption ('infectiousness'), homogenous information and behaviour of adopting individuals, thus entailing an equal chance for all of 'catching' the novel item, constant population, and everything else remaining unchanged. In reality, all of these and further conditions change and fluctuate to greater or lesser degrees all the time. So each empirical diffusion curve looks different, and can possibly change momentum and direction on its way (II/7.8.2). All the same, all such curves will be similar in that they represent an S-shape or bell-shape.

The degree of diffusion of an innovation has traditionally been identified with the stages of system structuration. This, however, obscures the fact that an S- or bell-curve representing a distribution of the adopters during diffusion of an innovation (emergence, take-off, saturation) is not the same as the structuration of an innovation (invention, development, prime unfolding, maturation, retention) – though these aspects are related to each other in the evolutive course of a life cycle, and can be described by applying the same or similar mathematical principles.

The important thing is not to confuse stages of diffusion and related groups of adopters with stages of structuration. The curve of an innovation's structural unfolding represents a learning curve, a curve of qualitative structural change of a system and its system-environment structures, whereas diffusion describes a growth curve in numbers, strictly speaking, a population growth curve. That is why different wording appears to be appropriate.

7.6 ACTOR FUNCTIONS: CREATION, PRODUCTION AND USE, REGULATION AND FINANCE

There are various actor functions that contribute to shaping the life course of a technological innovation, in principle as many as there are societal subsystems or subfunctions. In a given case, their degree of influence can vary with their degree of importance. Starting from the categories used in this text, a first distinction can be made between formative and effectuative functions (II/6.2). Effectuative functions in turn can be broken down into operative functions and controls functions. Particular operative functions are those of

a. creation, i.e. conceptualisation, design and re-design, development and re-engineering
b. production
c. use/consumption.

The functions of creation, production and use, as well as further functions such as regulation and finance, are at work synchronously rather than successively. Although the stages of structuration and diffusion determine some sequence, there is always some feedback as well, for example when it comes to steps of re-inventing or re-engineering. Furthermore, such retroconnective and readjusting mechanisms are actuated by the typical actors involved, i.e. creators (inventors, designers, developers), producers, regulators, investors, and others. As indicated in Figure 7.5, the relationship between creation, production and use represents a continuous feedback chain rather than a linear succession.

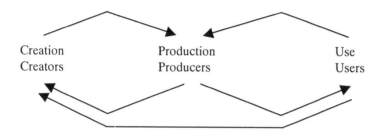

Creation	Production	Use
Creators	Producers	Users

Figure 7.5   Operations feedback between actor functions

The synchronous actor functions of a life cycle, though, are still poorly differentiated at the beginning of a life course as long as an innovation in its early formative stage is the business of an inventive vanguard. Soon, however, as the unfolding of an innovation passes through its different evolutive stages along the learning curve, specialised groups come in: teams of researchers, designers, developers, pioneer producers, test and lead users, and others. So the functions that make a life cycle soon take shape, and various groups of specialised actors equally soon fall apart. The operative functions of creation, production and use are existent at any time of a technological life cycle, albeit to different degrees. Creation is clearly the pre-eminent function during formation (invention and development), whereas production and use come to the fore with further unfolding and diffusion.

In more complex systems that can be broken down into a number of components with many elements, there are accordingly many life cycles at different levels at different stages. And accordingly, throughout the life cycle of the whole system re-invention, re-design, re-engineering may occur in components and elements, and, in consequence, in the system as such. As pointed out already in Rogers and Adhikarya (1979), re-invention or re-design can potentially occur at any stage of a life cycle. Re-invention or re-design needs itself to be classified according to its structural importance. Any re-invention that is implemented to a certain degree, and which influences the practical performance of suppliers and users, will alter the diffusion function, i.e. the empirical S-shape of a diffusional growth curve.

Any management of technological innovation will have to systematically combine the structuration and diffusion of an innovation with various actor functions (Dodgson 2000), particularly those of creation, production and use, regulation and finance. These functional categories of actors, and further ones such as opinion leaders, or promoters versus defenders (II/9), constitute an arena or social space where all aspects and questions relating to an innovation process at some moment are communicated and discussed, resulting in ongoing interaction surrounding the innovation process, thereby furthering or hindering it.

Not amalgamating the aspects of structuration and diffusion helps to maintain awareness of the fact that different groups of actors are of different importance for the advancement of an evolvement during the different structurational and diffusional stages of a life cycle. One could think of actor-specific diffusion curves, because the adoption of an innovation among researchers, developers, investors, producers, regulators and users is in each of these cases a different process with characteristics and problems of its own. The spreading of new ideas among inventors and developers is necessarily different from diffusion of a commercialised mass-product among consumers. Thus, an innovation's life cycle would entail a number of special diffusion curves, each representing specific and partial niche penetration. It can be left open here whether this is a sustainable idea and how it may be possible to integrate actor-function-specific diffusion curves into an overall measure of generalised niche penetration. Research practice so far has for the most part concentrated on the function of use.

Regulators and investors represent the most important actors relating to control functions, i.e. the ordinative function of regulation and administrative or managerial procedures, and the economic function of investment and financing. Quite often, though not necessarily, producers are also the main investors. Investment and regulation are interest-driven functions, be they of national, public, individual or pecuniary interest. They support, or do not support, the prime operations functions. The effects of finance on innovation are fairly clear, simply through the extent of funding. By contrast, the supportive or abortive effects of different ways of regulation are more difficult to grasp (Klemmer et al. 1999, Hemmelskamp et al. 2000).

Operative functions and control functions are in turn embedded in and conditioned by formative functions, i.e. political and cultural factors, such as support or non-support by politicians in government positions, parliaments, parties, confederations and unions, people's attitudes and lifestyles, needs and interests, as well as the predominant inspirations and visions in accordance with the basic values of a culture. Looking into these political and cultural factors too is outside the scope of the present work. But the existence and the defining impact of formative functions on effectuative functions should not be underestimated, and should be kept in mind, because ultimately it is neither machines nor money nor money-seeking lawyers who drive things ahead but people's minds and cultural conditions.

Table 7.3 gives an overview of some of the categories of life cycle analysis developed so far.

Table 7.3   Categories of life cycle analysis of technological innovations

Stages of structuration	Learning curve of structural and functional unfolding: 　　Formation through 　　　　Invention, discovery, exploration 　　　　Organised development 　　Prime unfolding 　　Maturation (late unfolding) 　　Retention; or decline, phase-out and replacement.
Structural impact	Major versus minor structural importance of an innovation (high vs. low impact), which may be sustaining or disruptive: 　　generic (in system or components) versus 　　incremental (in components) versus 　　modificational.
Degree of diffusion	Growth curve of adoption and niche penetration: 　　Emergence with vanguards and pioneers (early adopters) 　　Take-off through early and late majority 　　Final saturation with laggards.
Actor functions throughout a life cycle	Operative functions of 　　Creation (conceptualisation, design and re-design) 　　Production 　　Use All complemented/conditioned by effectuative controls functions of 　　Lawmaking, regulation, administration, management 　　Investment and finance and manifold formative functions of culture and politics.

7.7 LIFE CYCLE PRINCIPLES

7.7.1 The Life Cycle as Learning Curve: Growth-Efficiency Principle, Marginal Utility, Rebound Effect

In engineering, economics and psychology the innovation life cycle is often referred to as a learning curve. Psychologists quite literally mean learning, for example the growth in vocabulary a child acquires. Engineers refer to improving the performance of a technology (it is the engineers who actually learn), whereas economists refer to reduced prices as a result of improved technical performance (productivity) and economies of scale.

A life cycle as a learning curve empirically also takes on the typical S-shape (cumulative learning results) or bell-shape (rates of learning progress). Learning along that course is part of structural unfolding as much as social diffusion. Furthermore, and coming together with the structural unfolding, the learning curve results in increasing efficiency, i.e. decreasing demand for input per unit of output, or increasing output per unit of input, for example gasoline consumption of a motorcar per 100 km. Intensity (resource or energy intensity) is just another word for the same thing, i.e. input of materials or energy per unit of product. Economists tend to prefer to speak of productivity, i.e. input of energy or working time per unit of output. These are just different words for the same sort of input-output ratio.

Most often it is the increase in efficiency which is meant by learning. As the heuristics of the learning curve suggest, increase in efficiency is slow in the formative stages of a life cycle, this then accelerates with prime unfolding during take-off, and slows down again after the inflection point and during later maturation and retention. In this sense, the learning curve represents a growth-efficiency principle, i.e. a development of efficiency which goes hand in hand with restructuring (at first unfolding, then lessening) and diffusional growth (at first increasing, then decreasing). Just as diffusion reaches its achievable niche penetration (for example market saturation), and structuration approaches its stage of retention by only minor incremental modifications, so efficiency simultaneously approaches marginal utility. Reaching marginal utility towards the end of a learning curve is just another aspect of one and the same process.

The growth-efficiency principle represents one of the few universal principles we know of. It applies to biological systems as much as to societal and particularly technological systems. In the growth of a tree, for instance, the energy demand per unit of tree biomass (E/B in Figure 7.6) is high in early life compared to later life stages of the tree. Energy demand per unit of biomass decreases during continued growth of the tree, i.e. efficiency increases, and this increase slows down with maturation when the tree reaches its final height and volume (curve of B). In absolute terms, and in a tree, the process results in a continual growth of total energy demand (curve of E), which is lower in early life and higher in the late state, because the stock of biomass builds up, and total reproductive energy demand is the result of units of biomass multiplied by the energy demand per unit. As a consequence of increased efficiency, though, the growth rate of biomass is higher than the growth rate of total energy demand.

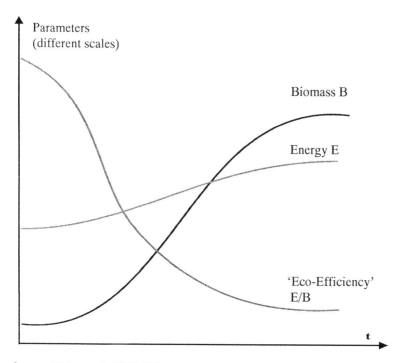

Source: Weber et al. (1988 107).

Figure 7.6 Growth of biomass, energy flow and efficiency in a tree

Another important aspect of the growth-efficiency principle is known as the rebound effect: growth and efficiency occur together, they induce each other. A typical example is rationalisation of production. At first it entails laying off some workforce; which enables enlarged restructured production that necessitates more workforce, usually with somewhat different skills, and results in an overall growth of production and employment. That is why the rebound effect has also been dubbed the echo effect and, with less acceptance, the boomerang effect with regard to phenomena of ecological interest such as population growth and growing demand for resources and energy.

Not only does efficiency increase with ongoing growth, at first accelerating then lessening, but growth comes as a consequence of increased efficiency too. In other words, the purpose of increasing efficiency is to increase, not to decrease, growth. The purpose of learning in the sense of increasing efficiency is to stabilise and to sustain the overall total growth of a system. Systems do not learn in order to stop growing, much less to shrink. If a learning curve is allowed to pursue and complete its life, growth will not come to a standstill until the niche potential of the system in consideration is exhausted, be that for internal reasons (a fully realised genetic or generic structural potential) or external reasons (limitation of living space and resources of any kind, in organisms as well as in machines).

A typical example of the rebound effect is the development of car traffic: Increasing efficiency of fuel consumption has not translated itself into absolutely shrinking fuel demand of a stagnant population of cars and car-drivers, but into bigger motors and more comfortable and faster cars, expanding their reach, thus the range of activity of their drivers. This has opened up more options within an enlarged action space, which in turn has made car-driving the best option to a growing number of travellers. Part of this rebound-growth process is ever more favourable per-unit costs and prices as discussed in the following passage. The rebound effect, as well as approaching marginal utility in later stages of the learning curve, are aspects of the growth-efficiency principle which are of particular importance when it comes to assessing the 'efficiency revolution' as a strategy of sustainable development (I/2.4.3).

7.7.2 Cost and Price Economics of a Learning Curve

In economic terms, and with regard to technological life cycles, increasing efficiency or productivity translates into decreasing unit costs and prices. As technologies, or products and their production processes, structurally evolve, and developers and producers learn to be more efficient, per-unit production costs come down, and so can selling prices. These come down because of market competition and also out of self-interest because higher turnover at somewhat lower prices is usually more profitable than low turnover at much higher prices. The process is reinforced by learning effects regarding market organisation (growing transaction efficiencies) and economies of scale because of growing volumes and relatively lower fixed cost.

As efficiency develops along the learning curve, so does the economic return on investment (expense-return ratio). Prospect of profit is certainly a main driving force behind innovative activities, albeit not the only one. The question arises about the stage at which such rewards can be expected to come. The stages of invention and development usually require investment with no return. In modern research companies, for example in biotech, this seems to be different; their return actually comes from investive R&D spending of larger corporate customers. The proof of profit starts with market launch and hoped-for take-off. If a life cycle continues to proceed successfully beyond that stage, break-even will be reached soon, and losses at the beginning turn into growing profits thereafter.

There is no commonly shared opinion, however, on how profits develop from that stage on, in absolute and per-unit terms. According to a model of the innovation cycle by Grossmann (2001 36), profits continue to rise and are highest when maturation is about to be accomplished, marginal utility reached and retention just beginning. This could indeed be the case in absolute terms, with larger volumes in turnover having so far compensated for decreasing per-unit margins caused by tightened market competition. Once this stage has been reached, however, profits do not seem to stay high. Instead, they decline because still shrinking margin profits, and if only slightly shrinking, cannot be compensated for by greater turnover any more. This is particularly true if new competitor products reach the market which cause turnover of older products even to decline. But there are different cases in which profits can be retained and companies are happy to have a 'cash cow' for many decades. Take the famous example of the medicine Aspirin, a remedy for headaches and fever which was introduced a hundred years ago and which has scarcely changed since then. Contrast this, say, with some washing detergents that still represent the same marketing brand but have been completely re-designed in chemical substance and represent 3G or 4G products.

Any successful innovation gives the same typical example of the growth-efficiency principle which holds true in technical and economic terms alike, for example, computers and telecommunications during recent decades. As the technical performance of computers and their production evolved and increased, piece costs and unit prices continued to come down. Ever new generations of memory chips, drives, screens or printers with ever more capacity have translated themselves into ever cheaper computers, or into more powerful computers with higher performance and better quality at a same price, or into a combination of both.

The history of the telephone throughout the 20th century gives the same picture. Transatlantic calls at their inception during the 1930s cost almost $100 a minute. In the 1940s the cost of a one-minute call had already come down to about $65, reaching $20 around 1950, further decreasing to 10, 5, 2 dollars and less today. In brief, as the telecom networks have structurally improved and grown in size and number of customers, per-unit costs for the telecom companies and per-unit prices for customers have correspondingly come down.

An important conclusion to be drawn from the economic learning curve is this: Never write off an invention or a new technological development simply because some experts say 'too expensive'. Any new technology is expensive at the beginning, and becomes ever more affordable throughout its life cycle as its evolves and grows. This certainly does not happen by itself. Someone has to do the learning – the different actors carrying on a life cycle, i.e. developers and designers, producers, users, regulators, investors. Someone has to finance it step by step, according to structural and diffusional stages. This poses corresponding financial necessities at every stage. But it does not make sense to ask whether future users of, say, hydrogen would be prepared to pay prices as high as the per-unit costs of today's development stages. Of course users are not prepared to pay such prices, and they certainly won't have to, because by the time that the fuel cycle of hydrogen has reached the take-off stage of being mass marketed in ever bigger volumes, its price will have come down to a small fraction of what it is today. The only relevant question is whether the next step in the evolvement of the system in consideration can be financed here and now, for example, going from a small demonstration prototype to a bigger one, then to larger developments, and so on. If there is a decisive will among the functional elites to get something done, money is never an insurmountable problem, neither at the beginning nor at later stages.

7.7.3 Shifting Differentials between Different Adopter Populations

Another general pattern in the evolvement of life cycles can be called the principle of shifting differentials. Differentials in consideration are development differences between actors within an adopter population. The differences can relate to any development indicator such as knowledge, productivity, volumes of input and output, profitability and real per-capita income. Kuznets investigated that phenomenon under the heading of 'U-curves' of income development. Scholars of industrial development and modernisation processes talk about unequal development, or deferred development respectively. These are just different expressions of the same general principle. This principle represents a characteristic inherent in any successful innovation and modernisation process: The differentials under consideration at first increase until a certain point at which differentials then decrease and converge again. Things may not necessarily be at the same level both at the beginning and at the end of the curve. Figure 7.7 shows the heuristics of that principle.

Figure 7.7.1 describes an actor A who enters into an innovation or upgrading process earlier than B who follows later. One can think of different social classes in a societal modernisation process, or different nations in train of industrialisation and ongoing high-tech modernisation, or different pioneers in new domains, or firms in an innovative market, or different households in a diffusion process of new consumer goods.

If there is no actor B following A in pursuing a particular life course, the differential between A and the rest of actors always gets bigger until retention, though at an accelerating rate until the inflection point, and at a slack-

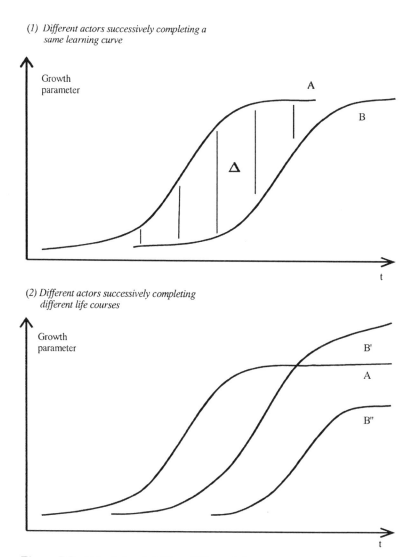

*(1) Different actors successively completing a
same learning curve*

*(2) Different actors successively completing
different life courses*

Figure 7.7 Principle of shifting differentials

ened pace thereafter. If there is a follower curve, it depends on the timing and shape thereof whether and when differentials Δ start to decrease again. As a rule of thumb one can say that differentials increase as long as A is taking off and B has not yet started to follow. Increased differentials tend to remain of the same size (i.e. no longer increasing) as long as A continues its take-off and B too has started its take-off. Differentials begin to decrease when A is maturating and reaching its retentive limits whereas B continues to unfold.

As symbolised in Figure 7.7.2, actors do not necessarily complete identical learning courses, or a learning course in the same way. As discussed below in II/7.8, life cycle trajectories in society can take on quite different shapes. They can be boosted or curbed and slowed down according to particular circumstances and interdependencies with co-cycles that can be co- or counter-directional. Cycles may even be redirected. And they possibly break down, or come to a standstill halfway. Quite a number of scenarios are conceivable and may come true. These include the model scenarios of B catching up to parity with A, as indicated in Figure 7.7.1; B′ in Figure 7.7.2 not only catching up but actually overtaking A; and B′′ completing some successive evolvement without catching up. Differentials will shift accordingly, with B′ running ahead of A, and with a lasting differential between A and B′′, though on a generally higher level.

Typically, shifting differentials can be studied in the diffusion of styles, practices and consumer goods among social classes. For example, market penetration for durable household goods in the 20th century is well documented. Its diffusional S-curves clearly display what was to be expected on the basis of income distribution, but also according to the sociological diffusion theory, which states that novel items diffuse top down, i.e. from higher to lower classes. Empirical data indeed show substantial diffusion of household durables among manager households within one decade, while that of normal employees follows with a 10 years delay and takes a longer time to reach saturation. Last comes the cycle of the non-active population of the time, again several years in delay and less steep. In the end, all reach very high penetration rates, with some small differentials remaining, i.e. the managers showing 95–100 per cent penetration, normal employees 85–95 per cent and the non-active population 80–90 per cent.

A quite similar picture can be seen in various statistical time-series comparing parameters of industrial development from different nations in the world system. Seen in a perspective of shifting differentials, growing gaps during certain times are no cause for alarm, but simply a confirmation of what is to be expected. This includes the subsequent stages of stasis (gap remaining the same) and later on also catching up (gap diminishes) – and this in fact is what ultimately happens in most cases. Despite harsh criticism and gloomy prophecies about the world development paradigm, social progress exhibited by shifting differentials over several decades is actually being achieved when measured by almost any standard, notably real per-capita income, energy demand, housing, education, life expectation, public infrastructure and provision of goods and services.

There are certainly exemptions to the rule, for example a number of countries, particularly in Africa, with repeated breakdowns in any attempt to modernise. And there are, as has always been the case, times of economic crisis and far-reaching structural change, depressed or chaotic periods, interim breakdowns of any kind, being reflected in interim periods of stagnant or declining parameters. This is not the place for a discussion of whether development crises happen of necessity or could and should be avoided. But the general learning curve picture when monitored over longer periods of time clearly enables a restatement of a prediction of modernisation theory: more or less all nations will finally become industrially developed and enjoy some decent living standard. Certain final differentials between the interdependent nations in the world system can, however, be expected to remain, though the final ranking will not necessarily be the same as it is today.

7.7.4 Co-related Life Cycles

No life cycle is an isolated single event. The systems whose existence is represented by their life cycle are all embedded in history and contemporary interaction. Every system has its origins and its destinations, and while evolving it is co-related with other systems in co- and counter-directional ways. Shifting differentials between subsequent cycles of adopter populations, as discussed in the previous chapter, are actually a first example of co-related cycles. They are not independent of each other, but, constituting social groups within the divisional structure of a nation, or nations within the divisional structure of the world system, they are part, actually subsystems, of the higher-level system which they belong to.

7.7.4.1 Continued cycles of a same system

Continued subsequent cycles of a particular system represent a special case of the co-relatedness of life cycles. Long waves of industrial innovation may again serve as the example of choice. Each of these can be empirically identified and distinguished from earlier and later such innovation cycles. Considered in isolation, for example the railway cycle of the 19th century or the recent IT cycle, they each start close to zero. Put into the bigger picture of overall industrial development, however, they turn out to be individual branches (cycles) of the tree of modern technology (II/7.3.5) which has so far unfolded in a transsecular cycle of several centuries. Hence, the final stage of the individual cycles is for the most part retention, and only possibly is it decline and phase-out. This does not exclude crises of retention, but these are rightly referred to as phases of consolidation, i.e. readaptation in order to maintain and carry on with what has been achieved.

In this sense, recurrent Schumpeter cycles of technological innovation represent a flight of S-shaped stairs climbing upwards, rather than representing waves going up and down again. The metaphor of long waves, associated with upswing and downswing, actually stems from a confusion between economic business cycles and technological innovation cycles (for which Kondratiev and Schumpeter cannot be said to be in no way responsible).

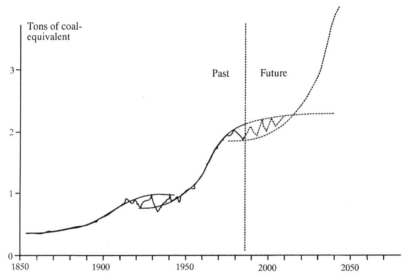

Note: This figure represents the per-capita annual energy consumption world-wide (data, fits, and a scenario for the future). It reproduces in part a drawing from J. Ausubel, A. Grübler, N. Nakićenović, 'Carbon Dioxide Emissions in a Methane Economy', *Climatic Change*, vol. 12 (1988): 245–63.

Source: Modis (1992 271).

Figure 7.8   Subsequent secular cycles of energy demand

Conversely, a long-term S-shaped cycle is usually built up of a number of shorter-term individual cycles. There is, for example, an impressive take-off curve showing the increasing velocity of vehicles and propulsion systems over a 200 year period (Bell 1973 203 after R.U. Ayres). In practice and empirical fact, this long-term development consists of an envelope curve made up of a number of subsequent overlapping innovation cycles, from the pony express in the mid-18th century via trains in the 19th and automobiles in the first half of the 20th century up to external combustion systems, propeller aircraft, gas turbines, magnetic levitation, jet engines, missiles, and whatever else the future will bring.

The long-term development of modern technology has been accompanied by a continued upswing of energy demand, in much the same way as the evolution of the modern economy has been accompanied by a continued upswing in per-capita income. Figure 7.8 shows how the transsecular growth in energy demand is composed of subsequent secular cycles. The first cycle in the chart represents the age of coal, succeeding cycles of wood and charcoal (which cannot be seen in the chart). The second cycle represents the age of mineral oil and nuclear energy. The next one, now emergent, seems bound to become the age of hydrogen and fuelless regenerative energy, possibly still in transitional combination with natural gas and coal in IGCC plants (with sequestration of CO_2). Figure 7.9, another example, illustrates how the periodic system was discovered in four successive cycles of about 60 years each from the mid 18th to the mid 20th century.

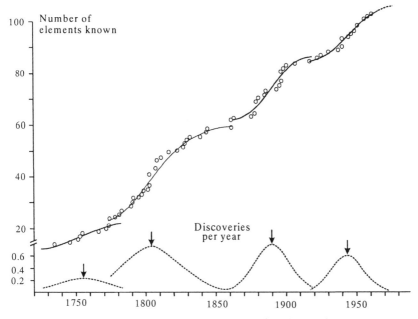

Note: The small circles indicate the number of stable chemical elements known at a given time. The S-curves are fits over limited historical windows, while the dotted lines below represent the corresponding life cycles. The arrows point at the centre of each wave hinting at some regularity.

Source: Modis (1992 130).

Figure 7.9 Discovery of the periodic system in four subsequent cycles

In empirical detail and reality, continued developments are not continual, neither steady throughout their life course nor always pleasantly smooth. All cycles – those of longer-term as much as the shorter-term cycles which the longer ones may be made of – come with the changeable dynamics of their own, and with a time of crisis when one cycle is approaching its end and a next new cycle is both taking over and taking off. Normally this results in some instability or even chaos, appearing as volatility of parameters as indicated in Figure 7.8.

7.7.4.2 Successive co-directional cycles

Another type of co-related cycles is successive co-directional cycles of different (sub)systems, or different factors or parameters of a same more complex system which they are part of. Successive co-directional cycles are interactively connected to each other, whereby the effects of their interdependence do not show up in a completely synchronous way but are staggered in time. All are nonetheless heading in the same direction. The principle is illustrated in Figure 7.10.

An example of such successive co-directional cycles is the interrelation between the investment cycle and the employment cycle within the entire complex of a business cycle. The bell curve of both rates tends to be closely correlated, with the investment rate coming first and the employment rate between about ½ and 1½ years later. Both factors are in turn just two out of many more co-evolving factors in an aggregate business cycle, some of which are actuated earlier, others later. Typical early co-cycles are those of producer orders and equity prices. Employment on the other hand is a typical late indicator of the entire business cycle. The co-cycles of GDP, industrial production and utilisation of capacity tend to be in-between and indicate where the overall aggregate business cycle roughly is at a given time.

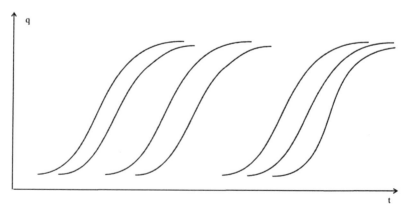

Figure 7.10   Successive co-directional cycles

Another example of succeeding related cycles is the epidemiological connection between the cycles of infection and outbreak of a disease. In the case of AIDS, the S-curve of accumulated deaths follows the S-curve of HIV infections at an interval of about 10 years. A similar co-cycle is constituted by the relation between the initial prospecting for and the eventual acquisition of natural resources such as metal ores and fossil fuels.

A further example is the bunch of recurrent social movements since late enlightenment and the beginning of the industrial revolution (Frank/Fuentes 1990, Huber 1989). They tend to start with the formation and upswing of an ideational current, anti-rationalist and anti-utilitarian (i.e. anti-capitalist, anti-industrial), such as Sentimentalism, Romanticism, existentialism, vitalism and flower power. This eventually finds expression in some kind of social critique, resulting in various movements targeted at different though co-related aims such as emancipation and self-actualisation, citizens' rights and liberties, democracy, participation and social justice. Often, it is a youth movement or student movement which represents such impulses. It may be connected to or followed by cycles of the peace movement, abolition movement, anti-apartheid movement, workers' movement, women's movement, search for authentic lifestyles, educational reform and not least movements for the protection of historic monuments, local culture and language, and movements for the conservation of nature and protection of the environment.

A hypothesis under debate on the recurrent emergence and withdrawal of social movements takes its own starting point in a combined cycle: Social movements of the kind listed above with a time-span of about three decades occur as a reaction to successive long waves of industrial modernisation of a total length of some more decades. The idea is that the tremendous take-offs of long waves create problem-laden structural change. Old social structures break apart, new ones establish themselves, thereby causing some tumultuous restructuring, hardship for those who lose out during these processes, alienation from hitherto familiar circumstances, and irritation in face of the new. As one or several of such problems accumulate, normally around the apex, i.e. the inflection point and the start of maturation of the innovative cycle, social movements arise as a response. They are aimed partly at dealing practically with those problems, and in part they simply react to them in some expressive and communicative way in order to come to terms with what is going on. While the social movement cycle is unfolding, the cycle of industrial innovation has reached retention and is consolidating for the time being. And while the social movement cycle is doing its job of societal readaptation and reintegration, and in the interim no new dramatic pressures build up, the movement cycle also comes to an end. This partly results in consolidation by retaining achievements of institutionalisation and professionalisation, and partly results in decline and withdrawal.

It seems that the entire transsecular cycle of societal modernisation, i.e. the transition from traditional to modern society such as it slowly started to emerge in the later middle ages, can be described as a sequence of co-related co-directional cycles of a length of several centuries, overlapping in time and interconnected by positive feedback-loops. The first of these transsecular cycles were those of formative modernisation of mind, knowledge, expressive styles, arts and literature, including what is known as rinascimento and the age of reformation (Protestantism and transition from theology to science) the focus of which was from the 15th to the middle of the 17th century. The 17th and 18th centuries were marked by an overlapping successive cycle of ordinative modernisation of state institutions, public administration and law. The next overlapping successive cycle was that of preoccupation with the economy from the mid-18th to the mid-20th century, relatively closely followed by the transsecular wave of industrialisation and technology in the 19th through to the present 21st century.

7.7.4.3 Countercycles

Countercycle is short for co-related counter-directional cycles. These may be recurrent, or happen only once. The case of continually recurrent, combined countercycles is known in systems theory as dynamic equilibrium.

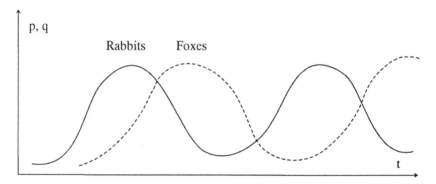

Figure 7.11   Combined countercycles (dynamic equilibrium)

A well-known example of such a dynamic equilibrium is the story of the rabbits and the foxes (Figure 7.11). It starts in a living space where there are originally no or few foxes, and good pastures for many rabbits that rapidly proliferate. This in turn is good food for foxes, coming in and growing in numbers as they feed on the rabbits which then start to decline in numbers as ever more of them are eaten by the growing population of foxes. From a certain point on, rabbits become scarcer, ever more foxes begin to starve, and the population of foxes declines. When the foxes have declined below a certain level of population density, the rabbits can recover in numbers again. And so will the foxes not very much later, and so the story goes on.

The interconnection between the cycles of the rabbits and the foxes may be interpreted as a successive co-cycle too, but the direction of the foxes' cycle – its impulse, or aim, or target – is overtly opposed to the rabbits' cycle: The rabbit population stops growing and goes down as the fox population rises; the rabbit population stops declining and starts to recover as the fox population falls.

If we reconsider the business cycle from the perspective of combined countercycles, a number of such counter-directional co-relations can be identified, resulting in the continual recurrence of upswing and downswing. Among such counter-directional factors in the business cycle are indebtedness and over-investment (growing capacities exceeding stagnant demand) with subsequent decline in investment and employment, and maybe also prices if there is price elasticity.

7.7.4.4 Substitutive countercycles
As regards the one-off countercycles, any example of disruptive innovation could be quoted. Figure 7.12 gives the example of the replacement of horse and carriage by the motorcar in the first half of the 20th century.

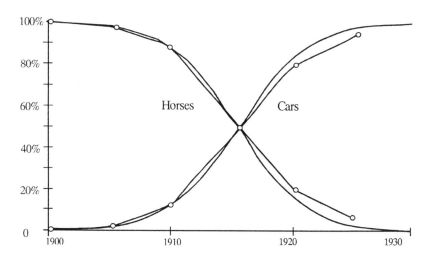

Source: Modis (1992 108), after drawings by Marchetti and Nakićenović, and Grübler (1990 186).

Figure 7.12   *Substitutive countercycles: replacement of horse and carriage by motorcars (disruptive innovation)*

Figure 7.13 shows subsequent substitutive countercycles relating to traffic
infrastructure. Further such examples include subsequent substitutions of
production processes, for example in the chemical industry where biotechnology of the 18th and 19th centuries was replaced by synthetic carbo- and
petrochemistry throughout the 20th century, which itself is now being replaced again by transgenic biotechnology. Comparable disruptive innovations can be traced back in steel metallurgy where the pre-industrial crucible
was replaced by the Bessemer process in the 19th century, itself replaced by
the open hearth process in the first half of the 20th century, the latter replaced by the basic oxygen process around the middle of the 20th century,
which in turn has now been replaced by the electric arc furnace. We owe
many of these examples to the works of Marchetti, Nakićenović and Grübler.

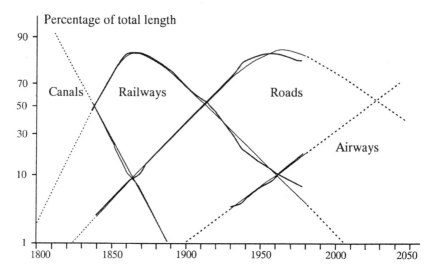

Note: The sum total in mileage among all transport infrastructures is split here among
the major types. A declining percentage does not mean that the length of the infrastructure is shrinking but rather that the total length is increasing. Between 1860
and 1900 the amount of railway track increased, but its share of the total decreased
because of the dramatic rise in road mileage. The fitted lines are projected forward
and backward in time. The share of airways is expected to keep growing well into
the second half of the 21st century.

Source: Modis (1992 263), adapted from a graph by Nakićenović.

*Figure 7.13   Subsequent substitutive countercycles relating to traffic
              infrastructure*

7.8 HISTORY MATTERS: OPENNESS AND CLOSEDNESS OF LIFE COURSES

7.8.1 Path Dependence, Lock-in and Resilience

Any system is selective in its connectivity and interactions: its structure and properties include a specific range of possibilities just as they simultaneously exclude others. There is no universal all-purpose system. Systems are particular, and the more they are advanced in their evolvement – from poorly specified fluidity and openness to mature specificity – the more their particularity results in path dependence and lock-in, excluding alternative options and related actors. No real system can be completely open and flexible. Instead, each system is exclusive to a certain degree. Lock-in as such is unavoidable, depending on the stage of structuration and diffusion. But there always remain certain degrees of freedom.

The thesis of path dependence and lock-in has been applied by its originators to technological innovation and industrial development. Lock-in, however, is not unique to the operative system of society. It seems to be a rather normal, actually universal process in the life course of any system that is gaining stability, i.e. consolidating its system structures, be they the structures of personality and human biography, the structures of organisms and technologies, or the structures of civilisation and society. Similarly, social milieu and class dynamics of closure, exclusiveness and outright exclusion are well known.

Path dependence can be understood as an autocatalytic self-stabilising process of formation of structure (Arthur et al. 1987, Arthur 1994, 1996). In a mathematical model of lock-in of technological paths, early random fluctuations determine which structure is selected and then kept (locked-in) at the expense of previously existent alternatives. In real life, this can be translated into an initial random walk of increasing returns to adoption which a group of adopters during small early events may experience. They adopt something, they learn about it, the more then this thing and its use is improved, the more attractive it becomes, the more it is adopted:

> The structure that emerges does not necessarily have to be the best or most efficient; events early on can lock in the system to an inferior technological path. ... Once a single technology structure emerges and becomes self-reinforcing, it is difficult to change it. If it were desirable to re-establish an excluded technology ... an ever-widening technical changeover gap would have to be closed (Arthur et al. 1987 302).

Historical examples include the selection of the internal-combustion engine for motorcar propulsion around 1890–1900, instead of electric power that seemed to be an equally possible option at the time, or FORTRAN as pro-gramming language during the 1950s, and the QWERTY typewriter key-board. These selections were non-predictable in the beginning, the results are non-superior, i.e. not representing the best of possible solutions, and the lock-in entails a considerable degree of structural rigidity – or stability, to look at it positively. Historical studies seem to suggest that first selections in favour of certain technological options and subsequently resulting lock-in and path dependence are not really subject to rational planning. Mensch (1979) had suggested that rapidity beats quality of innovation ('Being faster is adverse to being better'). But this isn't sure either.

Kemp/Soete (1992 442) have summarised the thesis of path dependence as 'a non-linear, evolutionary process ... with many bifurcations and possi-bilities of locked-in development, in which past history and historic events are important'. They suppose, however, the original choice in favour of a particular technology as being 'primarily driven by short-term economic benefits instead of longer-term optimality'. Selections that create technologi-cal path dependence can in any case not be seen to be the result of random events, not even in the initial steps, but of specifically motivated learning processes that represent progressive re-entries of results which reinforce and feed on themselves. They also include, beyond technological learning sensu strictu, self-stabilising actor constellations (networks), ever more vested in-terests, formal regulation of regime rules and finally economies of scale, all resulting in a typical learning curve.

Path dependence can empirically be studied in processes of the continued structural unfolding of a system on its learning curve, or alternatively, and more impressively, in discontinuities and phenomena of resilience. Resili-ence describes the ability of a system to absorb shocks and to keep to or to return to its original path when it has temporarily been perturbed and forced to change course.

Figure 7.14 shows long-term path dependence and resilience of patenting in Germany. The growth curve of first publication of patents, representing activities of research, development and restructuring of products and proc-esses, has repeatedly been deflected from its path by economic crises and war, as well as by temporary overshooting, but every time it returned to its original curve dynamics.

An example of resilience is recovery from a disease. In Marchetti (1986, 1988), to give another kind of example, there are charts showing resilience of air travel and car traffic to the two oil price shocks of 1973/74 and 1979/80. Sales of cars slumped twice for a relatively short period of time but then bounced back to higher than pre-slump volumes pursuing their growth path as if nothing had happened (also Modis 1992 46). Parallels to the slump in air travel as a consequence of the shock of September 11th 2001 invite themselves. The European slump in beef consumption during the BSE crisis of 2000–01 is of the same nature, with the general path of per-capita beef and meat consumption, which was resumed in 2001–02, heading slightly downwards in any case.

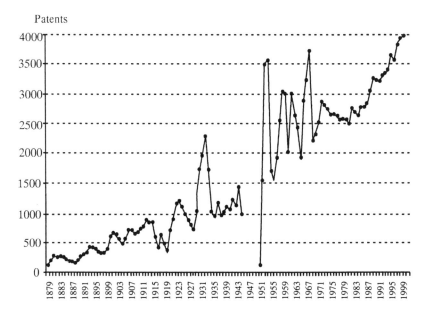

Source: Grupp et al. (2002 42).

Figure 7.14 Number of first publication of patents in Germany since 1879

7.8.2 Can Locked-in System Trajectories be Redirected?

If not properly put in a perspective of systems and evolution dynamics, the concept of path dependence could be misinterpreted in a deterministic way. From a voluntaristic point of view, fatalism or pessimism with regard to the role of active policies seems to ensue from path dependence and lock-in. Similar to simplistic theories of personal development – which postulate that a person's destiny is completely determined during their first months or even prenatal conditions – some scholars of technology studies also take the view that literally everything in a technology life cycle is determined during the early formative stage.

A similar interpretation of path dependence can be found in certain applications in forecasting. If a life cycle can empirically be represented by a time series of certain system parameters, such as growth in numbers or adoption rate, then the beginning and early course of the trajectory can indicate where the course will lead to at later times. This is the idea underlying the technique of mathematical curve-fitting (Modis 1992). Curve-fitting represents a non-linear approach to extrapolation. According to Marchetti, the method is applicable within a range of 10–90 per cent realisation of a niche potential. One of the comments on his findings is as follows:

> I was philosophically shocked. The diffusion runs with an astonishing stability and precision, insensitive to any initiative coming from the industry or sudden changes in economics and politics on the outside. ... My personal interpretation, based on hundreds of evident examples, is that people talk and the system does its business (Marchetti 1991 585).

The forecasting potential of curve-fitting is all too obvious. It is, however, not unconditional. The extrapolation of a curve holds in so far as there is constancy of conditions. The condition of 'ceteris paribus' is familiar from economic analyses and forecasts: From the variables m, n, o we can conclude x if everything else remains unchanged. Things, however, tend to be permanently in flux. It is simply a question of speed and view to the horizon. Extrapolation by curve-fitting can be a powerful tool in the hands of planners and policy makers if they are able to make sure of underlying conditions.

Marchetti refers to the examples of nuclear power and car traffic. Ongoing green criticism and public controversy did not change at all the continued and predictable trajectories of the diffusional course of these systems. There is, however, a simple explanation to this which Marchetti did not take into consideration. Firstly, green opposition to nuclear power, car and air traffic built up rather late, only after take-off and the inflection point of these innovation life cycles. Secondly, opposition remained on a purely cultural and political level without leading to effectuative consequences in terms of user behaviour, market demand, lawmaking, regulation and investment. The reason is, thirdly, that opposition, fierce as it may have appeared, was not strong enough. It either remained too weak, with the proponents exhausting themselves in discussion, or opposition got stuck in a stalemate vis-à-vis the supporters of nuclear power, car and air traffic. If cultural and political forces neutralise each other at the formative system level, this represents constancy of formative conditions, and the result is stability and continuity at the effectuative system level.

If, by contrast, there is a strong general political will to change things, based on reliable knowledge, then things undoubtedly can be changed – certainly not 100 per cent and overnight, but to a certain degree within a certain period of time, depending on structural constraints inherent in existing systemic interrelations.

In principle, the course of any technology life cycle can be redirected, by external as well as internal factors. Internal factors would encompass anything within the scientific and technical potential inherent in a particular technology. A classical external factor is availability of the resources which a system is made of and lives on. If there is abundance, system parameters are likely to follow a rather steep and long-lasting curve. If there is scarcity, the parameters of the system will develop more slowly at much lower levels. If available resources are next to nothing, there will be no system development at all. If resources begin to run out, a system and its trajectory will break down. And if resources can be made available again, the system has some chance of being able to recover.

Further external conditions of technological development include all factors relating to the economic and ordinative system, particularly regulation, investment and expected returns.

Beyond these effectuative factors, there are the formative ones: culture and politics in sociological terms, the human mind in psychological. The human mind is reflexive and can think something over again. People are capable of changing their mind, and human society is capable of changing course, of correcting itself by re-deciding matters, re-designing means and re-directing the course it follows. Certainly, to repeat the fact, not 100 per cent and overnight. History matters, or to put it differently, attained system structure matters. But there are degrees of freedom.

In fact, there can be no whether or not controversy about path dependence and predictability. The real question concerns the weight of each formative and effectuative factor which contributes to a life course trajectory. Identifying those factors in a given case is no easy task, the more so since life cycles of any kind are always part of higher-level further-reaching cycles. They may themselves consist of a number of lower-level shorter-term cycles, and these cycles are always woven into a web of co- and counterdirectional cycles as discussed above in II/7.7.3.

So, in spite of degrees of freedom and possibilities of redirecting course, the basic consensus of authors who have written on technological path dependence can be confirmed: The more the evolvement of an innovation has become path-dependent and the more its potential is unfolded and locked-in, the more difficult or even impossible it becomes to change direction as long as most of the framing conditions prevail. Changing path, i.e. a fundamental system change in the course of a well established trajectory, particularly during take-off and maturation, is possible only under extraordinary high-cost conditions. Most people would not voluntarily accept the prohibitive cost and turmoil of radical change at the halfway stage before reaching the later and final stages of marginal utility and niche saturation, where sunk costs in a literal and transferred sense are much lower. (More in II/8.3 on timeliness and windows of opportunity.)

8. Selective Dynamics: Connectivity and Timeliness

8.1 ADOPTION CRITERIA REVISITED

That which decides innovative niche competition is treated in models of evolution as selective environment or selective mechanisms. Diffusion theory emphasises adoption by actors as representing the positive side of selection. A similar notion is that of acceptance by actors, for example market demand-side acceptance by actual purchase. Systemic sociology prefers the term connectivity, encompassing any selective aspect under which an innovation may or may not become connected to existing structures.

Rogers (1995 210) summarised his empirical findings on selection of innovations in a set of five criteria explaining why certain actors who have got to know about an innovation (precondition of being informed) do or do not adopt it:

1. Advantage, i.e. whether an innovation promises to bring about certain improvements such as new choices, new sources of income, higher productivity or better quality.
2. Observability, i.e. whether the results of an innovation are visible to others.
3. Compatibility, i.e. whether an innovation is consistent, not only with existing interests, needs, desires, but also with existing values, normative correctness, legitimisation, or similar.
4. Complexity, i.e. whether an innovation is perceived to be difficult to understand and use. That is to say, simplicity fosters, complexity impedes diffusion.
5. Trialability, i.e. whether an innovation could be experimented with on a limited basis, or be adopted by a series of small, tentative steps. Trialability helps to remove barriers of uncertainty and aversion to risk.

Rogers' set of adoption criteria is derived from the empirical study of the diffusion of various consumer products, partly also management innovations. It is certainly useful, though neither complete nor unquestionable. For example, the degree of complexity, rather than being an adoption criterion is a general property of innovations. Complexity of an innovation, in that it represents a choice of low or high cost, turns out to be part of the advantage criterion. For the rest, however, complexity also remains an aspect of the compatibility of structural change as well as of effectuative feasibility.

8.2 CONNECTIVITY THROUGH ADVANTAGE, COMPATIBILITY AND CAPACITY

For systematic reasons, I prefer to regroup Rogers' adoption criteria into four categories of connectivity of an innovation: (1) Advantage of an innovation, (2) innovative capacity, (3) compatibility with existing structures and developments, and (4) timeliness, representing a particular co-evolutive aspect of compatibility, as discussed in the following.

Ashford (2001), similarly at first glance, holds the view that for a successful innovation three factors need to coincide: (1) the willingness to innovate as well as (2) the capacity and (3) the opportunity to do so. It will be seen at the end of this chapter how far there is actual overlap.

8.2.1 Advantage

In Rogers, the criterion of advantage is not represented to its full meaning. Relative advantage also implies disadvantage, for instance disadvantages of rejecting an innovation. Chances of adoption have to be measured against possible risks of adoption as well as a possible risk of rejection. Rogers' criterion of observability seems to be ambivalent in this respect. It partly means seeing the advantage others gain by adopting something, but also entails avoiding fears of falling behind the competition, or fears of becoming marginalised and isolated. Such aspects were at the basis of the epidemic diffusion model as mentioned in II/7.5. Mansfield (1961), following Griliches (1958), stated that adoption feeds adoption, i.e. the share of remaining non-adopters is influenced by the share of actors or competitors who already have adopted. This was explained as competitive pressure and called the bandwagon effect. Those mechanisms, however, seem to be part of the interactive dynamics of organisational osmosis, as outlined below in II/9.6, rather than in themselves representing a substantial selection criterion.

When the question is whether or not to adopt, some rational choice approach suggests itself. In Hedström et al. (2000 151) advantage is referred to as usefulness of an innovation, and in Hamblin/Miller (1976) as the experienced or simply the expected benefit of something. Utility and benefit are directly related to economic cost-benefit calculation and, in a generalised utilitarian way, to any rational choice reasoning, i.e. generalised low/high-cost reasoning. Both approaches are part of a broad field of theories of decision-making. They overlap with psychological theories of motivation that operate with the notion of incentive and disincentive. If we want to keep advantage, whether expected or experienced, as a central selective factor, it must be given a much broader meaning to encompass all such decisional and motivational aspects, including more complex cost-benefit evaluation, chance-and-risk assessment as well as incentive-disincentive balance.

Financial risk compared to the prospect of sustained financial profit is not the only decisive criterion in the rational choice calculation of, say, a potential producer. Alternatively, or in addition to that of making money, such criteria can also be achieving social esteem within one's peer community, the ambition to become famous, or some kind of power motive such as deriving satisfaction from being influential enough as to 'move things ahead' and shape realities. With inventors and pioneers it may also be self-actualisation through the pure joy of creation, or some idealistic sense of mission, or thirst for the adventure that can nowadays be found on the journey from the laboratory to the stock exchange. All of these possibilities can be variables with low or high load in an actor's individual utility function. They can determine the scoring of their priorities and preferences, and thus the critical threshold for decision in favour of or against adoption of an innovation.

Furthermore, the structure and outcome of such calculations depend on the specific function of an actor, i.e. creator, producer/investor, user or regulator. All of them possibly exist in different, partly competing groups, and perhaps involve a number of additional parties such as stakeholder groups, policy makers, professional milieus and occupational groups. The advantage outlook of different actors can obviously differ or even contradict each other depending on their positional role-set and the function they fulfil. There is at present no single model capable of dealing with all possible recombinations.

8.2.2 Compatibility and Impact

The criterion of compatibility refers to structural as well as functional compatibility, on macro levels as well as micro levels. It is similar to Rogers' definition of compatibility with regard to formative factors, such as compatibility with actors' basic values, beliefs and goals, with actors' lifestyles and working patterns, or compatibility with conventional wisdom (established paradigms) and hence support by the majority of experts.

Moreover, compatibility includes all aspects of effectuative connectivity to the existing legal, economic and technological structures, such as compliance with or impact on regulations and standards, compatibility with existing management structures, or with vested capital (problem of sunk costs) and predominant market structures, or operative compatibility with or impact on industrial infrastructure and skills.

Compatibility raises the question of structural impact (II/7.3). From which perspective can an innovation be regarded as having high or low impact? How easy or difficult would it be, when seen from a technical and organisational point of view, to implement the new thing under consideration? How many interfaces and resulting interdependencies does one item have with others? How many repercussions on these other items and how many necessities of re-arrangement would result from changing the status quo? A low-impact innovation is certainly more easily compatible with what exists than a high-impact innovation. If a high-impact innovation is to succeed, it must correspond to compatible structures on deeper, more basic system levels – otherwise it would not be connectable. New disruptive technologies, for example, are by definition incompatible with existing like technologies, but they represent a new manifestation of the deeper, more general paradigm of human development by scientific and technical progress. If they credibly promise whatsoever additional advantage in that direction, they have every chance of being adopted in spite of their being incompatible with existing predecessor industries.

Questions of impact may differ according to micro- or macro-perspective. For example, introducing new paper machinery based on a new chemical process has a considerable microstructural impact on the innovating paper mill. In principle, however, it does not much change the status and the markets of the pulp and paper industry on the whole. By contrast, the motorcar was not explicitly designed to replace horse and cart, nor did the rise of motorised traffic in its beginnings much change the microstructure of the horse-related businesses. The same holds true today when regarding clean energy technologies and their relation to fossil-burning technologies. Macrostructurally, however, they constitute a complex of completely new systems paths of their own, being functional niche competitors of old technologies, with the potential to undermine their existence and ultimately to replace them.

8.2.3 Innovative Capacity

Innovative capacity includes feasibility, learning capabilities and change-ability. The criterion of feasibility refers to practicability, in the sense of having the means at one's disposal, or being able to acquire the means to successfully adopt something, or to have the means to successfully pursuing an innovation as researcher, inventor, producer, etc. Of particular importance here are manageability, fundability and workability, i.e. managerial abilities, financial resources, technical and manpower capacities, especially regarding the knowledge base involved. Rogers' criteria of complexity and trialability are part of these aspects of feasibility.

An innovative project may appear to be incompatible with given structures, and to be unfeasible within given capacities and means available. Nevertheless it may become more compatible and feasible in the future, and can actively be made more compatible and feasible to a certain extent by learning and changing certain things. The extent to which this may occur is not easily determined, but undoubtedly there are degrees of freedom.

So learning capabilities and the ability to change are an aspect of particular relevance with regard to connectivity. Learning capabilities and change-ability include a willingness and an ability to acquire new knowledge and skill, including willingness to give up obsolete knowledge. More generally speaking, it requires the qualities of relative openness, responsiveness, flexibility or adaptivity. If there is such thing as innovativeness, particularly with regard to technological capabilities, as postulated in Dosi et al. (1990), then this refers to learning capabilities and the ability to change or to adequately react to changes.

Aspects of compatibility and feasibility, as well as costs of re-arranging and readjusting, of acquiring knowledge and means in order to make something more compatible and feasible, directly contribute to the advantage balance of an innovation.

The question of how knowledge is transferred, and how companies and other organisations acquire skill and knowledge relevant to being innovative, has attracted much attention. Audretsch (1996) identifies the following channels of innovative knowledge transfer:

-   by in-house learning, including R&D through in-house brain-power, in-house knowledge formation through teaching and training by internal and external personnel, including the study of any available kind of document
-   by hiring new competent brains
-   by imitation and reversed engineering from the adopters' side
-   by the import of goods and services from the adopters' side, known as equipment-embodied technological knowledge (OECD 1996 29–52)
-   through export, licensing and FDI from the first mover's side
-   through networking and joint ventures in order to acquire complementary technological assets, usually in R&D, but also in production methods and management.

According to Cohen/Levinthal (1989), R&D efforts serve a dual purpose. They generate new knowledge and recombine otherwise existing knowledge for one's own developments. To that end, scientific and technological knowledge from any source outside an organisation is likely to be exploited and assimilated (called absorptive capacity in Cohen/Levinthal).

8.3   TIMELINESS (WINDOW OF OPPORTUNITY)

The category of timeliness or opportunity can be understood as putting connectivity criteria, especially the criterion of compatibility, in a perspective of evolutive dynamics. The advantages of an innovation may be obvious, and system compatibility as well as feasibility may basically be given, but an innovation may nevertheless not be adopted because 'its time hasn't come yet', or in the terminology of systems evolution, a window of opportunity is not open yet.

With regard to Ashford's above-mentioned categories of willingness, opportunity and capacity, the notion of capacity in Ashford and in this work (feasibility, learning capability and changeability) clearly correspond to each other. Also his category of willingness seems to cover by and large what is described here as decisiveness and motivation to innovate on the basis of a positive balance of expected (dis)advantages and (dis)incentives. What remains to be settled is the question of opportunity. From my point of view it would add clarity if we translated the opportunity to innovate into the connectivity criteria of system compatibility on the one hand, and timeliness (evolutive window of opportunity) on the other.

An innovation without sufficient connectivity and without an open window of opportunity is impossible. It cannot be enforced by mere will. Failed examples of development endeavours called cathedrals in the desert are rooted in non-adequate connectivity, particularly through missing elements of compatibility. They represent attempts to implant high-tech, high-skill, high-complexity structures into an environment characterised by low-tech, low-skill and low-complexity, resulting in low connectivity.

Timeliness or evolutive opportunity of innovations can be described in terms of an innovation's life course. A system's life cycle in full upswing and unfolding is very unlikely to be cut short in favour of a possible competitor system. During take-off and initial maturation, sunk costs in terms of capital, skills and careers, user arrangements and of further vested interests appear to be very high. This only changes when marginal utility and niche saturation are approached. With full maturity and retention then the chances of new paradigms and competitor systems connecting and becoming part of an innovation and succession contest improve. The chances of success will also naturally be improved by the imminent decline of an obsolescent system, when a learning curve reaches marginal utility and perhaps enters into readjustment difficulties, or when performance difficulties arise because of stagnant productivity.

A general thesis on timeliness of innovations can be derived from this: The window of opportunity for future competitor systems opens up to the degree to which established like systems have accomplished their stage of maturity, thereby reaching marginal utility and diffusional niche saturation. A new paradigm combined with the ongoing unfolding of an established paradigm results in a no-go constellation. A stagnant or declining old paradigm, however, combined with a new paradigm which, on the whole, offers a net advantage, but which entails a high impact of change and learning to an uncertain degree, represents a maybe constellation. A stagnant or declining old paradigm, combined with a new paradigm which not only offers a positive advantage outlook, but also coming with clearly feasible perspectives of change and learning, results in a go-ahead constellation.

The next question then is how long a time window will remain open before it starts to close again. Normally, this is decided by the outcome of successor competition between rival innovations (II/9.4). As soon as one of the candidates emerges as the dominant design which is being locked in, rival candidates are simultaneously being locked out and their window of opportunity is closing accordingly. Rival innovations may, however, be adopted by different adopter populations (for example broad- versus narrow-gauge railways in different nations in the 19th century, or different telecommunications standards in different world regions today). In this case, although mutually exclusive, both paradigms are able to become established. If a rival system fails to find its niche, the failure may not be once and for all, and candidates may get second chances, though in technological successor competition they usually get just one. Strike while the iron is hot.

In those cases in which an innovation of major impact represents a technological regime shift, we can hint at certain conditions that have to be fulfilled in order to pave the way for such a generic system innovation. To begin with, no innovative regime shift comes down to earth as a set of unheard of ideas. Invention and development of an innovation must be advanced enough as to make an obvious candidate to be seriously adopted as a competitive paradigm. Being advanced entails having come a good bit further along the growth-efficiency learning curve of decreasing per unit costs (II/7.7.1).

The chances of inventions and developments becoming successful are much amplified if they belong to a particular group of innovations that occur within a long wave of generic system innovation which add to the tree of modern science and technology. Within the frame of a long wave, ever more inventions are induced, not here in the sense of cost theory, but in the sense of being technologically problem-induced or industrially issue-induced, wave-induced indeed. Once biotech and nanotech are broadly seen as the next big long-wave things to come, then of course many inventors and developers of biotech and nanotech will show up and try to join the race. This was the case with information and communications technology in the long wave we are currently still in – in its later stage of maturation and consolidation, which is probably just about time for the next long-wave candidates to start their rise.

Second, the advantages of novel systems must appear highly promising to certain groups of lead adopters, and the compatibility of the novel systems must not violate too many existing structures, which depend more specifically on the actor groups relevant during subsequent stages. Hurdles to an innovation are all the higher, and succession contest is all the harder, the more a novel system threatens established mentalities, knowledge base, practices and the vested interests of the latter system's creators, producers/investors and users. Their sunk costs, literally or in a transferred sense, must not be too high; otherwise there is no opportunity for innovation.

Advantages of innovations carry more weight, and any relative lack of compatibility of new systems may lose importance, the more the established industrial regimes no longer appear to be particularly promising (marginal utility), or the more there are recognised new problems that need to be dealt with, or new needs to be serviced, and where established technological regimes and markets demonstrate that they are incapable of adequately responding to those new challenges.

In the literature, these aspects have partly been discussed under the topic of bottlenecks or reverse salients (Hughes 1989), for example lack of affordable manpower, or climate change brought about by burning of fossils, or traffic congestion and other overcrowding brought about by outdated system components or infrastructure, or non-optimal modal split in a functional cluster. It should be kept in mind, however, that many inventions and developments are not problem-induced. They are created sui generis and emerge in their own right, and then have to look for problems and needs in order to become communicable and connective.

9. Interactive Dynamics: Cooperation and Competition

9.1 SOCIAL EMBEDDEDNESS AND COMMUNICATION

In his book *The Great Transformation* (1944) Karl Polanyi has described the transition from traditional mercantilism to modern industrial capitalism as a process of disembedding from old structures and reembedding in new ones. Sociology has used the term social embeddedness ever since when it addresses the fact that economic and industrial activities are part of a wider societal system. Such an understanding used to be self-evident to economists and other social scientists. It was also the basis of the socio-technical approach (II/6.2). Only in hermetically modelled neoclassical economics was that understanding lost. In diffusion research, however, embeddedness has always been a core element. As Coleman, Rogers and others have emphasised, any structural unfolding and diffusion of an innovation represents a socially embedded process (Coleman et al. 1966, Rogers 1995, of late Weyer et al. 1997).

Social embeddedness includes quite a number of aspects, for example interpersonal relations as well as further divisional aspects of social structure such as the institutional structure and the structure of social groups, milieus and classes. Furthermore, there are effectuative controls functions such as regulation and finance, inter/organisational dynamics and market competition. Finally, there are formative, political and cultural factors, particularly knowledge base and value base, not the least of which are national particularities as condensed into national innovation systems (II/10.2).

In some recent articles by Uzzi, the topic was taken up again. The fact 'that economic action is embedded in social structure' and relations (1997 35) is seen to be 'useful for understanding the sociological failings of standard neoclassical schemes' (1996 674). Uzzi, though, simply wants to combine some aspects of business organisation with some aspects of social network theory, saying that embeddedness 'promotes economic performance through interfirm resource pooling, cooperation, and coordinated adaptation but that also can derail performance by sealing off firms in the network from new information or opportunities that exist outside the network' (1996 675).

What Uzzi more specifically refers to are interpersonal ties. They 'vary between arm's-length and embedded' (1999 482). Arm's-length ties, according to Hirschman, who introduced this term, are sporadic and concentrate on carrying out some price-related transaction without any personal commitment. Embedded ties are closer, on a regular basis, and include some personal commitment. Embedded ties 'perform unique functions and have three features: trust, fine-grained information transfer, and joint problem-solving arrangements' (1996 677). Obviously, Uzzi's notion of embeddedness is highly selective. This is questionable because it only relates to interfirm networks in the rather narrow sense of regular informal interpersonal contacts accompanying interorganisational cooperation. As he remarks himself this represents just 'one of multiple specifications of embeddedness' (1997 36). Indeed we need to remain aware of the above listed multiple dimensions of social embeddedness.

Diffusion models of different sociological and economic origins converge in that they regard communication as paving the way for diffusion. Whatever advantage, compatibility, capability and timeliness there may be, all of these aspects, as well as information on a novel item, need to be communicated, i.e. socially transmitted. The most relevant channels of communication are mass-media (print, radio, TV), institutional publications (vocational, corporate, government) and word-of-mouth in interpersonal groups, special communities and other face-to-face contacts. In its beginnings in the 1940s, research on social diffusion, on mass communication and on group dynamics was all closely linked, and it identified what it called opinion leaders as being the decisive pivotal actors promoting diffusion.

According to a typology by Tushman/Rosenkopf (1992 325–37) the influence of non-technical 'social' factors in technological evolution increases with the degree of complexity of a technology or product. They distinguish four types of products, with the influence of 'social' factors supposed to be lowest at level 1 and highest at level 4:

1. non-assembled products such as cement, glass, paper, fibres, steel, screws, petroleum
2. simple assembled products such as stoves, containers, guns, escapements, balance wheels
3. closed systems (assembled, more complex), such as watches, TVs, VCRs, CD players, automobiles, aeroplanes
4. open systems (complex infrastructures connecting closed systems), such as telephone networks, railroad system, power grid.

9.2 NETWORKS

In the research on diffusion of innovations it has been known from the beginning that the decisive communication flows in innovation processes are part of interpersonal contact networks, be these institutionally formalised or informal. This involves primary and secondary opinion leaders, unilateral and multilateral influences, which lead to network patterns quite similar or equal to those known from group dynamics. That, by the way, is why groups and networks do not constitute two different realities as is sometimes stated in organisation theories. Groups and networks come together. A network constitutes a group, a group consists of a network, though the network may not yet completely describe all relevant aspects of a group structure.

During the 1980s the category of network was rediscovered as a generic term for any structured inter-actor constellation, including traditional group structures of kinship or peers, or in describing mutual personal relations in households, neighbourhoods, associations, volunteering and self-help groups as social support networks (Leinhardt 1977, Holland/Leinhardt 1979).

Another application of the network approach was the analysis of policy networks in describing the interorganisational dynamics of government agenda-setting and decision-making. Policy networks are constituted by a combination of strong/weak government agencies with strong/weak groups in society (industrial confederations, unions, citizens' initiatives, scientific experts). They may correspond to a model of dirigisme, concertation, corporatism, clientelism or pressure group pluralism (Knill 2000, Atkinson/Coleman 1989, Dunn/Perl 1994).

The network metaphor was soon applied to literally all sorts of cooperation among individuals or corporations. Overviews of network theory can be found in Scott (1992), Wassermann/Faust (1994), Jansen (1999) and Weyer (2000). It has become common practice by now to analyse the formation and further development of new technologies also in terms of network analysis. Seen from this angle, the actor constellation and its dynamics are of central importance to the fate of innovations and their life cycles.

It remains to be seen whether the attribution of network properties to anything that was formerly also analysed under topics like group, cooperation, communication, interaction or organisation represents an inflationary phenomenon rather than being an appropriate generalisation. In any case it has created a necessity to distinguish different kinds of networks with different functions and different groups of actors.

Many authors in the field start from the classical dichotomy of hierarchy versus market, and see networks either as a recombination of both or as a third basic form of organisation: a self-organising structure of interaction without a dominant actor, i.e. without an organising centre. In contrast to self-organisation, models of institutionalised interaction are thought to be founded by dominant actors who act as system-builders, i.e. as organising centre, and thus represent models of organisation by administration/authority or market. True social networks, according to Weyer (1997 61, 2000 7), would proceed on an informal basis, although these may be relatively enduring. They are built upon interpersonal trust. The actors are autonomous and enter into cooperation voluntarily with clear intentions of realising extra advantages that would otherwise appear to be unattainable. Network cooperation is expected to be mutually beneficial, even if it creates interdependencies of its own and a certain need to coordinate the different action lines of participants (1997 64).

Such a characterisation of networks can be said to mix adequate with unspecific criteria. It may be helpful to remember the origin of the traditional dichotomy of hierarchy versus market, i.e. a preoccupation with industrial production, and the question of whether it was preferable to have production that was centrally planned by administrative authorities and executed by nationalised corporations, or to have a competitive market economy of individually autonomous enterprises. Abstracted from that historical context there remains a generalised bipolar spectrum of organisation that extends from the one pole of directive authority (= hierarchy among actors, i.e. unequal standing, according to established rules) to the opposite pole of voluntary mutual agreement (= supply-demand complementary transactions on a negotiable basis among autonomous actors on equal standing). Real markets seldomly come close to this ideal latter state. Normally, they range somewhere in between because of positional power asymmetries based on cash, capital, information, knowledge and market position, and also on constraints of routinised chain-cooperation and habitualised patterns of behaviour.

A network is clearly not a 'third kind' of cooperative organisation that is neither hierarchical nor market-like. The decisive criterion is a different one. While hierarchy (central control) and market (self-organisation) represent cooperation within the boundaries of hierarchical or market-like role settings, a network represents a cross-boundary cooperation among actors in different roles, institutions and functions, or among different actors in an equivalent function (like-actor cooperation in mutual service and support networks). Networking is another word for cross-boundary cooperation, it may be interfunctional, interorganisational, intergroup, interpersonal, or any combination of these.

Networks are made of members of distinguished social units such as individuals, groups, communities and organisations, with whom they connect within and/or outside those social units. The relationships in networks are considered to represent weak ties rather than the strong ties of traditional social units such as family, kinship and company membership (Granovetter

1973, 1974). Networks nevertheless constitute cooperation among the participants. All networks are cooperative, though one purpose of networking may be to gain advantages over competitors outside the network. This represents a special case with regard to the general advantage of any networking, which is to gain benefits that would otherwise be unattainable.

Networks can be said to be intermediary, not however in the sense of being a half-way house between hierarchy and market, but in the sense of combining elements of formal and informal sectors. Also they are intermediary in the sense that they bridge otherwise closed settings, such as corporate organisations, or the role division between either producer or user/customer. Cross-boundary and like-actor network cooperation may as much include formalised elements based on law or individual contract as they include informal elements. Participants may include corporate and other institutional actors as much as actors ad personam.

Among the distinctions of networks to be found in the literature more specifically focussing on industry and technology there are:

- Strategic networks. These are usually set up around a core corporation that is the dominant actor in the network. They are aimed at synergies in gaining or maintaining a superior market position through joint distribution and sales systems, or improved services to the customer, not least through possible cost advantages resulting from such joint activities.
- Industrial networks or production networks (Grabher 1993). These are sectoral cooperations among industrial producers of different scale (large, medium, small), typically organised around supply-chain relationships, but sometimes also including the retail system, or, if relevant, scientific consultancy and research, financial institutions and public authorities on a local and state level. Depending on their specific shape and purpose they are called vertical in the case of cooperation along the chain of supplier-producer-user. In contrast to this they are called horizontal if the cooperation is, for instance, among industrial firms and consultancy firms, or in the case of so-called regional networks among a multitude of SMEs who pursue joint market research, product design policies and marketing activities.
- Innovation networks. These can be vertical or horizontal industrial networks the purpose of which is innovation, be it product, service or process innovation. Naturally, feedback cooperation between the carriers of different life-cycle functions (research and development, production, use, regulation, finance) is of particular importance here. Innovation networks provide access to advanced knowledge and skills, help to orient and make strategic decisions, create synergies and trust, and share costs and risks of innovation.

According to Biemans (1998), innovation networks represent a fourth level of inter-organisational cooperation after (1) customer involvement in product development, (2) buyer-seller relationships, and (3) strategic alliances. Fifth would come the virtual organisation. It can be said that all types of networks have to do with jointly creating or sharing knowledge in processes known now as organisational learning, reflecting a context of permanent structural change and an attendant necessity to constantly readjust thereto. The main reasons for entering into network activities are intensified market competition, technological complexity of products, shortened product life cycles and more demanding customers, all of which make it difficult even for well-established large companies to go it alone all the time.

9.3 TYPICAL NETWORKS DURING THE STAGES OF A LIFE CYCLE

Actor constellations carrying an innovation through its different stages of structuration and diffusion change over time. Key actors during early formation and emergence are creative individuals and small groups of vanguards or pathfinders. These people almost never belong to the establishment. Sometimes they are retired or retiring members of the establishment, but mostly they are newcomers, maybe persons with a marginal status. A few of them eventually become part of the establishment after having succeeded.

Creative and inventive persons tend to be outsiders quasi by functional necessity because searching on unknown territory and finding something new requires critical distance to (or ignorance of) anything which is part of conventional wisdom on common ground. Some do not care about being outsiders. One of the difficulties for creative persons is to make themselves distinguishable from mavericks and other breeds of strange persons.

Inventors are seldomly happy. Most inventions are ignored. Conventional wisdom is usually incapable of distinguishing the one invention among the many that is worth the future millions of investment. So vanguards and pathfinders quite often have their joy of cognition and discovery dampened by lack of recognition and perhaps even social isolation. The few who finally get some recognition are rarely still around to enjoy it. If they are awarded prices and given honours during their lifetime they have probably come up with some incremental and modificational add-on rather than having invented or discovered anything fundamentally new. There are a few famous exceptions to the rule. Maybe their function is to mislead creative newcomers about the gross existential risks they are taking.

During organised development pathfinders are followed by pioneers. The pathfinders either become one themselves, or, as is mostly the case, they are replaced by them. Pioneers in organised research and development are usually well integrated into scientific communities and industrial research and development institutions. They are productive team-workers rather than creative kick-driven individuals. Depending on whether things are at an earlier or later stage of development, the pioneers cooperate in typical R&D networks or innovation networks: Research and development projects, enjoying individual responsibility, or in a collective effort of a number of coordinated development projects, funded and perhaps also managed by government or industry agencies, and carried out by industrial development divisions and academic research institutes, or any combination of both.

Industrial innovation networks typically represent some form of research and development network in that they set up (self-)organised cooperation among academic, industrial and government agencies, the aim of which is to bring innovative ideas and designs from laboratory to practice. This latter includes re-design according to planned applications, adapting to existing conditions, testing, patenting, producing first prototypes, then pre-series in ever larger batches or volumes, until the threshold to market launch and regular production is reached.

Start-up companies, independent of or in cooperation with large corporations, play an important pioneering role during organised R&D. Start-ups are pioneering ventures by definition. Many of the start-ups of the latest 'new economy' have indeed grown from campus to the marketplace by making their way through the successive stages of research and development into regular production; this often in close cooperation with lead users, state or corporate, who guarantee effective demand to a certain extent.

From one R&D step to the next, two kinds of co-actors and controls actors become ever more important: administrative authorities and regulators on the one hand, donors of seed money and investors in venture capital on the other. Seed money normally comes from government sources or from foundations, whereas venture capital comes from investment agencies specialising in this field, which often make use of equally specialised segments of the stock exchange. Commercial banks, as creditors of external capital, do not take research and development risks. The earliest point where they come in is during advanced stages of take-off. Investment banking is about mergers and acquisitions of already established companies, except in the case of initial public offerings of newly exchange-registered start-ups, but here too the banks act as brokers rather than investors.

During the take-off stages of structural unfolding and the beginning of maturation, actor networks change again. Developing pioneers are replaced by experts in everyday routines in production, marketing and finance. This is where the above-mentioned industrial production networks and strategic networks come in, notably among producers and users, as well as producers and suppliers or, more generally speaking, producers and co-producers, also producers and distributors. In this way some sort of chain management comes into existence. Further networking is necessary with regulators and investors, by now, however, on an increasingly regular and now firmly established basis of law and procedures, and increasingly built-up stocks of capital.

The pivotal actors during this stage obviously are the producers, also including competition among different producers, even though producers act in cooperation with investors, designers, developers, users and regulators. The central role of production for take-off and maturation of a technology life cycle should not be underestimated. Any science and knowledge base of the economy, any customer-orientation, any turn from technology push to demand pull, and any social embeddedness of production of goods and services in network-settings and institutional arrangements cannot deny the central importance production has for the advanced development, prime unfolding, maturation and retention of a successful innovation life cycle. Among all of the actors involved, key manufacturers are in the best position to exert some control along the product chain.

9.4   NICHE CONTEST FOR INNOVATION AND SUCCESSION

As Schumpeter was among the first to stress, diffusion is not just about accepting and adopting. It also is about rejecting, about selection in its negative sense. The characteristics of structural change apply to any diffusional process, with winners on the side of successful innovation and losers whose positions are rendered obsolete. Any innovation comes hand in hand with some replacement and release, with opportunity and risk. Most major innovations are a challenge to the status quo and a threat to those whose stakes are vested in that status quo. That is why conflicts between challenging innovators, or proponents of the new, and challenged defenders, or opponents to the new, are a matter of central concern in diffusion theory.

Any innovation finds itself in a situation of niche contest, of competition and assertion. It is ignored, or ridiculed, or disputed and fought. Possibly it is recognised, adopted here, but rejected there. A novel thing may be equal, superior or inferior to rival like items. Niche contest arises from the simple fact that anything truly new constitutes a double challenge. It challenges anything old that may be affected, anything not able or not willing to readapt to structural change, thus constituting a struggle between new and old; and it challenges any new alternative that is on a similar track and wants to emerge from the competition as the dominant design, thus constituting a struggle between new and new, between innovative alternatives. One can refer to these two basic forms of niche contest as innovation contest (new-new competition) and succession struggle (new-old competition).

Typical examples of succession struggles in technology are situations of niche competition such as the following:

- between precious metal currencies (old) and fiat currencies (new) from the mid-19th to the mid-20th century; today, on a lower system level, between payment in cash (old) and cashless payment (new)
- between historicising and modern architecture around 1890–1930
- between horse-driven and engine-driven vehicles around 1890–1930
- between valve tube, and transistor and semi-conductor from the late 1940s to the 1970s
- between record disc, tape and compact disc from the 1950s to the 1990s
- between fossil and renewable energy today.

Typical examples of innovation contests aimed at becoming the dominant design include the following:

- between German PAL and French SECAM, as technical standards for colour TV
- between American and South-Korean CDMA 2000, and European and Japanese W-CDMA (also known as UMTS), as technical standards for 3G mobile telephony
- between mainframes, PCs and handhelds
- between telecoms services such as Teledata (or Teletel, BTX, Datavision, Viewdata respectively) and the Internet
- between aluminium and carbon-fibre reinforced plastics, as construction materials in aeroplanes and ground vehicles
- between internal-combustion engines that run on hydrogen, and fuel cells burning hydrogen, as alternative systems of car propulsion
- between insulating materials and thermo paints, as eco-efficient methods of maintaining indoor temperature.

The distinction does not always seem to be clear-cut. Is the choice between record disc, tape and compact disc an example of innovation contest or of succession struggle? Another example which in a certain sense includes both types of niche competition is the expert conflict between followers of hot and cold fusion. The supporters of hot fusion are not only an innovation rival representing one of the future energy alternatives, but they also represent the challenged establishment which upholds the paradigm of thermo-nuclear physics, in opposition to the supporters of cold fusion who would tear down the wall between physics and electrochemistry, and who accept empirical facts that cannot yet be explained within the framework of the traditional paradigm.

9.5 INTERACTIVE CHALLENGE-DEFENCE DYNAMICS (PROMOTERS, OPINION LEADERS)

The challenge-defence dynamics of niche contest for innovation and succession comes with the interactive role complement of challengers and defenders, proponents and opponents. Depending on the context of discourse, there are quite a number of terms describing that constellation, particularly with different names for the innovative leaders such as vanguards, pathfinders, pioneers, promoters, opinion leaders, change agents or sponsors. Typically, they are related to a specific stage of a life cycle.

The adoption curve of diffusion theory as shown in Figures 2.3, p.266, and 2.4, p.275, actually identifies its stages by the defining groups of actors involved. First come the original innovators, the creators and inventors, visionary vanguards and territory-mapping pathfinders. They are joined then by early adopters, the well-organised developing pioneers, who definitively set up an innovation's life course and give it hold. Innovators and early adopters have to stand firm against opposition and inertia or lack of interest by those who will eventually become adopters later on. The important part of this happens during prime unfolding and diffusional take-off, with the majority of early producers and users becoming the dominant group of actors. Thereafter, winning over the rest of the adopter population, including the laggards, is more a question of time and silent accommodation rather than being part of a true challenge-defence dynamics of niche contest.

Since everything starts with certain formative aspects of knowledge and communication, the focus of competition during the formative stages of invention and early development is on the new paradigms challenging the old ones, and on discourses revolving around the pros and cons of old and new approaches. Almost everything of what Thomas Kuhn (1962) has written on the competitive succession of scientific paradigms and academic schools also holds true with regard to the basic steps of technological innovation, which in modern society is actually just a special case of scientific innovation. Most notably, and according to all experience, the defenders of old paradigms usually display no willingness to be persuaded to become converts, however rational and sensible discourses may be. Either they have too much to lose in terms of influence, money, social standing and personal identity, or their mental learning curve has come to a standstill and they lack the will or the capacity to embrace a follow-up cycle of continued learning. Equally, peers and the broader public tend not to accept when persons start to become somehow different from what they have hitherto been known for, i.e. different from the role-labelling they were given. Defenders of established paradigms thus have to be bypassed and left behind, or wrestled down in some sort of power struggle. But if the defenders are able to maintain their dominant position, the challengers will have to wait until the defenders have died off. Mannheim's sociology of knowledge is in a similar vein when asserting that there are limits to the learning capacity of individuals, and that cultural evolution therefore proceeds on a biological basis of succession of generations.

When it comes to organised development and early adoption within an industrial company, the antipodal groups can be identified as prospectors and defenders (Table 9.1).

Table 9.1   Prospectors and defenders

Adaptive sequence	Defenders	Prospectors	Analysers	Reactors
1. Selection of areas for future innovations	Minimal search, low priority	Regular searching for opportunities	Combine stable and innovative domains	Unfocussed awareness of uncertainty
2. Choice of technology/ product and market domain	Narrow search process	Broad search process	Two domains strategy	Narrow and unstable
3. Choice of production processes and distribution	Incremental evolvement of existing processes	Varied	Two processes	Inconsistent
4. Reconfiguring existing processes and structures	Focus on efficiency	Focus on innovation	Co-executives of efficient and innovative	Unfocussed reactions

Source: Clark (1987 132), after Miles/Snow (1978).

The defenders focus on keeping and marginally improving the existing structures, mostly on efficiency or on appearance, whereas the prospectors focus on continually searching for new innovative opportunities within a broad spectrum of possibilities. Furthermore, there is an intermediary group of 'analysers' who would prefer to combine elements of new and old while keeping a close watch on what competitors are doing. There is yet another group of unfocussed 'reactors' with unstable and inconsistent views, politically speaking the unreliable, unpredictable minority that may tip the scales. The model seems to be good news for innovators in that it suggests that, supposing an innovation is really competitive, it is just a question of time until innovative prospectors form a coalition with analysers and reactors.

The model of promoters by Witte and Hauschildt represents another approach specifically referring to the formative stages of a life cycle (Witte

1973, Hauschildt 1997). From studies of innovation processes in industry and medium-sized businesses they concluded that to become successful an innovation needs to be pushed ahead, i.e. promoted, by a typical pairing of actors: expert promoters (Fachpromotoren) and line promoters (Machtpromotoren). In the recent American literature on intrapreneurship that pairing of actors reappeared as the relationship between intrapreneurs (expert promoters) and their intra-organisational sponsors (line promoters). Expert promoters contribute professional credibility based on the necessary technological knowledge and know-how. They are the technology promoters in the first instance. Line promoters, in contrast, are those who are influential through their having a say in the command line of an organisation, including the capacity to make funds and manpower available. They promote from the side of managerial controls measures, i.e. from the ordinative and economic side. Only by acting as a pair can expert promoters (intrapreneurs) and line promoters (sponsors) satisfy what is required to move an innovation ahead.

During the stage of prime unfolding and diffusional take-off through market launch, mass marketing and regular production, an innovation needs to be promoted by all members of an organisation, i.e. all of the management and employees of a company or a whole branch or sector. The challenge-defence dynamics now become mostly inter-organisational, particularly in the form of market competition between different first movers, or between innovative and conservative companies. The first movers, lead users as well as lead producers, may eventually become the new market leaders.

Throughout the stages of a life cycle, structuration and diffusion are catalysed by communication. That is why the centrepiece of sociological diffusion theory has been the analysis of the communication flows furthering or hindering an innovation. One-sided as this may seem to be from a comprehensive systems point of view, it is nevertheless trivial to say that everything in an innovation life cycle includes and relies on communication at any time. Communication flow may not be everything in innovation, but without communication there would be no innovation processes at all.

In its beginnings, diffusion theory was closely related to emerging theories of mass communication, both relying on insights from group dynamics, equally new at the time, and the related analysis of communication networks. The influential nodes in the networks ('Whom would you address to if ...') were identified as group leaders, or as informal opinion leaders in more formalised settings where formal and factual authority may drift apart.

In principle, anybody can be an opinion leader and persuade others to adopt an innovation by credibly personifying the arguments in favour of it. In practice, and with regard to the innovative challengers, the role of being an opinion leader is for the most part carried out by those who act as expert promoters and line promoters. They are backed in the process by 'deputy'

320   *Innovation Life Cycle Analysis*

opinion leaders of second and third order so to speak, i.e. by people who become stakeholders in the process and speak out and act in favour of an innovation. On the side of the defenders and opponents there exist much the same communication networks built around opinion leaders, though, of course, representing the counter-side of the innovation process. They are quite often successful, or successful at least for a long period of time, in preventing new things from happening.

An innovation of general importance, reaching far beyond individual organisations, is subject to mass communication and political processes. So far, diffusion theory has transposed its findings from microstructural group dynamics to the macrolevel of mass communication and institutional dynamics as well as market dynamics. This may not basically be wrong, though things on macrolevels are certainly more differentiated and complex. In principle, however, the mechanisms at work seem to be similar. For instance, persuasive charismatic opinion leaders as described by Hedström et al. (2000 152) are not necessarily confined to being political agitators who constitute party networks on a meso and macrolevel. On an everyday face-to-face microlevel, charisma does exist as well and can be equally important.

9.6 INTERACTIVE INSIDE-OUTSIDE DYNAMICS: ORGANISATIONAL OSMOSIS

Niche contest and challenge-defence dynamics occur externally as well as internally – external meaning the relationships between different individuals, groups, communities, organisations; internal referring to diverging processes inside a company, or industrial sector, or a scientific community or social milieu. External and internal contest occurs at any system level (micro and macro, individual and collective). Individual persons too have their inner dialogues and conflicts about novel things.

Understanding the inside-outside dynamics of innovation processes is of particular importance, and a particularly intriguing object of study. The majority of a status quo sees innovation as a process of outside assimilation by those on the inside, thereby changing themselves and their organisation. In the beginning, innovators are outsiders, and many remain so, being unable or unwilling to become part of difficult and often cumbersome follow-through processes. To most of the active participants in these processes, the innovation they assimilate and promote initially comes from the outside.

In a current thesis, Müller (2004) analyses innovative assimilation as a process of organisational osmosis. Osmosis of substances from outside a cell to its inside requires certain conditions of external and internal pressure and differences in concentration of certain substances. The analogy to physical pressure is a common feature in political science, as in public pressure, pressure groups, putting someone under pressure, or similar. Less obvious are the aspects corresponding to physical concentration of substances. The analogy here would be a situation in which a novel issue has reached a certain high degree of 'enrichment' outside an organisation, in contrast to a poor degree of 'enrichment' inside that organisation, the issue in question being relevant to that organisation. Concentration or enrichment can be understood as the amount of available information and knowledge about an issue as well as the evaluative load of an issue, i.e. the place it has been given within a ranking order of priorities and preferences. One could refer to this as the specific issue load of something. If an issue load grows ever higher outside an organisation, to which the issue is relevant, but does not grow inside that organisation, then a gap of load differentials between inside and outside builds up, resulting in tensions of a psychological as much as of a political nature. The situation takes on the character of cognitive and political dissonance which needs to be overcome – either by readjusting to the challenge and adopting the issue in some limited way, or by an explicit and continued rejection of the issue, resulting in escalating confrontation. This would create opposite camps, make adversaries or even enemies of each other and increasingly block communication and cooperation with each other.

Such a situation of escalating confrontation cannot go on forever. Existing load differentials, since they represent tension and stress, have a tendency to release themselves. In society, if it does not end up in violent eruptions, this is done by creating differentiation both within inside camps and outside camps, commonly known as factional grouping into hawks and doves, hardliners and compromisers, fundamentalists and realists, traditionalists and modernisers, or, as we could say in this context, rejecters and adopters. It is the latter who begin to ease tension by looking for opportunities to make contact, start talking, build bridges, encourage themselves and others to adopt and enter into cooperation. Psychologically, the adopting compromisers may appear to be 'weaklings' who lack the strength to stand the tension and fighting. It may equally be, however, that the rejecting hardliners are simply inaccessible rigid characters, who lack a proper understanding of what is going on, whereas the coalition-builders and networkers are the brighter brains with the wider horizons.

An organisation of former defenders cannot assimilate an innovation without such an internal differentiation, because this is the basic precondition for inside promoters to get a chance. If the diffusion of an innovation is to succeed on its way from representing an outsider minority to a greater and respected inside minority and later majority, than the previously existing homogeneity of challenging and defending actor groups has to be dissolved or broken up into factions. Inside promoters themselves contribute in turn to creating a difference between rejecters and adopters.

Early adopters who act as cooperators to the outside and promoters from within established institutions – they could be called bridgeheads – are of particular importance to the whole process of diffusion of an innovation. Greater insight into the dynamics of organisational osmosis would thus be an important additional element of innovation policies, particularly on the questions of how to identify and approach these persons, and how to contribute to creating internal action space for them from the outside. Inside promoters in organisations are dependent on co-directional outside pressure, the more so the more a process is still in its beginnings. Outside pressure can be created by formative currents in peer milieus (especially vocational milieus, social movements and political campaigns, involving celebrities of various kinds) and also by investors, stakeholders and government agencies. Their formative impulses are transmitted by direct communication and by mass media.

An ironic example of applied osmosis theory is the 'dual strategy' pursued by the students' movement during the 1960s–70s. The idea was to be a militant social movement outside the established institutions (extra-parliamentary) or to set up parallel counter-institutions of their own, while at the same time sending activists on a 'long march' into working inside established institutions as internal promoters, hereby furthering inside-outside communication and internal assimilation of the movement's revolutionary ideas. It did not work, however, in the sense of bringing about revolution, whatever this was meant to be, but it did bring many 'revolutionaries' onto a path leading to position and authority within the establishment.

9.7 DECENTRAL MARKET DIFFUSION AND CENTRALLY ADMINISTERED IMPLEMENTATION

Most diffusion processes, such as launching a consumer product or adopting a particular verbal usage, are seen from the perspective of a decentral individuals-related process. Someone takes the initiative, others follow by voluntary spontaneous adoption, or by going through some controversial process of adoption, and so an 'epidemic' innovation spreads. Such a pattern of decentral diffusion is fully in tune with the model of an ideal market process. Sometimes it is also called a bottom-up process. This can be misleading. The rhetoric of bottom-up versus top-down seems to be appropriate within the setting of a hierarchy of some kind such as the hierarchy of higher and lower ranks within an organisation. As cultural sociologists know, the direction of diffusion of innovations tends to be top-down rather than bottom-up, starting with the more educated or the better-off, then trickling down the social hierarchy. That is why the axis in consideration here is decentral (market) diffusion versus centrally administered implementation, which is different from bottom-up versus top-down.

Decentral diffusion has the advantage of not requiring too much administrative planning and implementation. On the other hand, many innovations in modern society actually do require just that and would not be feasible without far-reaching long-term planned-for cooperation between government and other state agencies, research institutions, companies, confederations, etc. Within a complex institutional setting, most things would not work without an organised institutional process of planning and implementation. And as we know according to all experience, decentral processes and market mechanisms, and centrally administered and collectively coordinated processes can be, and actually are, combined in manifold ways.

So it is necessary to be more precise about the settings in which individual decentral processes do better, and the cases that require a coordinated, administered process which is step by step conceptualised, implemented, evaluated and reconceptualised. Part of a general answer to this is certainly the difference between complex generic system innovations and minor incremental and modificational ones. Examples of the latter are new drugs or new consumer brands, generally coming on a relatively low level of impact to the adopter. These were the examples from which Rogers, Coleman and others drew their conclusions. By contrast, generic system innovations – such as the transition from carbon to hydrogen, or (inter)national currency reform, necessitating collective risk-sharing, shared learning and complex restructuring – clearly necessitate coordinated processes in addition to and complementing decentral individual action. Central planning does not need to be autocratic dirigisme carried out by exclusive elites, but can equally be done in a participative way involving those affected by it. Either way, central planning cannot replace and should not be designed to replace decentral individual action. The latter does, however, need to be coordinated and oriented by the former.

Compared to decentral diffusion, which is appealing to liberal minds and faithful free marketeers, centrally administered implementation has much less appeal, particularly under contemporary conditions of the collapsed eastern planned economies and the crises-struck democratic statism of continental Europe. Moreover, centrally administered implementation could appear to come with a higher risk of wrong decisions, bad planning and abortive developments because of a lack of self-correcting mechanisms. A closer, preferably empirical look into this reveals it in many cases to be a prejudice, the more so since the actors involved in today's innovation processes tend to be corporations whose microstructure is in fact bureaucratic: industrial and financial bureaucracies, academic research bureaucracies, government bureaucracies, media bureaucracies, even civil action groups bureaucracies.

Decentral diffusion of the formerly innovative consumer habit of tobacco smoking has presumably done as much harm to society as a centrally administered failure such as nuclear fission. Hayek was of course right in saying that exclusive elitist bureaucracies never represent as much knowledge as open inclusive big markets. But that comparison creates a rather sketchy picture. Administrative planning processes can be inclusive and participative, just as real market processes have a tendency towards closure, exclusiveness and oligopolist rule. Ultimately, it is neither markets nor bureaucracies that are knowledgeable, or less so, but rather the individual market participants and bureaucrats involved in a process.

9.8 PROTOTYPE COMPETITION AMONG PIONEERS, MARKET LEADERSHIP COMPETITION AMONG FOLLOW-THROUGH MASS PRODUCERS

The early formation of a new paradigm usually generates a variety of prototypes. These compete thereafter, during organised development, in an innovation contest to become the dominant design for market launch, or one of the two or three dominant designs in the field (II/9.4). Examples include the choice between internal or external combustion engines or electro motors in automobiles at the beginning of the 20th century, or in the 1960s the choice between PAL and SECAM as the two prototype technical standards of colour TV, or, during the 1980s–90s, alternative design approaches to computer operating systems and software applications.

Since the turn of the new century there is, for instance, the prototype competition between Bluetooth and 802.11b, also known as Wi-Fi, to become the dominant technical standard for wireless data transmission between IT devices such as laptops, mobiles, handhelds and printers. Another contemporary battle to become established as the dominant design, in this case in 3G mobile telephony, is between the telecom systems 1xRTT from South Korea and FOMA from Japan. The one relies on the technical standard CDMA 2000 developed and implemented in South Korea and America, the other one on the Japanese and European W-CDMA (or UMTS) standard. The winners of such a contest are bound to emerge as market leaders during market launch and take-off with mass production. If, however, the choice in favour of one of the two systems is decided at government level for reasons of national interest, then both standards may be adopted as the dominant design in different world regions.

Tushman/Rosenkopf (1992) have designed a generalised four-stage cycle of how new technological options are selected:

1. Technological discontinuity, i.e. new variants or opportunities emerge. If these disrupt rather than enhance established structures they usually originate with new entrants or some other outsiders rather than with veterans.
2. Era of ferment, i.e. a stage of experimenting and testing under conditions of uncertainty.
3. Selection, i.e. decision-making processes, also including the dynamics of will-building, bargaining and contracting in different networks, finally resulting in adoption of a new dominant design.
4. Retention, i.e. a state of technological closure and incremental change, in which the dominant design is implemented and diffused until new competitive alternatives eventually emerge when their window of opportunity begins to open.

The model certainly is about selective dynamics, but there is some ambiguity as to whether it relates to selective microstructural decision-making processes that repeatedly occur in many forms throughout any life cycle, or whether it is meant to represent the macrostructural evolvement of an entire technology life cycle as such. It seems that, rather than representing a life cycle model, the Tushman/Rosenkopf model takes a merely microstructural perspective and represents a kind of policy cycle which consists of the steps of problem definition and agenda-setting, working out and planning for measures to be taken, followed by implementation, then evaluation, and da capo.

Prototype competition during the stage of development up to early take-off certainly entails such policy cycles within the microstructure of actors participating in the contest. But the macrostructural outcome, even if it has been centrally planned for, cannot fully be subject to such rational proceedings. There are just too many intervening variables. As discussed in II/7.8.1, prototype selection, i.e. lock-in of what is thereupon becoming the dominant design, does not seem to be a particularly rational process, though it is decisive to the future fate of an innovation. It holds the key to whether there will be a successful transition from invention to production take-off, or whether a particular life cycle comes to its end at this stage.

Later on, having entered the stage of prime unfolding and diffusional take-off, there is a different competition: the competition about who will be market leader. This is where hypotheses about first-mover advantages come in. Controversial arguments rage over the question of whether early movers or first movers, called pioneers in Schumpeter and Sombart, gain a lasting advantage over the competition. Such a position is strongly supported in Porter/van der Linde (1995), also by the broader public with regard to legendary success stories of innovative companies pioneering the new technology of their time. No doubt, the diffusion of any key industrial technology has seen the rise of new companies which have become long-standing market leaders. Rising new-economy stars of the recent past, mostly in the USA, have been

Texas Instruments, Fairchild, Intel in semiconductors, DEC, Apple, Dell, Cray, Sun Microsystems, Compaq in non-mainframe computers, Hewlett-Packard in printers, Microsoft in software, Cisco in Internet systems, Airtouch and Vodafone in mobile telecommunications, Genentech, Cetus, Amgen, Integrated Genetics and Applied Biosystems in biotechnology.

One of the advantages of first movers in comparison to old businesses – if not hampered by market entry barriers – is the fact that they face no problems of internal substitutional restructuring with all its attendant friction and cost. All of the pioneers are promoters of the new thing, there are no threatened vested interests, no sunk costs, and there is no in-house opposition, at least not on the strategic level (Dosi et al. 1990 129).

Many innovators at first sought cooperation with established actors, but were denied it and had to set up their own business. In this way Xerox came into being because Kodak rejected Ch. Carlsson's idea of copy machines. Later on Apple was founded because now it was Xerox that rejected St. Jobs' idea of personal computers.

Romer (1986) and Arthur (1996), in a similar vein as Porter/van der Linde, have argued that the shift from traditional 19th century smokestack bulk-processing industries to 21st century high-tech industries also implies a shift from Marshallian economies of diminishing returns to new economies of increasing returns, i.e.

> the tendency for that which is ahead to get further ahead, for that which loses advantage to lose further advantage. ... Increasing returns to adoption generate not equilibrium but instability. If a product or a company or a technology – one of many competing in a market – gets ahead by chance or clever strategy, increasing returns can magnify this advantage, and the product or company or technology can go on to lock in the market (Arthur 1996 100).

A case in point is the story of Microsoft. The focus is on knowledge as the basic form of capital. Knowledge, as Romer put it, 'is an input in production that has *in*creasing marginal productivity' (1986 1002). Among the most cited reasons for the competition not catching up but losing ground is the fact that advanced technologies of today are embodied in very specialised high-level knowledge, sophisticated organisation and skilful complex production processes, all of which are difficult, time-consuming and expensive to transfer or to acquire.

The thesis of first-mover advantage represents the positive view on taking the risks of innovation. It has been contradicted by various authors, pointing out that only a few of the rising stars survive in the long run, whereas others fail at one or other of the successive steps of ongoing market and company growth. In biotech, for instance, most innovative start-ups had to concede to formal cooperation or be taken over by some big chemical or pharmaceutical corporation. Of late, bio-companies are now also taking over other bio-companies that did not manage to grow big enough.

Another aspect of this was put forth by Florida/Kennedy (1990) in their much debated thesis on America's high-tech companies who were considered to be successful innovators (pioneers). They broke through with new technological systems, but then lost ground, at the time especially to the Japanese, when it came to following through the next steps of mass production or mass marketing and continual incremental improvement with regard to the involved industrial production processes as well as certain steps of middle-range product re-design, often related to mass customisation. Breakthrough companies capture return on their investment only at the beginning of take-off, whereas substantial profits are to be yielded during continued take-off and later stages of maturity, hopefully leading into an ideal retentive 'cash cow' constellation (II/7.7.1.2).

The hypothesis on the sequence of increasing returns may be true. A general hypothesis on pioneers, however, being replaced by follow-through mass producers cannot be supported. There are certainly many cases in which pioneers have been taken over by bigger, established players, but there are equally many cases in which pioneers have managed to follow through and set up themselves as the leading mass-producers.

9.9 TECHNOLOGY PUSH VERSUS DEMAND PULL

Another way of approaching the interactive dynamics of innovation, apparently economic in nature, is to ask whether decisive impulses come from the demand side or the supply side. This is to ask in this context whether the dominant force is science-driven technology push or market demand pull.

Technology push refers to the development preferences of researchers, inventors, developers and producers, whereas demand pull refers to information from marketing people about user and consumer preferences, or inquiries, thus not really effective purchasing demand. Two-thirds of innovations can be classified as demand-oriented, whereas one-third is technology-driven (Schmalholz/Penzkofer 1997 31). One can assume, however, that most of the two-thirds of demand-oriented innovations represent incremental changes of minor importance.

It is obvious that demand pull, where it becomes apparent, strongly determines the process of market penetration (diffusional take-off and saturation). Less obvious is the influence of demand pull on earlier formative stages of structuration and diffusion. It should be noted that there is a terminological parallel to the industrial paradigms of Taylorism/Fordism (i.e. technology push) and Toyotism (i.e. demand pull). Aside from contemporary myths of pro-active marketing, it will remain a basic truth that new supply items, i.e. major innovations in systems, components and infrastructures, are introduced by risk-taking entrepreneurial initiative; the entrepreneur may even be a huge corporate actor. The above mentioned examples of Xerox and Apple are perfect evidence of this. Supply either creates or it does not create resonance on the demand side. Demand cannot be a substitute for supply side trial-and-error learning. Demand, however, gives an effective selection feedback. In this way demand is of considerable importance to optimising products and increasing efficiency of processes during the structural unfolding and diffusion of innovations.

There is a plausible and empirically proven link between company size and innovativeness, particularly concerning the structural impact of an innovation, the ability to create or adopt innovations, and the speed of adoption. Small and medium-sized companies (SMEs) tend to innovate less, and if they do, the innovations are late incremental and modificational rather than generic systems or component innovations; they are entirely market- or customer-oriented, not technology-driven or vision-driven (Audretsch 1996 113, Schmalholz/Penzkofer 1997). By contrast, it is the big companies and research institutions who have the necessary means and capacities, or risk-taking new start-ups with the necessary knowledge and expertise, who dare to pioneer in generic innovations and paradigmatic regime shifts. These, in turn, are more technology-driven, orienting at new opportunities that will have to be worked out and communicated step by step to an ever broader audience, and finally to the public in general as the potential customers who need to be convinced that the new thing would serve a good use. So science-driven technology push is more important in the beginning and during the first stages of an innovation life cycle, whereas demand pull gains increasing selective influence during market introduction, take-off and later stages of niche penetration (Coombs et al. 1987 103).

A special case of technology push and demand pull feeding back into each other is cooperation between smaller high-tech start-ups and established corporations who play the role of lead users; the latter not only being the main applicants of certain innovations but also having institutionalised experience in experimenting, product development, getting things registered or licensed, and in marketing. Typically, such cooperation between innovators and lead users can be found in most of the IT sectors as well as in biotechnology (Casper 1999).

10. Location Matters: Regional Clusters and National Innovation Systems

As Dosi et al. (1990) have observed, the capabilities and capacities necessary to develop, produce and sell new products, are closely related to location and agglomeration economies which exercise competitive exclusion on other locations. With regard to nations and national innovation systems, different capabilities and capacities result in international technology gaps.

Seen from a geographic point of view, all of the actors contributing to an innovation could in principle be widely scattered on the map. Normally, however, innovative actors and innovation networks are locally or regionally concentrated. The reason is advantages of close personal contacts. Knowledge spills over among people. Spillover occurs as persons communicate and cooperate, the more so the more they do it face-to-face. That is why location matters, geographic proximity, and why also in the knowledge economy there are regional industrial clusters where innovative activities and new developments are most intensive.

10.1 REGIONAL CLUSTERS OF COMPETENCE

Throughout history, any form of production and trade have created typical territorial concentrations of activities. Before industrialisation, centres used to be trading places and, related to this, financial centres. Obvious candidates had to be ports, or be located at a crossroads of important trade routes. Typical examples include Venice, Antwerp, London or the locations connected to the Hanseatic League.

Industrialisation then created typical industrial towns and whole industrial districts as described by Alfred Marshall. Well-known examples of traditional industrial districts include Manchester, the English Midlands, the Ruhrgebiet in Germany, the Ohio basin in the US, home to the American car industry, also the rust belt, the old heavy industry region in the American northern middle east. The important factors constituting a traditional industrial region were local access to raw materials, particularly coal, and availability of a numerous industrial labour army.

329

Today, the transition to the high-tech knowledge economy is creating similar regional clusters of a new variety. The crucial factor now is expertise and services revolving around high technology, such as research, education and training, finance, high-speed traffic, life-style and living environment. The key factor to all of it seems to be technological, economic and legal expert knowledge, i.e. professional skill and experience. That is why such centres are also called clusters of competence. The competence involved is usually productional rather than scientific, although in the evolutionary course of the new knowledge and high-tech industries these aspects become ever more integrated. A productional cluster of competence is 'a territorial concentration of elements in a supply chain' (Rehfeld 1995 190). A regional cluster of competence can be expected to include a critical mass of research institutions as well as established companies and start-ups, specialising in innovative sectors such as nowadays information technology, telecommunications, e-business, media, financial services, new materials, microtech, nanotech, biotech, clean energy, air and spacecraft, high-end vehicles, thereby attracting capital and highly qualified brains, and gaining networking advantages based on local proximity (Jacobs 1997, Braczyk et al. 1998).

Recent examples of innovative regions, particularly in IT, include California's Silicon Valley south of San Francisco and Boston's Route 128. Others were more or less successful in trying to join in, among them the science city of Tsukuba in Japan, London's M4, the east-west strip along motorway 4, the Silicon Glen in Scotland, regional clusters in France such as Grenoble-Isère Tetrapole, Sophia-Antipolis, South Paris Scientific City and the technopole of Toulouse, in Italy the triangolo between Torino, Ivrea and Novara/Milano, in Germany the regional clusters around Munich and Stuttgart. As another claimant to the league the Silicon Fen (Cambridge) has of late introduced itself. In biotechnology there are Biotech Bay (which is identical with Silicon Valley), Biotech Beach in San Diego, Cambridge, Uppsala and also Israel.

Most of the European examples lie within the so-called Eurobanana (DATAR 1989). It circumscribes a roughly banana-like broad ribbon stretching from London and South East England via Benelux and the Rhine territories to the Alpine territories of Germany, Austria, Switzerland and France, to Northern Italy and Southern France, into connecting Spanish regions. The regions within the range of the Eurobanana are said to be home to the most advanced of Europe's industries. The banana metaphor was, of course, sharply criticised by Scandinavian countries. They proposed a bunch of grapes, with the bigger ones in northern latitudes. Paris, although outside the banana too, has not made a protest since to the French, who invented the term Eurobanana, Paris is the sun at the centre of the orbiting banana planets.

Centres of traditional industry may be able to manage a transition to high technology. For instance, the greater regions of London and Paris have so far managed to keep their place at the frontline of any new development in occidental economic history. They were already important centres during the middle ages, later on becoming important places of trade and finance. Both have by now innovative regional clusters of the 'new economy'. London, alongside New York, has become the outstanding example of a global city.

With the shift in economic and production factors from raw materials to brain power, the question being discussed is whether people and their amenities still follow industrial companies, or whether, just the other way round, companies follow brains and amenities. It is certainly not just by chance that new high-tech centres are quite often located in beautiful landscapes with high standards of living and much infrastructure serving mobility, housing, education, cultural animation and leisure (Andersson 1990a and b).

The rise of the new high-tech regions comes with strong interregional competition: for corporate investment, for advanced science and technological research, for high-tech specialists and other high potentials; a contest fought with the weapons of tax reliefs, subsidies, good schools and universities, close access to airports and fast rail stations, various amenities and attractive environment and living conditions. The new competition among regions carries traits of social Darwinism. It contributes to changing the pattern of centre-periphery structures within the world system at the national level, possibly also creating new or deepened welfare disparities between central and peripheral regions.

If local clusters are seen as a phenomenon of decentralisation then there would appear to be a contradiction between 'decentral' local clusters of high-tech industries and corporate networks of 'central' global players. This, however, is misleading. Regional high-tech clusters *are* 'nodes in global networks' (Amin/Thrift 1995 115).

Furthermore, high-tech clusters are, in principle, not really different from traditional industrial districts or mercantile centres of commerce. The phenotypes certainly change in appearance over time. For instance, the qualifications of 19th century industrial workers were certainly not the same as the skills and expertise needed in a high-tech environment of today. But in principle the factors at work always include, as described by Marshall, (1) patterns of mutually complementing specialisation, (2) a large enough pool of adequately skilled and experienced people, (3) a common cross-organisational flow of communication providing information and orientation to the benefit of all participants (networking), and finally (4) a factor of atmosphere, as Marshall put it, i.e. a conviction of being at the forefront of current developments, and a motivational readiness to go for it. Continual invention of new buzzwords in business literature, and constant renaming of the same things, obscures these far-reaching continuities in transsecular modernisation processes (Cooke 1995).

10.2 NATIONAL INNOVATION SYSTEMS

Geographic proximity as well as coherence of national institutions and culture are thought to contribute to the emergence of national systems of innovation (Nelson 1992). In the same way as there are typical patterns of national character, or at least stereotypes thereof, typical national patterns of innovation were also said to have been identified for the UK, the USA, Germany, Japan, the Netherlands, the Soviet Union, Brazil, South Korea, as discussed in Freeman/Soete (1999) or in Clark (1987). Also included were some general comparisons between the world regions of Europe, North America, Latin America and East Asia.

National innovation systems, according to Lundvall (1988, 1993), represent the idea of vertical and horizontal networks extended to the national level, i.e. the national web of cooperative activities aimed at innovation targets, or contributing to conditions favourable to innovation. There have been a number of attempts to identify in more detail the specific factors interrelated in a national innovation system. Economists, of course, tend to stress economic factors including industrial infrastructure as well as human capital, while other social scientists also include regulation and governance structures as well as certain patterns of mentality such as cultural connectivity and open-mindedness, or attitudes towards hard working, or towards risk and uncertainty.

At the national level, or any macrolevel, a couple of factors come to the fore that tend to remain in the background as long as things are considered from the micro-perspective of individual actors. Among those factors are the national institutions of education, training and research, be these public or private, as well as the role of regulation, and patterns and style of governance at the national, transnational and regional state level. These are aspects of the institutional structure of nations. Hence one could define a national innovation system, as Freeman/Soete (1999) put it, as 'the network of institutions in the public and private sectors whose activities and interactions initiate, import, modify and diffuse new technologies'. In the words of Nelson (1992), a national innovation system 'doesn't need to be conscious or planned for. It is constituted by the interplay of government, administration, confederations, industry, finance, academia, and other R&D institutions'.

As Freeman/Soete have stated, the concept of national innovation system can be traced back to Friedrich List. His preoccupation was with Germany catching up with England in terms of industrial development and wealth. In a book from 1841 titled *The National System of Political Economy*, List criticised the idea that the market would do it all. He argued in favour of well-targeted national government policies, aimed at regulating and conditioning, though not directly intervening in firms and markets. His conviction was that economic success depends on non-tangible factors such as national culture and work ethics, support for science and technical invention, good education and training, reliable administration and other similar factors, all of which he considered to be of the same importance as the accumulation of tangible capital in a conventional sense.

Pavitt and Patel have established a connection between knowledge intensity, or R&D intensity respectively, and nationally clustered innovativeness. As they observe, most of a company's R&D, even within transnational corporations, is done 'at home'. As a consequence, companies' innovative activities are significantly influenced by their home country's national system of innovation: the quality of basic research, workforce skills, systems of corporate governance, the degree of competitive rivalry, or persistent patterns of private investment or public procurement (Pavitt/Patel 1996 143).

With regard to the patterns in question, many authors, among them Nelson, identify two basic alternatives: a free market model where government does not intervene but provides favourable framing conditions, and a corporate model of coordination and negotiated cooperation. Soskice (1994 273) thus distinguishes between 'liberal market economies' including the USA, the UK, Ireland, Australia, New Zealand, and 'coordinated market economies' including Germany, Scandinavia (both labelled industry-coordinated) and Japan (which is called group-coordinated). These different institutional settings would explain why

> in a broadly similar external environment in terms of world markets and access to modern technology national innovation systems continue to produce different micropatterns of behavior. The line of reasoning is that different national systems of institutions act to refract differentially the external framework onto domestic actors. Different national institutional systems generate different advantages – in terms of skill production, technology diffusion, access to long-term finance, cooperative systems of industrial relations, attitude to risk, and so on – which lead to companies engaging in world markets according to their *comparative institutional advantage* (Soskice 1994 271).

The liberal pattern would explain why the USA is supposed to be good at radical generic innovation, less good at incremental improvement, and bad at efficient production processes, whereas, in complete contrast, the coordinated patterns of Germany and Japan are said to be bad at innovating new systems, but good at incremental product innovation (Germany) and process perfection (Japan).

But does it correspond to realities? Why and how precisely would it be so? Couldn't rather the liberal pattern be suspected of lacking the coherent framing and organised cooperation which is required if the complex conditions of modern innovation processes are to be satisfied? In contrast, the coordinated pattern could be expected to be particularly good at delivering just this. Remember MITI's myth during the 1970s–80s. A liberal pattern, in turn, primarily driven by market competition, could actually be expected to be very good at increasing efficiencies in incremental ways, whereas it is the coordinated system, with its tendency towards corporatism and bureaucracy, that could be suspected of being more inefficient in microstructural detail.

Stylised national differences are actually less pronounced in practice. The US certainly is devoted to a liberal market economy, but at the same time and depending on the issue, it also tends to be rather interventionist and well co-ordinated. In the US, regulation and government demand, particularly with regard to military planning, have often made a significant innovative impact. Without planning for 'smart bombs', the microchip would hardly have been developed with the rapidity and to the extent that it was. With telecommunications and GPS it was similar, and at present it is the same with micro-electro-mechanical systems.

The French and German backlog in last generation innovations, notably IT and applied biotech, is of a recent nature. In previous decades, France and Germany used to contribute considerably to generic systems change in technology, despite a continued tendency towards statism and corporatist institutional settings throughout different political periods.

Appropriate regulation, as well as some kind of national coordination, is necessary in any relevant field of action, although appropriate regulation must not be confused with ever more regulation, extending down to the nitty-gritty of every detail. Appropriate national regulation and coordination further innovation because they create certainty regarding the law, a reliable framework of rules, standards and targets, a common perspective and individual planning horizons. Over-regulation and heavy bureaucratisation, by contrast, strangle initiative and innovation – at the national level as much as at the level of companies and other corporate actors. The existence of characteristic national differences can of course not be denied, but it remains questionable whether 'liberal versus coordinated' offers an appropriate frame of reference. Civil markets and national coordination by public bodies need not be adverse to each other, and can both perform best when they come in the right combination (II/9.2).

Differences in national innovation systems are also rooted in differences in national culture – unless distinctive features of culture are conceived of from the beginning as being an integral part of a national innovation system. The co-relatedness and co-directionality between formative and effectuative factors within nations and nation-states are familiar to sociologists with a Weberian historical-institutional filiation of sociology of knowledge. They continue to be regarded with suspicion, however, by neoclassical economists who, favouring 'homogenous' market conditions and 'equilibriums' of convergence, tend to externalise, vulgo ignore, some of the more important facts of life. There is a wide body of comparative cross-country studies which relate differences in industrial and economic structure and performance to differences not only in institutional settings, but also to differences in culture (Albach 1993, Putnam et al. 1993, Soskice 1994, Thompson et al. 1990).

With globalisation and a restructured role for the nation-state, the concept of national innovation system may need some reformulation or enlargement. In the EU – particularly in monetary and financial institutions, policies of market regulation and competition, agriculture, manufacturing, technology and environment – European law is increasingly being superimposed over national lawmaking and regulation. Standards regarding schooling and higher graduation are equally subject to europeanisation, i.e. assimilation into EU-set standards. Many citizens in individual EU member states still do not realise this fact because EU directives are enacted in the form of national law, the EU origin of which cannot be seen to those not in the know. Brussels tends to avoid directly recognisable lawmaking through the use of EU ordinances which take immediate effect in every member country.

Another example of transnationalisation of hitherto national patterns regards high potentials and research. In today's knowledge economy brain power is of the utmost importance. Since people are becoming ever more mobile, there is now stronger international competition to attract those high powered brains. Brain drain has long been a problem to those countries where high potentials have been educated and paid for, but where working and living conditions are not competitive and attractive enough to keep them. The USA has become the biggest net importer of brain power (OECD 2001a, 2002b) as it has the biggest regional innovation clusters and by far the biggest slice of the world's R&D. Any table of patents is led by US laboratories, be these corporate or public, as any list of Nobel prizes is also led by Americans or foreign nationals working at US universities or corporate laboratories. As a reaction to this, non-US companies have begun setting up research subsidiaries in the US or buying American start-ups.

If such tendencies prove to represent a stable secular trend, the notion of 'national system of ...' would need to be specified by putting it into a world system perspective. Many national innovation networks are actually of multinational and in some sense of global reach, though dominated by certain national players and their still prevailing national interest.

Appendix: Systematics of Technological Environmental Innovations

Energy

Safer nuclear
   Pebble bed reactor
   CAESAR (Clean and environmentally safe reactor; just uranium-238)
   Transmutation of nuclear waste to shorten half-life to a few decades
   Nuclear fusion (unclear)
   Cold fusion (unclear)

Cleaner carbon
   Beneficiation of coal (artificial ageing, desulphuration)
   Synfuels (synthesis gas, liquefied coal)
   Pressurised-fluidised bed combustion
   Co-generation of heat and power, e.g. district heat and power stations
   Combined cycle power plants (steam turbine plus hot gas turbine)
   Integrated gasifier combined cycle IGCC, and
   CO_2 sequestration in central power stations, in connection with steam reformation (production of H_2)

New fuels
   Methane hydrate
   Biofuels
      Fuel crops
      Biodiesel
      Combustion of biomass
      Biogas from fermenters
   Silane (unclear)
   Hydrogen production
      Steam reformation
      Electrolysis
      Gravitational electrolysis
      Solar hydrogen
      Brown's gas

Hydrogen storage
  Cryotanks
  Absorption into sodium hydride (Borax)
  Absorption into metal hydride
  Nanocubes
Fuel Cells
  SOFC (solid oxide fuel cell) i.e. high temp, large, stationary
  MCFC (melted carbonate fuel cell) i.e. high temp, large, stationary
  PAFC (phosphoric acid fuel cell) i.e. medium temp, medium-sized,
    stationary
  PEM-FC (proton-exchange membrane, or polymer-electrolyte
    membrane fuel cell) i.e. relatively low temp, small and miniature,
    stationary and mobile
  DMFC (direct methanol fuel cells) i.e. low temp, small and miniature,
    mobile and portable
  Flow cell batteries (large plants for interim power storage)

Latest generation batteries and accus

Micro-electro-mechanical systems (MEMS)
  Free piston micro engines
  Miniaturised gas turbines
  Nuclear MEMS
Photovoltaics. Next generation solar cells made of
  ultra-thin-layer silicon
  special metals (copper-iridium-gallium, cadmium-telluride)
  flexible organic polymers with inorganic nanorods
Solar thermal
  Solar thermal heat
  Concentrating power
    Parabolic trough plants
    Parabolic dish reflectors
    Solar towers
Wind
  Up-current towers
  Wind turbines, stand-alone and in wind farms
    Offshore wind farms
    Lightweight wind turbines with hinged blades
Hydroelectric power
  Running-water power stations (river barrages)
  Storage power stations (large dams)

Tidal power
    Tidal power plants
    Tidal underwater turbines (watermills)
Underwater-current turbines
Wave power
    Power buoys
    Limpet (land-installed marine-powered energy transformer)
    Energy ship (hybrid of wind and hydropower)
    Ocean thermal energy conversion (OTEC)
Geothermal
    Heat in combination with heat pumps (about 100 m depth)
    Combined heat and power plants (hot water from 4–5,000 m depth)

Distributed power generation (micropower)
Integrated two-ways-flow grid management

Natural Resources

Cascadic retention management of water and groundwater
    Sustainable groundwater regime
    Drainage management of surface waters
    Flood prevention regime (dams and restoration of flooding zones)
    Percolation management of rainwater and purified effluent water
Rainmaking in arid regions
Decentral water supply with small plants
Drinking water by desalination of sea water

Low-impact mining (coal, metals)
Low-impact mining (oil)
    Single shaft recovery
    Multi-directional wells, aided by seismic imaging of oil fields
    Multi-phase pumping
Mining biotechnology: phytomining, bioleaching

Sustainable forestry
    Sustainable forestry regimes adapted to local conditions
    Licensed logging and certified wood industry
Fisheries
    Scaling down of open sea fisheries
    Aquaculture
        Fish farming, naturally bred and transgenic varieties
        Offshore ranching, naturally bred and transgenic varieties

Agriculture

Organic farming

Ecological modernisation of conventional agroindustry
Fermentation of manure (biogas plants) in intensive animal farming
External auditing and brand licensing in animal and field farming
Adoption of a number of green regime rules in animal farming concerning, e.g., animal medication, animal density, animal transport
Phytase (enzymatic additive to feedingstuffs)
Biological pest control
Integrated pest management
CleanWeeder in combination with new way of growing rice

Precision farming in the field
Sensor-controlled automated rake
Camera-assisted weed-detection
Tractor tyres with variable pressure
Computer-aided and satellite-based systems (global positioning) on and off the field which control precise dosage of seeds, fertiliser, pesticides, water, exactly to the square meter

Closed-loop greenhouses
Closed-loop agrofactories on several floors

Crop and animal identity verification (produce tracking)

Industry crops (natural fibres and high-content crops as feedstocks for industrial production)

Transgenic crops and animals
Conventional cross-breeding enhanced by genetic engineering, such as
Genetic markers
DNA fingerprint technology
Automated genetic screening
GM crops tolerant of pesticides and herbicides
GM crops tolerant of climate stress and hostile environments
GM crops resistant to special pests and immune to special disease
GM high-yield crops of high quality and content-value
Apomixis-stimulated seeds, i.e. self-replicating ('self-cloning') genderless crops
Molecular farming in crops (nutrients, pharmaceuticals and chemical specialty substances)
Molecular farming in animals (secretion of useful extractable substances in the milk of female animals)

Transgenic animals with optimised useful properties
  EnviroPig (more complete absorption of phosphate nutrients)
  Medaka fish and salmon with higher level of growth hormones
Transgenic spider silk

Chemistry and Chemicals

Biofeedstocks (vegetable oil, fat, starch, sugar, cellulose, fibres, etc.), natural and transgenic

Phytochemistry partly replacing carbo- and petrochemistry

Biotechnological processes
  Microbial biosynthesis
  Microbial photosynthesis (photolytic production of hydrogen)
  Biocatalytic processes
    GM-enhanced fermentation
    Enzymatic catalysis

Biosensors
  DNA-chip technology
  Enzymatic
  Microbial
  Biooptical

Benign substitution of harmless or low-impact substances for
  Heavy metals
  Persistent organic pollutants
  Halogenated compounds
  Organic solvents
  Concentrated acids or alkaline solutions

Low-impact chemistry (inherently safe products, short-range chemistry, green chemistry)

New Materials

Secondary materials
  Secondary raw materials from reprocessing of waste materials
  Cycleware
Biotic compound materials (made of natural fibres and plant proteins)
New plastics and synthetic fibres
New metal alloys
New ceramics
Crystal breeding

New composite materials
  Composite materials on the basis of carbon fibres, e.g. carbon fibre
      reinforced plastics, carbon fibres reinforced by synthetic resin
  Metal-like materials such as organic metals (e.g. Polyanilin), aerogels,
      metallic glass or transparent glass-like metals
  Polytronics, i.e. semiconducting materials made of organic polymers
      and metals
Nanomaterials
  Carbon nanotubes (as working material in electronics, chemistry
      and medicine)
  Nanosensors
  Nanofluids
  Thermoelectric materials (direct conversion of heat into electricity, or
      vice versa)
  Artificial zeolites (nano-porous minerals)

Materials Processing

Nanotechnical surface treatment
  Self-protecting, self-healing, self-cleaning surfaces through
      protein/enzyme nanocoatings and micro-rough bristly surfaces
  Light-refracting nanostructured surfaces (colours without pigments)
  Ultra-thin extra-strong coatings by colloidal (electrostatic) self-
      assembly of layers
  Ultra-thin extra-strong coating by thermal gaseous spraying
  Inductive decoating of surfaces
  Plasma treatment of surfaces (among others, plasma treatment of
      plastics, vacuum-sputtering of solar-thermal surfaces, dry
      decontamination)
Dry powder coating technology
Regenerative combustion
Clean-burn technology on the basis of porous burners (also flameless
      oxidation)
Advanced membrane technology
High-performance lasers
Sonic devices, high and low frequency
Powder metallurgy, foamed metals, dry metal-working
Sulphur- and chlorine-free pulping
Biotechnological processing of biotic materials (e.g. in pulping and paper
      making, leather tanning, textiles manufacturing, food processing)

Building

Re-densification of urban space, spatial re-integration of operative functions

Multi-floor buildings in closed street blocks/carées preferable to high-rise buildings, as both are preferable to small stand-alone houses

Building materials
>   Design for reusability of construction elements and recyclability of building materials
>   Elimination of hazardous substances in building materials
>   High-quality building materials made of mineral and wooden cycleware
>   Polymer matrix composites
>   Foamed minerals

Energy design
>   North-south alignment of new buildings
>   Daylighting instead of artificial lighting
>   Automated lighting control
>   Transparent heat insulation (windows, façade glass)
>   Sandwich walls
>   High-performance insulation materials
>   Thermo paints
>   Photovoltaic panels roof-top or façade-integrated
>   Solar thermal collectors roof-top, maybe in combination with
>   Calorific boilers
>   Solar air and ventilation system
>   Heat recirculation in pipes beneath the floor
>   Heat pumps
>   Geothermal heating and cooling

Water design
>   Roof-top turf
>   Minimisation of soil sealing

Facility management and contracting

Vehicles

Optimum modal split of transport
Logistic traffic optimisation
Soft car driving

Internal-combustion engines
>   Diesel engines
>>   Diesel soot microfilters
>>   Plasma Diesel-exhaust cleaning

      Diesel common rail injection

      Elsbeth motors

   Incremental improvements of internal combustion in Otto motors

      Three-way exhaust catalyst

      Lean-burn motor

      Fast starter-generator

      Pressure wave turbocharger/supercharger

      Direct fuel injection

      Cranked connecting rod

      Variable valve system

      Camless engine

      Cartronics (interconnecting and fine-tuning controls of
         components)

      Mechatronics (similar, including substitution of electromecha-
         nical actuators for conventional mechanics)

Hybrid propulsion (combination of two different propulsion systems in one
      car, e.g. electric motor combined with internal combustion)

Hydrogen-fuelled and fuel cell-powered cars

      Hydrogen-fuelled internal-combustion motor car

      Hybrid car combining hydrogen combustion with electric motor

      E-cars powered by fuel cells (various designs of FCs, fuel, and
      fuel storage)

Materials

      Product stewardship. Take-back of disused cars

      Recycling of car parts and materials

      Design for disassembly and recyclability

      Increased share of aluminium and new compound materials

      Increased share of cycleware and biotic materials

FC-powered ships and submarines

FC-powered propeller machines

Hydrogen-fuelled jet planes

Fuel-efficient redesign of aircraft components, including new materials

Airship cargolifter

Micro-electro-mechanical propulsion systems in small, unmanned airplanes

Magnetic levitation train (Maglev)

Utility Goods: Office and Household Appliances

Teaming-up in use

Leasing/renting instead of owning

Modular design

Power-efficient office and household appliances and devices
    Power-efficient refrigerators, washing machines, TV sets, etc.
    Power-efficient light bulbs and fluorescent tubes
    LEPs (light emitting polymers) or LEDs (light emitting diodes)
New refrigeration technology
    Closed-cycle air refrigeration
    Thermo-acoustic refrigeration
    Magnetic refrigeration
Preference for high-quality pure materials
Design for disassembly and recyclability
Increased share of cycleware and biotic materials
Environment-oriented chain management

Materials Reprocessing and Waste Management

Refurbishing and reuse of machines, plants, and product parts
Industrial symbiosis (combined processes, i.e. cascadic re-entry of waste
    products or byproducts in subsequent production processes)
Dry-stabilate process (one-bin principle in municipal waste collection and
    separation with integrated recovery of valuable content materials)
Automated detection and separation in sorting plants
    High-pressure waterbeam crushing
    Sonic and optical (colour reflection) sensors detecting materials
    Software-and-actuator systems controlling mechanic separation
Disassembly of complex products in semi- or fully-automated plants
Materials recycling
    In-process recycling (within a factory or production line)
    In-process recovery of valuable and hazardous waste content
    Macrostructural recovery and recycling (secondary raw materials)
        particularly metals such as steel, aluminium, platinum, gold,
        silver, copper, and building materials, plastics, textile fibres and
        organic base materials (e.g. amino acids)
Biological treatment of biotic final wastes
    Composting in closed plants with recovery of biogas
    Co-fermentation of various wastes in biogas plants
Thermal treatment of final waste and hazardous waste
    Incineration plants with heat and power co-generation
    Waste incineration in industrial blast furnaces
    Solar and mechanical drying of wet sludges for incineration
    Smouldering-burning process (pyrolysis)
    Thermoselect process (pyrolysis)
Closing down of all landfills, except a few for slag and ash and selected mineral construction waste, as far as these are not recyclable yet

Emissions Control

Airborne emissions
 CO_2 sequestration (as under energy)
 Car exhaust treatment (as under vehicles)
 Electric filtration of airborne particles
 Aerosol micro filtration
 Dry desulphuration (in contrast to conventional wet desulphuration)
 Catalytic removal of NO_x
 Removal of VOCs and CVOCs by
  Catalysts
  Flameless oxidation
  UV oxidation
  Biocatalytic waste air treatment with microbes
 Advanced membrane-type filtration of gases

Sewage water purification
 3-phase-sewage water purification plant with integrated ammonia-
  stripping
 Advanced membrane-type filtration of effluent
  Ceramics membranes and cellulose-stainless steel membranes
  Ultrafiltration with various diaphragms
  Gel stripping of high-molecular substances
 Electrolysis of brine (replacing disinfection of water with chlorine)
 Biocatalytic purification
  New breeding generations of bacteria (GM)
  Combining bacteria with coke
  Microfiltration through biomembranes
 Sequential reactor plants
 Plant-growth purification stations
 Small purification units in decentral applications
  Catalytic
  Electrolytic
  Microfiltration through biomembrane

Site remediation
 Enshrining contaminated soil in a concrete coffin
 Thermomechanical treatment in rotary skilns, off- and on-site
 Chemical decontamination
  Use of neutralising agents
  Bauxsol™ (red mud)
 Phytoremediation

Microbial bioremediation
    Various cases of GM-enhanced bacteria
    High-pressure aeration with biological and mineral additives
Restoration of damaged habitats, i.e. recultivation and reforestation of
    dug-over, abandoned or desertificated land, e.g. newly developed
    sites, traffic ways, shut-down mining areas and slag heaps, fields
    devastated by agroindustrial use, formerly forested areas

Environmental Measuring and Monitoring

Spectroscopic detection and measurement
    Ion mobility spectrogram synthesiser
    Mobile case-sized mass spectrometer
Chromatographic detection and measurement
    Chemical fingerprinting (isotope markers)
Laser-based metrology
    Laser-based quality/consistency probing
    Laser-based sewage water meter and analyser
    LIPAN (laser induced plasma analyser)
Optical, photo-acoustic and sonar methods
    32-channel array detector (UV light)
    Photo-acoustic sensors
    Combined sonar and radar remote detection of pollutants
Membrane-type dialytical analysis
Electrolytic analyses (nitrate monitoring)
Electrical analyses (measuring resistance, e.g. mercury monitoring)
DNA-chip technology, optical biosensors, enzymatic and cell biosensors
Natural bioindicators (lichens, indicator plants, small organisms)
Ultra-fine soot monitoring
Wireless environmental sensor networks (interconnected motes, i.e. remote
    sensors and radio transceivers)
Satellite-based earth observation (environmental monitoring)

References

Abernathy, William J. / Clark, Kim B. (1985), 'Innovation. Mapping the Winds of Creative Destruction', *Research Policy*, **14**, 3–22.

Albach, Horst (1993), *Culture and Technical Innovation. A Cross-Country Analysis and Policy Recommendations*, Berlin/New York: Walter de Gruyter.

Allenby, Braden R. (1999), 'Earth Systems Engineering. The Role of Industrial Ecology in an Engineered World', *Journal of Industrial Ecology*, **2** (3), 73–93.

Allenby, Braden R. / Cooper, William E. (1994), 'Understanding Industrial Ecology from a Biological Systems Perspective', *Total Quality Environmental Management*, Spring 1994, 343–354.

Amin, Ash / Thrift, Nigel (1995), 'Neo-Marshallian nodes in global networks', in Krumbein (ed.), 115–139.

Anastas, Paul T. / Warner, John C. (1998), *Green Chemistry. Theory and Practice*, Oxford University Press.

Anastas, Paul T. / Williamson, T.C. (1998), *Green Chemistry. Frontiers in Benign Chemical Synthesis and Processes*, Oxford University Press.

Andersen, Mikael Skou / Massa, Ilmo (eds) (2000a), *Journal of Environmental Policy and Planning*, Special issue on *Ecological Modernization*, **2** (4), Wiley & Sons.

Andersen, Mikael Skou / Massa, Ilmo (2000b), 'Ecological Modernization. Origins, Dilemmas and Future Directions', in Andersen/Massa (eds), 337–345.

Andersson, Ake E. (1990a), *Creation, Innovation and Diffusion of Knowledge*, Stockholm: Institute for Future Studies.

Andersson, Ake E. (1990b), *Infrastructure and the Metropolis*, Stockholm: Institute for Future Studies.

Angel, David (1994), *Restructuring for Innovation. The remaking of the U.S. semiconductor industry*, New York: The Guilford Press.

Arthur, W. Brian (1994), *Increasing Returns and Path Dependence in the Economy*, Ann Arbor: University of Michigan Press.

Arthur, W. Brian (1996), 'Increasing Returns and the New World of Business', *Harvard Business Review*, July/August 1996, 100–109.

Arthur, W. Brian / Ermoliev, Yu. M. / Kaniovski, Yu. M. (1987), 'Path-dependent processes and the emergence of macro-structure', *European Journal of Operational Research*, **30**, 294–303.

Ashford, Nicholas A. (1993), 'Understanding Technological Responses of Industrial Firms to Environmental Problems. Implications for Government Policy', in Fischer/Schot (eds), 277–307.

Ashford, Nicholas A. (2001), 'Government and Environmental Innovation in Europe and North America', in Proceedings...

Ashford, Nicholas A. / Ayers, C. / Stone, R.F. (1985), 'Using Regulation to Change the Markets for Innovation', *Harvard Environmental Law Review*, September, 419–466.

Atkinson, Malcolm / Coleman, William D. (1989), 'Strong States and Weak States: Sectoral Policy Networks in Advanced Capitalist Economies', *British Journal of Political Science*, **19**, 47–67.

Audretsch, David B. (1996), 'International Diffusion of Technological Knowledge', in Koopmann/Scharrer (eds), 107–135.

Ausubel, Jesse H. / Marchetti, Cesare (1996), 'Elektron. Electrical Systems in Retrospect and Prospect', *The Liberation of the Environment, Daedalus,* special issue, **125** (3), 139–169.

Ayres, Robert U. (1993), 'Industrial Metabolism. Closing the Materials Cycle', in Jackson, T. (ed.), 165–188.

Ayres, Robert U. and Leslie W. (1996), *Industrial Ecology. Towards Closing the Materials Cycle*, Cheltenham: Edward Elgar.

Ayres, Robert U. / Simonis, Udo Ernst (eds) (1994), *Industrial Metabolism. Restructuring for Sustainable Development*, Tokyo: United Nations University Press.

Baumol, William (2002), *The Free-Market Innovation Machine. Analysing the Growth Miracle of Capitalism*, Princeton University Press.

Beck, Ulrich (1992), *Risk Society. Towards a New Modernity*, London: SAGE Publications.

Beck, Ulrich / Giddens, Anthony / Lash, Scott (1994), *Reflexive Modernization. Politics, Tradition and Aesthetics in the Modern Social Order*, Cambridge: Polity Press.

Becker, Frank / Englmann, Frank C. (2001), 'Public Policy, Voluntary Initiatives and Water Benign Process Innovations', in Proceedings...

Beise, Marian et al. (2003), 'The Emergence of Lead Markets for Environmental Innovations', in Horbach, Jens / Huber, Joseph / Schulz, Thomas (eds), *Nachhaltigkeit und Innovation*, Munich: oekom, 11–53.

Bell, Daniel (1973), *The Coming of Post-Industrial Society*, New York: Basic Books.

Bennett, John William (1976), *The Ecological Transition. Cultural Anthropology and Human Adaptation*, New York: Pergamon Press.

Biemans, Wim G. (1998), 'The Theory and Practice of Innovative Networks', in During, W. / Oakley, Raymond P. (eds), *New Technology-Based Firms in the 90s*, Vol. IV, London: Paul Chapman, 10–26.

BMU (ed.) (1999), *Erneuerbare Energien und Nachhaltige Entwicklung*, Berlin: Bundesministerium für Umwelt.

Bolter, David (1984), *Turing's Man*, London: Duckworth.

Bouma, Jan Jaap / Jeucken, Marcel / Klinkers, Leon (eds) (2001), *Sustainable Banking. The Greening of Finance*, Sheffield: Greenleaf Publishing.

Bourg, Dominique / Erkman, Suren (eds) (2003), *Perspectives on Industrial Ecology*, Sheffield: Greenleaf Publishing.

Braczyk, Hans-Joachim / Cooke, Philip / Heidenreich, Martin (eds) (1998), *Regional Innovation Systems*, London: UCL-Press.

Braungart, Michael (2002), *Brot für die Welt, Fleisch für die Müllverbrennung*, ed. by EPEA, Hamburg.

Braungart, Michael / McDonough, William (1998), 'The next Industrial Revolution', *The Atlantic Online*, October 1998.

Braungart, Michael / McDonough, William (2002), *Cradle to Cradle. Remaking the way we make things*, New York: North Point Press.

Calleja, Ignacio / Lindblom, Josefina / Wolf, Oliver (2002), 'Clean Technologies in Europe. Diffusion and Frontiers', *The IPTS Report*, No.5, November 2002, 13–22.

Casper, Steven (1999), 'National Institutional Frameworks and High-Technology Innovation in Germany. The Case of Biotechnology', *Wissenschaftszentrum Berlin discussion paper*, FS I 99 – 306, March 1999.

Charles, Daniel (2003), 'Corn that Clones Itself', *MIT Technology Review*, **106** (2), March 2003, 32–41.

Christensen, Clayton (1997), *The Innovator's Dilemma: When new technologies cause great firms to fail*, Harvard Business School Press.

Clark, Peter A. (1987), *Anglo-American Innovation*, Berlin/New York: Walter de Gruyter.

Clear Tech (2002), 'Free Energy Technologies Overview', www.free-energy.cc/background.html by Clear Tech Inc., Metaline Falls, WA, as of December 2002.

Clift, Roland / Longley, A.S. (1995), 'Introduction to Clean Technology', in Kirkwood, R.C. / Longley, A.S. (eds), *Clean Technology and the Environment*, London: Chapman & Hall.

Cohen, Wesley M. / Levinthal, Daniel A. (1989), 'Innovation and Learning: The two Faces of R&D', *Economic Journal*, **99** (3), September, 569–596.

Coleman, James S. / Katz, Elihu / Menzel, Herbert (1966), *Medical Innovation. A Diffusion Study*, Indeanapolis: Bobbs-Merrill Comp.

Cooke, Philip (1995), 'Innovation networks and regional development. Learning from European experience', in Krumbein, W. (ed.), 233–247.

Coombs, Rod / Saviotti, Paolo / Walsh, Vivien (1987), *Economics and Technological Change*, Totowa, N.J.: Rowman & Littlefield.

DATAR (1989), *Les villes européennes*, Paris: DATAR (i.e. Délégation à l'Aménagement du Territoire et à l'Action Régionale).

Diebolt, Claude / Monteils, Marielle (2000), 'The New Growth Theories. A Survey of Theoretical and Empirical Contributions', *Historical Social Research*, **25** (2), 3–22.

Dodgson, Mark (2000), *The Management of Technological Innovation*, Oxford University Press.

Dosi, Giovanni (1982), 'Technological Paradigms and Technological Trajectories', *Research Policy*, **11**, 147–162.

Dosi, Giovanni / Pavitt, Keith / Soete, Luc (1990), *The Economics of Technical Change and International Trade*, Hertfordshire: Harvester/Wheatsheaf.

Dunn, James A. / Perl, Anthony (1994), 'Policy Networks and Industrial Revitalization. High Speed Rail Initiatives in France and Germany', *Journal of Public Policy*, **14**, 311–343.

Dürkop, Jutta / Dubbert, Wolfgang / Nöh, Ingrid (1999), *Beitrag der Biotechnologie zu einer dauerhaft umweltgerechten Entwicklung*, Umweltbundesamt Berlin, Texte 1/99.

Ehrlich, Paul R. and Anne H. (1990), *The Population Explosion*, New York: Simon & Schuster.

Erdmann, Georg (2001), 'Innovation, Time, and Sustainability', in Proceedings...

EU Commission Community Research (2001), *Sustainable Production. Challenges & Objectives for EU Research Policy*, published by the European Commission.

Fairley, Peter (2002), 'Solar on the cheap', *MIT Technology Review*, **105** (1), January/ February 2002, 48–53.

Fairley, Peter (2003), 'Recharging the Power Grid. Huge batteries', *MIT Technology Review*, **106** (2), March 2003, 50–56.

Fischer, Kurt / Schot, Johan (eds) (1993), *Environmental Strategies for Industry*, Washington, D.C./Covelo, Ca.: Island Press.

Fischer-Kowalski, Marina (1997), 'Society's Metabolism', in Redclift/ Woodgate (eds), 119–137.

Florida, Richard / Kennedy, Martin (1990), *The Breakthrough Illusion. Corporate America's Failure to move from Innovation to Mass Production*, New York: Basic Books.

FMEL (2002), *Innovation Policy*, published by the German Federal Ministry of Economics, Technology and Labour, Berlin.

Frank, André Gunder / Fuentes, Marta (1990), 'Social Movements in recent World History', in Amin, Samir / Arrighi, Giovanni / Frank, Andre Gunder / Wallerstein, Immanuel (eds), *Social Movements and the World-System*, New York: Monthly Review Press, 139–180.

Freedman, David H. (2002), 'Fuel Cell vs. the Grid', *MIT Technology Review*, **105** (1), January/February 2002, 40–47.

Freeman, Christopher (1987), *Technology Policy and Economic Performance. Lessons from Japan*, London: Frances Pinter Publishers.

Freeman, Christopher / Soete, Luc (1999), *The Economics of Industrial Innovation*, 3rd edition (1st in 1974), London: Frances Pinter Publishers.

Frosch, Robert A. (1996), 'Toward the End of Waste. Reflections on a New Ecology of Industry', *The Liberation of the Environment, Daedalus*, special issue, **125** (3), 199–212.

GlobeScan (various years), *Surveys of Sustainability Experts*, published by Environics International Ltd., Toronto.

Grabher, Gernot (ed.) (1993): *The Embedded Firm. On the Socioeconomics of Industrial Networks*, London/New York: Routledge.

Graedel, Thomas (1994), 'Industrial Ecology. Definition and Implementation', in Socolow et al. (eds), 23–42.

Graedel, Thomas / Horkeby, Inge / Norberg-Bohm, Victoria (1994), 'Prioritizing Impacts in Industrial Ecology', in Socolow et al. (eds), 359–370.

Granovetter, Mark (1973), 'The Strength of Weak Ties', *American Journal of Sociology*, **78**, 1360–1380.

Granovetter, Mark (1974), *Getting a Job. A study of contacts and careers*, 2nd ed. 1995, University of Chicago Press.

Griliches, Zvi (1958), 'Research Costs and Social Returns. Hybrid Corn and Related Innovations', *Journal of Political Economy*, **66** (3), October 1958, 419–431.

Grossmann, Wolf-Dieter (2001), *Entwicklungsstrategien in der Informationsgesellschaft*, Berlin/Heidelberg: Springer.

Grübler, Arnulf (1990), *The Rise and Fall of Infrastructures. Dynamics of Evolution and Technological Change in Transport*, Heidelberg: Physica Verlag.

Grübler, Arnulf (1994), 'Industrialisation as a Historical Phenomenon', in Socolow et al. (eds), 43–68.

Grübler, Arnulf (1996), 'Time for a Change: On the Patterns of Diffusion of Innovation', *The Liberation of the Environment, Daedalus*, special issue, **125** (3), 19–42.

Grupp, Hariolf (1998), *Foundations of the Economics of Innovation. Theory, Measurement and Practice*, Cheltenham: Edward Elgar.

Grupp, Hariolf / Dominguez-Lacasa, Icíar / Friedrich-Nishio, Monika (2002), *Das deutsche Innovationssystem seit der Reichsgründung*, Heidelberg: Physica Verlag.

Hamblin, Robert L. / Miller, Jerry L.L. (1976), 'Reinforcement and the Origin, Rate and Extent of Cultural Diffusion', *Social Forces*, **54** (3), March 1976, 743–759.

Handbook of Vegetable Oil Esters (2002), *Innovative Products in Metal Cleaning*, publ. by European LIFE Project, Cooperation Office Hamburg, May 2002.

Hauschildt, Jürgen (1997), *Innovationsmanagement*, München: Vahlen.

Hedström, Peter / Sandell, Rickard / Stern, Charlotta (2000), 'Mesolevel Networks and the Diffusion of Social Movements. The Case of the Swedish Social Democratic Party', *American Journal of Sociology*, **106** (1), July 2000, 145–172.

Hemmelskamp, Jens (1999), *Umweltpolitik und technischer Fortschritt. Eine Untersuchung der Determinanten von Umweltinnovationen*, Heidelberg: Physica.

Hemmelskamp, Jens / Rennings, Klaus / Leone, Fabio (eds) (2000), *Innovation-Oriented Environmental Regulation*, Heidelberg: Physica Verlag.

Henderson, Rebecca M. / Clark, Kim B. (1990), 'Architectural Innovation. The Reconfiguration of Existing Product Technologies and the Failure of Established Firms', *Administrative Science Quarterly*, **35**, 9–30.

Herold, J. / Roland, W. (1996), *Henkel Automotive's Environmental Strategy*, Heidelberg/Düsseldorf: Henkel KG internal paper.

Hirschhorn, Joel / Jackson, Tim / Baas, Leo (1993), 'Towards Prevention. The emerging environmental management paradigm', in Jackson (ed.), 125–142.

Holland, Paul W. / Leinhardt, Samuel (eds) (1979), *Perspectives on Social Network Research*, New York: Academic Press.

Hoogma, Remco / Weber, Matthias / Elzen, Boelie (2001), 'Integrated long-term strategies to induce regime shifts to sustainability: The approach of strategic niche management', in Proceedings ...

Huber, Joseph (1989), 'Social Movements', *Technological Forecasting and Social Change*, 35 (1989), 365–374.

Huber, Joseph (1995), *Nachhaltige Entwicklung*, Berlin: Edition Sigma.

Huber, Joseph (2000), 'Towards Industrial Ecology. Sustainable Development as a Concept of Ecological Modernization', in Andersen/Massa (eds), 269–285.

Huber, Joseph (2001), *Allgemeine UmweltSoziologie*, Opladen: Westdeutscher Verlag.

Hughes, Thomas P. (1989), 'The Evolution of Large Technological Systems', in Bijker, Wiebe E. / Hughes, Thomas P. / Pinch, Trevor J. (eds), *The Social Construction of Technological Systems*, Cambridge, Mass.: MIT Press, 51–82.

Jackson, Tim (1993), 'Principles of Clean Production', in Jackson (ed.), 143–164.

Jackson, Tim (ed.) (1993), *Clean Production Strategies. Developing Preventive Environmental Management in the Industrial Economy*, Lewis Publishers.

Jacobs, Dany (1997), 'Knowledge-Intensive Innovation. The Potential of Regional Clusters', *The IPTS Report*, No.16, July 1997, 24–31.

Jänicke, Martin / Weidner, Helmut (eds) (1995), *Successful Environmental Policy*, Berlin: Edition Sigma.

Jänicke, Martin / Blazejzak, Jürgen / Edler, Dietmar / Hemmelskamp, Jens (2000), 'Environmental Policy and Innovation', in Hemmelskamp/Rennings/Leone (eds), 126–152.

Jansen, Dorothea (1999), *Einführung in die Netzwerkanalyse*, Opladen: Leske und Budrich.

Jansson, AnnMari / Hammer, Monica / Folke, Carl / Costanza, Robert (eds) (1994), *Investing in Natural Capital. The Ecological Economics Approach to Sustainability*, Washington: Island Press.

Jensen, Peder (2003), 'Unmodified Vegetable Oil as an Automotive Fuel', *The IPTS Report*, No. 74, May 2003, 18–23.

Kazazian, Thierry (2003), 'The Ecodesign Process', in Bourg, Dominique / Erkman, Suren (eds), *Perspectives on Industrial Ecology*, Sheffield: Greenleaf Publishing.

Kemp, René (1997), *Environmental Policy and Technical Change. A Comparison of the Technological Impact of Policy Instruments*, Cheltenham: Edward Elgar.

Kemp, René / Soete, Luc (1992), 'The Greening of Technological Progress. An evolutionary perspective', *Futures*, June 1992.

Kemp, René / Schot, Johan / Hoogma, Remco (1998), 'Regime Shifts to Sustainability through Processes of Niche Formation. The approach of strategic niche management', *Technology Analysis and Strategic Management*, **10** (2), 175–195.

Kemp, René / Rotmans, Jan (2001), 'The Management of the Co-Evolution of Technical, Environmental and Social Systems. The Case for Transition Management', in Proceedings…

Kemp, René / Arundel, Anthony / Smith, Keith (2001), 'Survey Indicators for Sustainable Development', in Proceedings…

Klemmer, Paul / Lehr, Ulrike / Löbbe, Klaus (1999), *Environmental Innovation. Incentives and Barriers*, Berlin: Analytica.

Knill, Christoph (2000), 'Policy-Netzwerke', in Weyer (ed.), 111–133.

Koopmann, Georg / Scharrer, Hans-Eckart (eds) (1996), *The Economics of High-Technology Competition and Cooperation in Global Markets*, Baden-Baden: Nomos.

Krishnan, Rajaram / Harris, Jonathan M. / Goodwin, Neva R. (eds) (1995), *A Survey of Ecological Economics*, Washington, D.C. / Covelo, Ca.: Island Press.

Krumbein, Wolfgang (ed.) (1995), *Ökonomische und politische Netzwerke in der Region*, Münster: LIT-Verlag.

Kuhn, Thomas S. (1962), *The Structure of Scientific Revolutions*, The University of Chicago Press.

Leinhardt, Samuel (ed) (1977), *Social Networks. A Developing Paradigm*, New York: Academic Press.

Leisinger, Klaus M. (2001), *Biotechnologie, Ernährungssicherheit und Politik*, Basel: Novartis Stiftung für Nachhaltige Entwicklung.

Lindemann, Peter A. (2001), *The Free Energy Secrets of Cold Electricity*, publ. by Clear Tech Inc., Metaline Falls, WA.

Litfin, Karen T. (ed.) (1998), *The Greening of Sovereignty in World Politics*, Cambridge, Mass./London: The MIT Press.

Lomborg, Björn (2001), *The Sceptical Environmentalist. Measuring the real state of the world*, Cambridge University Press.

Lopes, Paul / Durfee, Mary (eds) (1999), *The Social Diffusion of Ideas and Things, The Annals of the American Academy of Political and Social Science*, special issue, **566**, November.

Lovins, Amory (2002), *Small is Profitable. The Hidden Economic Benefits of Making Electrical Resources the Right Size*, London: Earthscan Publ.

Lowe, Ernest (1993), 'Industrial Ecology. An Organising Framework for Environmental Management', *Total Quality Environmental Management*, Autumn 1993, 73–85.

Lucas, Robert E. Jr. (1988), 'On the Mechanics of Economic Development', *Journal of Monetary Economics*, **22**, 3–42.

Lundvall, Bengt-Åke (1988), 'Innovation as an Interactive Process: From User-Producer Interaction to the National System of Innovation', in Dosi, Giovanni / Freeman, Christopher / Nelson, Richard (eds), *Technical Change and Economic Theory*, London/New York: Pinter, 349–369.

Lundvall, Bengt-Åke (1993), 'User-Producer Relationships, National Systems of Innovation and Internationalisation', in Foray, Dominique / Freeman, Christopher (eds), *Technology and the Wealth of Nations. The Dynamics of Constructed Advance*, London/New York: Pinter.

Mansfield, Edwin (1961), 'Technical Change and the Rate of Imitation', *Econometrica*, **29** (4), October, 241–318.

Marchetti, Cesare (1980), 'Society as a Learning System: Discovery, Invention, and Innovation Cycles Revisited', *Technological Forecasting and Social Change*, **18**, 267–282.

Marchetti, Cesare (1986), 'Stable Rules in Social and Economic Behavior', *IIASA-Paper*, Institute for Applied Systems Analysis, Laxemburg near Vienna.

Marchetti, Cesare (1988), 'Kondratiev Revisited. After One Kondratiev Cycle', *IIASA-Paper*, Institute for Applied Systems Analysis, Laxemburg near Vienna.

Marchetti, Cesare (1991), 'Branching out into the Universe', in Nakićenović/ Grübler (eds), 583–592.

Mensch, Gerhard (1979), *Stalemate in Technology. Innovations overcome the Depression*, Cambridge, Mass.: Ballinger Pub.

Miles, R.E. / Snow, C.C. (1978), *Organisational Strategy, Structure and Process*, New York/ London/ Tokyo: McGraw-Hill.

Milieudefensie (1992), *Action Plan Sustainable Netherlands. A perspective for changing northern lifestyles*, published by Friends of the Earth Netherlands, Amsterdam.

Modis, Theodore (1992), *Predictions*, New York: Simon & Schuster.

Mol, Arthur P.J. (1995), *The Refinement of Production. Ecological Modernization Theory and the Chemical Industry*, Utrecht: van Arkel.

Mol, Arthur P.J. (1997), 'Ecological Modernization. Industrial Transformations and Environmental Reform', in Redclift/Woodgate (eds), 138–149.

Mol, Arthur P.J. / Sonnenfeld, D.A. (2000), 'Ecological modernisation around the world. An introduction', *Environmental Politics*, **9** (1), 3–14.

Müller, Axel (2004), *Organisational Osmosis*, Thesis to be submitted at Martin-Luther University, Halle an der Saale.

Nakićenović, Nebojsa (1996), 'Freeing Energy from Carbon', *The Liberation of the Environment, Daedalus*, special issue, **125** (3), 95–112.

Nakićenović, Nebojsa / Grübler, Arnulf (eds) (1991), *Diffusion of Technologies and Social Behavior*, Berlin: Springer.

Nakićenović, Nebojsa / Grübler, Arnulf / McDonald, Alan (eds) (1998), *Global Energy Perspectives*, Cambridge University Press.

Nelson, Richard R. (1992), *National Innovation Systems. A Comparative Study*, Oxford University Press.

Nelson, Richard R. / Winter, Sidney G. (1982), *An Evolutionary Theory of Economic Change*, London/Cambridge, Mass.: The Belknap Press of Harvard University Press.

OECD (1992), *OECD-proposed Guidelines for Collecting and Interpreting Technological Innovation Data (Oslo Manual)*, Paris: OECD Publications Service.

OECD (1996), *Technology and Industrial Performance*, Paris: OECD Publications Service.

OECD (1998), *Biotechnology for Clean Industrial Products and Processes. Towards Industrial Sustainability*, Paris: OECD Publications Service.

OECD (2001a), *Innovative People. Mobility of skilled personnel in national innovation systems*, Paris: OECD Publications Service.

OECD (2001b), *OECD Environmental Data 2001*, Paris: OECD Publications.

OECD (2001c), *The Application of Biotechnology to Industrial Sustainability*, Paris: OECD Publications.

OECD (2001d), *Towards Sustainable Household Consumption? Trends and Policies in OECD countries*, Paris: OECD Publications.

OECD (2002a), *Sustainable Development. Critical Issues*, Paris: OECD Publications.

OECD (2002b), *International Mobility of the Highly Skilled. From statistical analysis to the formulation of policies*, Paris: OECD Publications.

Ökopol (2001), 'Inherently Safe Chemicals?', in *Substitution of Hazardous Chemicals*, Conference Proceedings, SubChem project, University for Applied Sciences, Hamburg, 8–9 Oct 2001.

Oertel, Dagmar / Fleischer, Torsten (2001), *Brennstoffzellen-Technologie: Hoffnungsträger für den Klimaschutz*, Berlin: Erich Schmidt Verlag.

Paton, Bruce (1994), 'Design for Environment. A Management Perspective', in Socolow et al. (eds), 349–358.

Pauli, Gunter (1998), *Upcycling. The Road to Zero Emissions*, Sheffield: Greenleaf.

Pavitt, Keith / Patel, Parimal (1996), 'What makes high technology competition different from conventional competition? The central importance of national systems of innovation', in Koopmann/Scharrer (eds), 143–171.

Perrow, Charles (1984), *Normal Accidents. Living with high-risk technologies*, New York: Basic Books.

Plichta, Peter (2001), *Benzin aus Sand. Die Silan-Revolution*, München: Langen.

Polanyi, Karl (1944), *The Great Transformation*, New York: Rinehart & Company.

Porter, Michael E. / van der Linde, Claas (1995), 'Toward a New Conception of the Environment-Competitiveness Relationship', *Journal of Economic Perspectives*, **9** (4), Fall, 97–118.

Proceedings of the International Conference *Towards Environmental Innovation Systems*, held 27–29 Sep 2001 in Garmisch-Partenkirchen, org. by J. Hemmelskamp and M. Weber on behalf of the German Ministry of Education and Research, the Austrian Ministry of Transport, Innovation and Technology, and the European Commission DG Research.

Putnam, Robert D. / Leonardi, Robert / Nanetti, Rafaella (1993), *Making Democracy Work. Civic traditions in modern Italy*, Princeton University Press.

Redclift, Michael / Woodgate, Graham (eds), *The International Handbook of Environmental Sociology*, Cheltenham: Edward Elgar.

Rees, William E. / Wackernagel, Mathis (1994), 'Ecological Footprints and Appropriated Carrying Capacity. Measuring the Natural Capital Requirements of the Human Economy', in Jansson et al. (eds), 362–391.

Rehfeld, Dieter (1995), 'Produktionscluster und räumliche Entwicklung. Beispiele und Konsequenzen', in Krumbein, W. (ed.), 187–205.

Rip, Arie / Kemp, René (1998), 'Technological Change', in Rayner, S. / Majone, E.L. (eds), *Human Choice and Climate Change*, Vol. II, *Resources and Technology*, Washington, D.C.: Batelle Press, 320–401.

Rogers, Everett M. (1995), *Diffusion of Innovations*, New York: The Free Press; completely revised version of the first edition of 1971.

Rogers, Everett M. / Shoemaker, F. Floyd (1971), *Communication of Innovations*, New York: Macmillan/The Free Press.

Rogers, Everett M. / Adhikarya, Ronny (1979), 'Diffusion of Innovations. An Up-to-Date Review and Commentary', *Communications Yearbook*, Vol. 3, 1979, New Brunswick, 67–82.

Romer, Paul M. (1986), 'Increasing Returns and Long-Run Growth', *Journal of Political Economy*, **94** (5), October, 1002–1035.

Romer, Paul M. (1990), 'Endogenous Technological Change', *Journal of Political Economy*, **98**, 71–102.

Rosenberg, Nathan / Landau, Ralph / Mowery, David C. (eds) (1992), *Technology and the Wealth of Nations*, Stanford University Press.

Rossi, Mark (1997), 'Moving beyond Eco-Efficiency: Examining the Barriers and Opportunities for Eco-Effectiveness', Paper given at the Greening of Industry Network Annual Conference, Santa Barbara, November 1997.

Scheringer, Martin (2002), *Persistence and Spatial Range of Environmental Chemicals*, Weinheim: Wiley-VCH.

Schmalholz, Heinz / Penzkofer, Horst (1997), 'Innovation, Wachstum und Beschäftigung', *ifo-schnelldienst*, Heft 17/18, 27–35.

Schmidt-Bleek, Friedrich (1994), *Wieviel Umwelt braucht der Mensch? MIPS, das Maß für ökologisches Wirtschaften*, Berlin: Birkhäuser.

Schmidt-Bleek, Friedrich / Weaver, Paul (eds) (1998), *Factor 10. Manifesto for a sustainable planet*, Sheffield: Greenleaf.

Schumpeter, Joseph (1934), *Theorie der wirtschaftlichen Entwicklung*, Berlin: Duncker & Humblot, first edition 1911.

Schumpeter, Joseph (1939), *Business Cycles*, 2 Vol., New York: McGraw-Hill.

Scott, John (1992), *Social Network Analysis. A Handbook*, London: SAGE.

Socolow Robert / Andrews, Clinton / Berkhout, Frans / Thomas, Valerie (eds) (1994), *Industrial Ecology and Global Change*, Cambridge University Press.

Soskice, David (1994), 'Innovation Strategies of Companies. A Comparative Institutional Approach of Some Cross-Country Differences', in Zapf, Wolfgang / Dierkes, Meinolf (eds), *Institutionenvergleich und Institutionendynamik*, Berlin: Edition Sigma, 271–289.

Spaargaren, Gert (1997), *The Ecological Modernization of Production and Consumption. Essays in Environmental Sociology*, Thesis, Landbouw Universiteit Wageningen.

Stahel, Walter / Giarini, Orio (1993), *The Limits to Uncertainty. Facing risks in the new service economy*, Dordrecht: Kluwer Academic Publishing.

SubChem (2002), *Substitution of Hazardous Chemicals*, SubChem research network, University of Applied Sciences, Hamburg.

Sustainable Production (2001), *Challenges and Objectives for EU Research Policy*, report of the expert group on 'Competitive and Sustainable Production in the Period to 2020', publ. by the EU Commission, Community Research.

Teece, David J. (1992), 'Strategies for Capturing the Financial Benefits from Technological Innovation', in Rosenberg et al. (eds), 175–206.

Thompson, Michael / Ellis, Richard / Wildavsky, Aaron (1990), *Cultural Theory*, Boulder, Co.: Westview Press.

Tischner, Ursula / Schmincke, Eva / Rubik, Frieder / Prösler, Martin (2000), *How to do EcoDesign*, Frankfurt: Verlag form.

Tushman, Michael / Rosenkopf, Lori (1992), 'Organisational Determinants of Technological Change. Toward a sociology of technological evolution', *Research in Organizational Behavior*, **14**, 311–347.

UNCED (1992), *Documents of the Rio-Conference on Environment and Development* [including Agenda 21, Rio Declaration on Environment and Development, Forest Principles, Climate Convention, and others], New York: United Nations Press.

UNEP/IE (eds) (1997), *Ecodesign. A promising approach to sustainable production and consumption*, Paris: UNO Publishers.

UNEP (1995–96), *Cleaner Production Worldwide*, Vol. I, II, III, Paris: UNEP Cleaner Production Programme.

UNEP (2000), *Global Environment Outlook 2000*, publ. by UNEP, Nairobi; also publ. by Earthscan Publ., London.

Utterbeck, J.M. / Abernathy, W.J. (1975), 'A Dynamic Model of Process and Product Innovation', *Omega*, **3** (6), 630–656.

Utterbeck, J.M. / Abernathy, W.J. (1978), 'Patterns of Industrial Innovation', *Technology Review*, **7**, 41–47.

Uzzi, Brian (1996), 'The Sources and Consequences of Embeddedness for the Economic Performance of Organisations. The Network Effect', *American Sociological Review*, **61** (8), August, 674–698.

Uzzi, Brian (1997), 'Social Structure and Competition in Interfirm Networks. The Paradox of Embeddedness', *Administrative Science Quarterly*, **42**, 35–67.

Uzzi, Brian (1999), 'Embeddedness in the Making of Financial Capital. How Social Relations and Networks benefit Firms seeking Financing', *American Sociological Review*, **64** (8), August, 481–505.

Vergragt, Philip J. (2000), *Strategies towards the Sustainable Household. SusHouse Project Final Report*, Faculty of Technology, Policy and Management, Delft University of Technology.

Wassermann, Stanley / Faust, Katherine (1994), *Social Network Analysis. Methods and Application*, Cambridge University Press.

WCED (1987), *Our Common Future (Brundtland-Report)*, ed. by World Commission on Environment and Development, Oxford University Press.

WDR (1992), *World Development Report. Development and Environment*, ed. by the World Bank, Washington, D.C., also published by Oxford University Press.

Weber, Bruce H. / Depew, David J. / Smith, James D. (eds) (1988), *Entropy, Information and Evolution*, Cambridge, Mass.: MIT Press.

Weyer, Johannes (ed.) (2000), *Soziale Netzwerke. Konzepte und Methoden der sozialwissenschaftlichen Netzwerkforschung*, München: Oldenbourg.

Weyer, Johannes / Kirchner, Ulrich / Riedl, Lars / Schmidt, Johannes F.K. (1997), *Technik, die Gesellschaft schafft. Soziale Netzwerke als Ort der Technikgenese*, Berlin: Edition Sigma.

Wildavsky, Aaron (1995), *But is it true? A citizen's guide to environmental, health and safety issues*, Cambridge, Mass.: Harvard University Press.

Williamson, Olivier E. (1975), *Markets and Hierarchies. Antitrust Analysis and Implications*, New York: The Free Press.

Witte, Eberhard (1973), *Organisation für Innovationsentscheidungen. Das Promotoren-Modell*, Göttingen: Vandenhoek & Ruprecht.

Witzel, Walter / Seifried, Dieter (eds) (2000), *Das Solarbuch. Fakten, Argumente, Strategien*, publ. by Energie- und Solaragentur Region Freiburg.

Wulff, Claudia (1999), 'Gentechnik in der Landwirtschaft, Nahrungsmittelverarbeitung und industriellen Produktion', in *Gentechnik*, ed. by Bundeszentrale für Politische Bildung, Bonn, 182–373.

Index